Stochastic Control and Mathematical Modeling

Applications in Economics

This is a concise and elementary introduction to stochastic control and mathematical modeling. This book is designed for researchers in stochastic control theory studying its application in mathematical economics and those in economics who are interested in mathematical theory in control. It is also a good guide for graduate students studying applied mathematics, mathematical economics, and nonlinear PDE theory.

Contents include the basics of analysis and probability, the theory of stochastic differential equations, variational problems, problems in optimal consumption and in optimal stopping, optimal pollution control, and solving the Hamilton–Jacobi–Bellman equations with boundary conditions. Major mathematical requisitions are contained in the preliminary chapters or in the appendix so that readers can proceed without referring to other materials.

Hiroaki Morimoto is a Professor in Mathematics at the Graduate School of Science and Engineering at Ehime University. His research interests include stochastic control, mathematical economics and finance and insurance applications, and the viscosity solution theory.

ENCYCLOPEDIA OF MATHEMATICS AND ITS APPLICATIONS

The titles below, and earlier volumes in the series, are available from booksellers or from Cambridge University Press at www.cambridge.org.

71 G. E. Andrews, R. Askey and R. Roy *Special Functions*
72 R. Ticciati *Quantum Field Theory for Mathematicians*
73 M. Stern *Semimodular Lattices*
74 I. Lasiecka and R. Triggiani *Control Theory for Partial Differential Equations I*
75 I. Lasiecka and R. Triggiani *Control Theory for Partial Differential Equations II*
76 A. A. Ivanov *Geometry of Sporadic Groups I*
77 A. Schinzel *Polynomials with Special Regard to Reducibility*
78 T. Beth, D. Jungnickel and H. Lenz *Design Theory II, 2nd edn*
79 T. W. Palmer *Banach Algebras and the General Theory of *-Algebras II*
80 O. Stormark *Lie's Structural Approach to PDE Systems*
81 C. F. Dunkl and Y. Xu *Orthogonal Polynomials of Several Variables*
82 J. P. Mayberry *The Foundations of Mathematics in the Theory of Sets*
83 C. Foias et al. *Navier–Stokes Equations and Turbulence*
84 B. Polster and G. F. Steinke *Geometries on Surfaces*
85 R. B. Paris and D. Kaminski *Asymptotics and Mellin–Barnes Integrals*
86 R. J. McEliece *The Theory of Information and Coding, 2nd edn*
87 B. A. Magurn *An Algebraic Introduction to K-Theory*
88 T. Mora *Solving Polynomial Equation Systems I*
89 K. Bichteler *Stochastic Integration with Jumps*
90 M. Lothaire *Algebraic Combinatorics on Words*
91 A. A. Ivanov and S. V. Shpectorov *Geometry of Sporadic Groups II*
92 P. McMullen and E. Schulte *Abstract Regular Polytopes*
93 G. Gierz et al. *Continuous Lattices and Domains*
94 S. R. Finch *Mathematical Constants*
95 Y. Jabri *The Mountain Pass Theorem*
96 G. Gasper and M. Rahman *Basic Hypergeometric Series, 2nd edn*
97 M. C. Pedicchio and W. Tholen (eds.) *Categorical Foundations*
98 M. E. H. Ismail *Classical and Quantum Orthogonal Polynomials in One Variable*
99 T. Mora *Solving Polynomial Equation Systems II*
100 E. Olivieri and M. Eulália Vares *Large Deviations and Metastability*
101 A. Kushner, V. Lychagin and V. Rubtsov *Contact Geometry and Nonlinear Differential Equations*
102 L. W. Beineke and R. J. Wilson (eds.) with P. J. Cameron *Topics in Algebraic Graph Theory*
103 O. J. Staffans *Well-Posed Linear Systems*
104 J. M. Lewis, S. Lakshmivarahan and S. K. Dhall *Dynamic Data Assimilation*
105 M. Lothaire *Applied Combinatorics on Words*
106 A. Markoe *Analytic Tomography*
107 P. A. Martin *Multiple Scattering*
108 R. A. Brualdi *Combinatorial Matrix Classes*
109 J. M. Borwein and J. D. Vanderwerff *Convex Functions*
110 M.-J. Lai and L. L. Schumaker *Spline Functions on Triangulations*
111 R. T. Curtis *Symmetric Generation of Groups*
112 H. Salzmann et al. *The Classical Fields*
113 S. Peszat and J. Zabczyk *Stochastic Partial Differential Equations with Lévy Noise*
114 J. Beck *Combinatorial Games*
115 L. Barreira and Y. Pesin *Nonuniform Hyperbolicity*
116 D. Z. Arov and H. Dym *J-Contractive Matrix Valued Functions and Related Topics*
117 R. Glowinski, J.-L. Lions and J. He *Exact and Approximate Controllability for Distributed Parameter Systems*
118 A. A. Borovkov and K. A. Borovkov *Asymptotic Analysis of Random Walks*
119 M. Deza and M. Dutour Sikirić *Geometry of Chemical Graphs*
120 T. Nishiura *Absolute Measurable Spaces*
121 M. Prest Purity *Spectra and Localisation*
122 S. Khrushchev *Orthogonal Polynomials and Continued Fractions*
123 H. Nagamochi and T. Ibaraki *Algorithmic Aspects of Graph Connectivity*
124 F. W. King Hilbert *Transforms I*
125 F. W. King Hilbert *Transforms II*
126 O. Calin and D.-C. Chang *Sub-Riemannian Geometry*
127 M. Grabisch et al. *Aggregation Functions*
128 L. W. Beineke and R. J. Wilson (eds.) with J. L. Gross and T. W. Tucker *Topics in Topological Graph Theory*
129 J. Berstel, D. Perrin and C. Reutenauer *Codes and Automata*
130 T. G. Faticoni *Modules over Endomorphism Rings*

ENCYCLOPEDIA OF MATHEMATICS AND ITS APPLICATIONS

Stochastic Control and Mathematical Modeling

Applications in Economics

HIROAKI MORIMOTO

Ehime University, Matsuyama, Japan

CAMBRIDGE
UNIVERSITY PRESS

CAMBRIDGE UNIVERSITY PRESS
Cambridge, New York, Melbourne, Madrid, Cape Town, Singapore,
São Paulo, Delhi, Dubai, Tokyo

Cambridge University Press
32 Avenue of the Americas, New York, NY 10013-2473, USA

www.cambridge.org
Information on this title: www.cambridge.org/9780521195034

First published 2010

Printed in the United States of America

A catalog record for this publication is available from the British Library.

Library of Congress Cataloging in Publication data

Morimoto, Hiraoki.
Stochastic control and mathematical modeling : applications in economics /
Hiroaki Morimoto.
p. cm. – (Encyclopedia of mathematics and its applications)
Includes bibliographical references and index.
ISBN 978-0-521-19503-4 (hardback)
1. Stochastic control theory. 2. Optimal stopping (Mathematical statistics)
3. Stochastic differential equations. I. Title. II. Series.
QA402.37.M67 2010
629.8′312 – dc22 2009042538

ISBN 978-0-521-19503-4 Hardback

To My Teacher M. Nisio

Contents

Preface

The purpose of this book is to provide a fundamental description of stochastic control theory and its applications to dynamic optimization in economics. Its content is suitable particularly for graduate students and scientists in applied mathematics, economics, and engineering fields.

A stochastic control problem poses the question: what is the optimal magnitude of a choice variable at each time in a dynamical system under uncertainty? In stochastic control theory, the state variables and control variables, respectively, describe the random phenomena of dynamics and inputs. The state variable in the problem evolves according to stochastic differential equations (SDE) with control variables. By steering of such control variables, we aim to optimize some performance criteria as expressed by the objective functional. Stochastic control can be viewed as a problem of decision making in maximization or minimization. This subject has created a great deal of mathematics as well as a large variety of applications in economics, mathematical finance, and engineering.

This book provides the basic elements of stochastic differential equations and stochastic control theory in a simple and self-contained way. In particular, a key to the stochastic control problem is the dynamic programming principle (DPP), which leads to the notion of viscosity solutions of Hamilton–Jacobi–Bellman (HJB) equations. The study of viscosity solutions, originated by M. Crandall and P. L. Lions in the 1980s, provides a useful tool for dealing with the lack of smoothness of the value functions in stochastic control. The main idea used to solve this maximization problem is summarized as follows:

(a) We formulate the problem and define the supremum of the objective functional over the class of all control variables, which is called the *value function*.
(b) We verify that the DPP holds for the value function.
(c) By the DPP, the value function can be viewed as a unique viscosity solution of the HJB equation associated with this problem.
(d) The uniform ellipticity and the uniqueness of viscosity solutions show the existence of a unique classical solution to the boundary value problem of

the HJB equation. This gives the smoothness of the viscosity solution of the HJB equation.

(e) We seek a candidate of optimal control by using the HJB equation. By using Itô's formula, we show the optimality.

This book is divided into three parts: Part I – Stochastic Calculus and Optimal Control Theory; Part II – Applications to Mathematical Models in Economics; and Part III – a collection of appendices providing background materials.

Part I consists of Chapters 1–5. In Chapter 1, we present the elements of stochastic calculus and SDEs, and in Chapter 2, we present the formulation of the weak solutions of SDEs, the concept of regular conditional probability, the Yamada–Watanabe theorem on weak and strong solutions, and the Markov property of a solution of SDE.

In Chapter 3, we introduce the DPP to issue (b). The verification of the DPP is rather difficult compared to the deterministic case. The Yamada–Watanabe theorem in Chapter 2 makes its proof exact. The supremum of (a) is taken over all systems in the weak sense.

Chapter 4 provides the theory of viscosity solutions of the HJB equations for (c). Using Ishii's lemma, we show the uniqueness results on viscosity solutions.

Chapter 5 is devoted to the boundary value problem of the HJB equations for (d) in the classical sense. Section 5.4 explains how to apply (a)–(e) in stochastic control.

Part II consists of Chapters 6–12. Here we present diverse applications of stochastic control theory to the mathematical models in economics. In Chapters 6–10, we take the state variables in these models as the remaining stock of a resource, the labor supply, and the price of the stock. The criteria in the maximization procedure are often given by the utility function of consumption rates as the control variables. Along (a)–(e), an optimal control is shown to exist.

Chapters 11 and 12 deal with the linear and nonlinear variational inequalities, instead of the HJB equations, which are associated with the stopping time problem. The variational inequality is analyzed by the viscosity solutions approach for optimality.

Part III consists of Appendices A–H. These provide some background material for understanding stochastic control theory as quickly as possible.

The prerequisites for this book are basic probability theory and functional analysis (see e.g., R. B. Ash [2], H. L. Royden [139], and A. Friedman [69]). See M. I. Kamien and N. L. Schwartz [80], A. C. Chiang [33], A. K. Dixit and R. S. Pindyck [46], L. Ljungqvist and T. J. Sargent [107], and R. S. Merton [114], for economics references.

Acknowledgments

I have greatly profited from the valuable comments and thoughtful suggestions of many of my colleagues, friends, and teachers, in particular: A. Bensoussan,

N. El Karoui, W. H. Fleming, H. Ishii, Y. Ishikawa, K. Kamizono, N. Kazamaki, S. Koike, H. Kunita, J. L. Menaldi, H. Nagai, M. Okada, S. Sakaguchi, Ł. Stettner, T. Tsuchikura, J. Zabczyk, and X. Y. Zhou.

I would like to thank I. Karatzas very much for reading the earlier drafts and his encouragement to publish this book. Thanks also go to K. Kawaguchi for many discussions on the mathematical treatment of economics and for his knowledge of the economic problems. I wish to thank T. Adachi, H. Kaise, and C. Liu for their careful reading of the manuscript and assistance at various stages.

H. Morimoto
Matsuyama
January 2009

Part I

Stochastic Calculus and Optimal Control Theory

1

Foundations of Stochastic Calculus

We are concerned here with a stochastic differential equation,

$$\begin{aligned} dX(t) &= b(X(t))dt + \sigma(X(t))dB(t), \qquad t \geq 0, \\ X(0) &= x \in \mathbf{R}^N, \end{aligned}$$

in N-dimensional Euclidean space $\mathbf{R}^N$. Here b, σ are Lipschitz functions, called the *drift term* and the *diffusion term*, respectively, and $\{B(t)\}$ is a standard Brownian motion equation defined on a probability space $(\Omega, \mathcal{F}, P)$. This equation describes the evolution of a finite-dimensional dynamical system perturbed by noise, which is formally given by $dB(t)/dt$. In economic applications, the stochastic process $\{X(t)\}$ is interpreted as the labor supply, the price of stocks, or the price of capital at time $t \geq 0$. We present a reasonable definition of the second term with uncertainty and basic elements of calculus on the stochastic differential equation, called *stochastic calculus*.

A. Bensoussan [16], I. Karatzas and S. E. Shreve [87], N. Ikeda and S. Watanabe [75], I. Gihman and A. Skorohod [72], A. Friedman [68], B. Øksendal [132], D. Revuz and M. Yor [134], R. S. Liptzser and A. N. Shiryayev [106] are basic references for this chapter.

1.1 Review of Probability

1.1.1 Random Variables

Definition 1.1.1. *A triple* $(\Omega, \mathcal{F}, P)$ *is a* probability space *if the following assertions hold:*

(a) Ω *is a set.*
(b) $\mathcal{F}$ *is a* σ*-algebra, that is,* $\mathcal{F}$ *is a collection of subsets of* Ω *such that*
  (i) $\Omega, \ \phi \in \mathcal{F}$,
  (ii) *if* $A \in \mathcal{F}$, *then* $A^c := \Omega \setminus A \in \mathcal{F}$,
  (iii) *if* $A_n \in \mathcal{F}, n = 1, 2, \dots,$ *then* $\bigcup_{n=1}^{\infty} A_n \in \mathcal{F}$.

(c) *P is a probability measure, that is, a map $P : \mathcal{F} \to [0, 1]$, such that*
   (i) $P(\Omega) = 1$,
   (ii) *if $A_n \in \mathcal{F}, n = 1, 2, \dots$, disjoint, then* $P(\bigcup_{n=1}^{\infty} A_n) = \sum_{n=1}^{\infty} P(A_n)$.

Definition 1.1.2. *A probability space $(\Omega, \mathcal{F}, P)$ is complete if $A \in \mathcal{F}$ has $P(A) = 0$ and $B \subset A$, then $B \in \mathcal{F}$ (and, of course, $P(B) = 0$), that is, $\mathcal{F}$ contains all P-null sets.*

Remark 1.1.3. *Any probability space $(\Omega, \mathcal{F}, P)$ can be made complete by the completion of measures due to Carathéodory. We also refer to the proof of the Daniell Theorem, Theorem 2.1 in Chapter 2.*

Definition 1.1.4. *For any collection $\mathcal{G}$ of subsets of Ω, we define a smallest σ-algebra $\sigma(\mathcal{G})$ containing $\mathcal{G}$ by*

$$\sigma(\mathcal{G}) = \cap\{\mathcal{F} : \mathcal{G} \subset \mathcal{F},\ \mathcal{F}\ \sigma\text{-algebra of } \Omega\},$$

which is the σ-algebra generated by $\mathcal{G}$.

Example 1.1.5. *On the set of real numbers* **R**, *we take $\mathcal{G} = \{$open intervals$\}$ and denote by $\mathcal{B}(\mathbf{R})$ the σ-algebra $\sigma(\mathcal{G})$ generated by $\mathcal{G}$, which is the Borel σ-algebra on* **R**.

Definition 1.1.6. *Let $(\Omega, \mathcal{F}, P)$ be a complete probability space.*

(a) *A map $X : \Omega \to \mathbf{R}$ is a random variable if*

$$X^{-1}(B) := \{\omega : X(\omega) \in B\} \in \mathcal{F}, \qquad \text{for any } B \in \mathcal{B}(\mathbf{R}).$$

(b) *For any random variable X, we define the σ-algebra $\sigma(X)$ generated by X as follows:*

$$\sigma(X) = \sigma(\mathcal{G}) = \mathcal{G}, \quad \mathcal{G} := \{X^{-1}(B) : B \in \mathcal{B}(\mathbf{R})\} \subset \mathcal{F}.$$

Proposition 1.1.7. *Let X, Y be two random variables. Then Y is $\sigma(X)$ measurable if and only if there exists a Borel measurable function $g : \mathbf{R} \to \mathbf{R}$ such that*

$$Y(\omega) = g(X(\omega)), \qquad \text{for all } \omega \in \Omega.$$

Proof. Since $Y = Y^+ - Y^-$, we will show the "only if" part when $Y \geq 0$.

(1) Suppose that Y is a simple random variable. Then Y is of the form:

$$Y(\omega) = \sum_{i=1}^{n} y_i 1_{F_i}(\omega),$$

where $y_i \geq 0$, $F_i \in \sigma(X)$ and the F_i are pairwise disjoint. By definition, there exists $D_i \in \mathcal{B}(\mathbf{R})$, for each i, such that $F_i = X^{-1}(D_i)$. Clearly, the D_i are pairwise disjoint. Define

$$g(y) = \begin{cases} y_i, & y \in D_i, \\ 0, & y \notin \bigcup_{i=1}^{n} D_i. \end{cases}$$

Then

$$Y(\omega) = \sum_{i=1}^{n} y_i 1_{\{X^{-1}(D_i)\}}(\omega) = \sum_{i=1}^{n} y_i 1_{D_i}(X(\omega)) = g(X(\omega)).$$

(2) In the general case, there exists a sequence of simple random variables Y_n converging to Y. Let g_n be the corresponding sequence of measurable functions such that $Y_n = g_n(X)$. Define

$$g(y) = \liminf_{n\to\infty} g_n(y).$$

Then g is $\mathcal{B}(\mathbf{R})$ measurable and

$$Y(\omega) = \liminf_n Y_n(\omega) = \liminf_n g_n(X(\omega)) = g(X(\omega)), \quad \omega \in \Omega. \quad \blacksquare$$

1.1.2 Expectation, Conditional Expectation

Definition 1.1.8. *Let X be a random variable. The quantity*

$$E[X] = \int_\Omega X(\omega) dP(\omega)$$

is the expectation of X, where $E[X^+]$ or $E[X^-]$ is finite.

Definition 1.1.9. *Let X, Y be two random variables on a complete probability space $(\Omega, \mathcal{F}, P)$.*

(a) *The expression $X = Y$ will indicate that $X = Y$ a.s., that is, $P(X \neq Y) = 0$.*
(b) *For $1 \leq p < \infty$, the norm $\|X\|_p$ of X is defined by*

$$\|X\|_p = (E[|X|^p])^{1/p}.$$

(c) *If $p = \infty$, then*

$$\|X\|_\infty = ess\ sup|X| = \inf\{\sup_{\omega\notin N}|X(\omega)| : N \in \mathcal{F}, P(N) = 0\}.$$

(d) *The L^p spaces are defined by*

$$L^p = L^p(\Omega) = \{X : random\ variable, \|X\|_p < \infty\}.$$

Proposition 1.1.10.

(i) *$L^p(\Omega)$ is a Banach space, that is, a complete normed linear space, for $1 \leq p \leq \infty$.*
(ii) *$L^2(\Omega)$ is a Hilbert space, that is, a complete inner product space, with inner product $(X, Y) = E[X \cdot Y], \quad X, Y \in L^2(\Omega)$.*

For the proof, see A. Friedman [69, chapter 3].

Definition 1.1.11. *Let $X_n, n = 1, 2, \ldots,$ and X be random variables.*

(a) *$X_n \to X$ a.s. if $P(X_n \to X$ as $n \to \infty) = 1$.*

(b) $X_n \to X$ *in probability if* $P(|X_n - X| \geq \varepsilon) \to 0$ *as* $n \to \infty$, *for any* $\varepsilon > 0$.

(c) $X_n \to X$ *in* L^p *if* $\|X_n - X\|_p \to 0$ *as* $n \to \infty$.

Proposition 1.1.12. *Let* $X_n, n = 1, 2, \ldots,$ *and* X *be random variables.*

(i) *If* $X_n \to X$ *a.s., then* $X_n \to X$ *in probability.*
(ii) *If* $X_n \to X$ *in* L^p $(p \geq 1)$, *then* $X_n \to X$ *in probability.*
(iii) $X_n \to X$ *in probability if and only if* $E[\frac{|X_n - X|}{1+|X_n - X|}] \to 0$.
(iv) *Let* $\varepsilon_n \geq 0$ *and* $\sum_{n=1}^{\infty} \varepsilon_n < \infty$. *If* $\sum_{n=1}^{\infty} P(|X_{n+1} - X_n| \geq \varepsilon_n) < \infty$, *then* X_n *converges a.s.*
(v) *If* $X_n \to X$ *in probability, then it contains a subsequence* $\{X_{n_k}\}$ *such that* $X_{n_k} \to X$ *a.s.*

For the proof, see A. Friedman [69, chapter 2].

Definition 1.1.13. *A family* $\{X_n : n \in \mathbf{N}\}$ *of random variables* X_n *on* $(\Omega, \mathcal{F}, P)$ *is uniformly integrable if*

$$\lim_{a \to \infty} \sup_n \int_{\{|X_n| \geq a\}} |X_n| dP = 0.$$

Proposition 1.1.14. *Assume that one of the following assertions is satisfied:*

(i) $E[\sup_n |X_n|] < \infty$,
(ii) $\sup_n E[|X_n|^p] < \infty$, *for some* $p > 1$.

Then $\{X_n\}$ *is uniformly integrable.*

Proof.

(1) We set $Y = \sup_n |X_n|$. Then, by (i),

$$P(Y \geq c) \leq \frac{1}{c} E[Y] \to 0 \quad \text{as } c \to \infty.$$

Thus

$$\sup_n \int_{\{|X_n| \geq c\}} |X_n| dP \leq \int_{\{Y \geq c\}} Y dP \to 0 \quad \text{as } c \to \infty.$$

(2) By Chebyshev's inequality,

$$\sup_n P(|X_n| \geq c) \leq \frac{1}{c^p} \sup_n E[|X_n|^p].$$

Thus, by (ii) and Hölder's inequality,

$$\sup_n \int_{\{|X_n| \geq c\}} |X_n| dP \leq \sup_n (E[|X_n|^p])^{1/p} (E[1_{\{|X_n| \geq c\}}])^{1/q}$$
$$\leq \sup_n E[|X_n|^p] (\frac{1}{c})^{p/q} \to 0 \quad \text{as } c \to \infty,$$

where $1/p + 1/q = 1$. ■

Proposition 1.1.15. *Let $\{X_n\}$ be a sequence of integrable random variables such that $X_n \to X$ a.s. Then $\{X_n\}$ is uniformly integrable if and only if*

$$\lim_{n\to\infty} E[|X_n - X|] = 0.$$

Proof. Let $Y_n = X_n - X$.

(1) Suppose that $\{X_n\}$ is uniformly integrable. Since

$$\begin{aligned} E[|X_n|] &= E[|X_n|1_{\{|X_n|\geq a\}}] + E[|X_n|1_{\{|X_n|<a\}}] \\ &\leq \sup_n E[|X_n|1_{\{|X_n|\geq a\}}] + aP(|X_n| < a), \qquad \text{for any } a > 0, \end{aligned}$$

we have $\sup_n E[|X_n|] < \infty$, taking sufficiently large $a > 0$. By Fatou's lemma

$$E[|X|] \leq \liminf_{n\to\infty} E[|X_n|] < \infty.$$

Also, by Chebyshev's inequality,

$$\begin{aligned} E[|Y_n|] &\leq E[|X_n|1_{\{|Y_n|\geq a\}} + |X|1_{\{|Y_n|\geq a\}} + |Y_n|1_{\{|Y_n|<a\}}] \\ &\leq E[|X_n|1_{\{|X_n|\geq c\}}] + 2cP(|Y_n| \geq a) + E[|X|1_{\{|X|\geq c\}}] \\ &\quad + E[|Y_n|1_{\{|Y_n|<a\}}] \\ &\leq \sup_n E[|X_n|1_{\{|X_n|\geq c\}}] + \frac{2c}{a}\sup_n E[|X_n| + |X|] + E[|X|1_{\{|X|\geq c\}}] \\ &\quad + (1+a)E[\frac{|Y_n|}{1+|Y_n|}]. \end{aligned}$$

Letting $n \to \infty$, $a \to \infty$, and then $c \to \infty$, we get $\limsup_{n\to\infty} E[|Y_n|] = 0$.

(2) Conversely, it is easy to see that

$$P(|Y_n| \geq a) \leq \frac{1}{a}E[|Y_n|] \to 0 \quad \text{ as } a \to \infty$$

and

$$\begin{aligned} E[|Y_n|1_{\{|Y_n|\geq a\}}] &= E[(|Y_n| - |Y_n| \wedge m)1_{\{|Y_n|\geq a\}}] + E[(|Y_n| \wedge m)1_{\{|Y_n|\geq a\}}] \\ &\leq E[|Y_n| - |Y_n| \wedge m] + mP(|Y_n| \geq a). \end{aligned}$$

Hence, letting $a \to \infty$ and then $m \to \infty$, we get

$$\limsup_{a\to\infty} E[|Y_n|1_{\{|Y_n|\geq a\}}] = 0, \qquad \text{for each } n.$$

Next, for any $k \in \mathbf{N}$,

$$\begin{aligned} \sup_n E[|Y_n|1_{\{|Y_n|\geq a\}}] &\leq \sup_{n\leq k} E[|Y_n|1_{\{|Y_n|\geq a\}}] + \sup_{n>k} E[|Y_n|1_{\{|Y_n|\geq a\}}] \\ &\leq \sum_{n=1}^{k} E[|Y_n|1_{\{|Y_n|\geq a\}}] + \sup_{n>k} E[|Y_n|]. \end{aligned}$$

Letting $a \to \infty$ and then $k \to \infty$, we have $\limsup_{a\to\infty} \sup_n E[|Y_n|1_{\{|Y_n|\geq a\}}] = 0$. By the same calculation as shown in (1), we have

$$\begin{aligned}\sup_n E[|X_n|1_{\{|X_n|\geq c\}}] &\leq \sup_n E[|Y_n|1_{\{|X_n|\geq c\}} + |X|1_{\{|X_n|\geq c\}}]\\ &\leq \sup_n E[|Y_n|1_{\{|Y_n|\geq a\}}] + \frac{2a}{c}\sup_n E[|X_n|]\\ &\quad + E[|X|1_{\{|X|\geq a\}}].\end{aligned}$$

Letting $c \to \infty$ and then $a \to \infty$, we obtain uniform integrability. ■

Definition 1.1.16. *Let $X \in L^1(\Omega, \mathcal{F}, P)$ and let $\mathcal{G}$ be a sub-σ-algebra of $\mathcal{F}$, that is, $\mathcal{G} \subset \mathcal{F}$ is a σ-algebra. A random variable $Y \in L^1(\Omega, \mathcal{G}, P)$ is the conditional expectation of X given $\mathcal{G}$ if*

$$\int_A Y dP = \int_A X dP, \quad \text{for all } A \in \mathcal{G}.$$

We write $Y = E[X|\mathcal{G}]$.

Proposition 1.1.17. *Let $X \in L^1(\Omega, \mathcal{F}, P)$ and $\mathcal{G}$ a sub-σ-algebra of $\mathcal{F}$. Then the conditional expectation $Y \in L^1(\Omega, \mathcal{G}, P)$ of X given $\mathcal{G}$ exists uniquely.*

Proof. Without loss of generality, we may assume $X \geq 0$. Define

$$\mu(A) = \int_A X dP, \quad \text{for } A \in \mathcal{G}.$$

Then μ is a finite measure, absolutely continuous with respect to P. By the Radon–Nikodým Theorem (cf. A. Friedman [69]), there exists, uniquely, $Y \in L^1(\Omega, \mathcal{G}, P)$, $Y \geq 0$, such that $\mu(A) = \int_A Y dP$, for $A \in \mathcal{G}$. ■

Remark 1.1.18. *We recall that $L^2(\Omega, \mathcal{G}, P)$ is a closed subspace of the Hilbert space $L^2(\Omega, \mathcal{F}, P)$. If $X \in L^2(\Omega, \mathcal{F}, P)$, then $E[X|\mathcal{G}]$ coincides with the orthogonal projection $\hat{X}$ of X to $L^2(\Omega, \mathcal{G}, P)$, that is,*

$$E[|X - E[X|\mathcal{G}]|^2] = \min\{E[|X - Y|^2] : Y \in L^2(\Omega, \mathcal{G}, P)\}.$$

Proposition 1.1.19. *Let $X_n, X \in L^1(\Omega, \mathcal{F}, P)$, $n = 1, 2, \ldots$, and $\mathcal{H}, \mathcal{G}$ be two sub-σ-algebras of $\mathcal{F}$. Then, the following assertions hold:*

(i) $E[E[X|\mathcal{G}]] = E[X]$.
(ii) $E[X|\mathcal{G}] = X$ *a.s. if X is $\mathcal{G}$-measurable.*
(iii) $E[aX_1 + bX_2|\mathcal{G}] = aE[X_1|\mathcal{G}] + bE[X_2|\mathcal{G}]$ *a.s.,* $\quad a, b \in \mathbf{R}$.
(iv) $E[X|\mathcal{G}] \geq 0$ *a.s. if $X \geq 0$.*
(v) $E[X_n|\mathcal{G}] \nearrow E[X|\mathcal{G}]$ *a.s. if $X_n \nearrow X$.*
(vi) $E[\liminf_{n\to\infty} X_n|\mathcal{G}] \leq \liminf_{n\to\infty} E[X_n|\mathcal{G}]$ *a.s. if $X_n \geq 0$.*
(vii) $E[X_n|\mathcal{G}] \to E[X|\mathcal{G}]$ *a.s. if $X_n \to X$ a.s. and $\sup_n |X_n| \in L^1$.*
(viii) $f(E[X|\mathcal{G}]) \leq E[f(X)|\mathcal{G}]$ *a.s. if $f : \mathbf{R} \to \mathbf{R}$ is convex, and $f(X) \in L^1(\Omega)$.*
(ix) $E[E[X|\mathcal{G}]|\mathcal{H}] = E[X|\mathcal{H}]$ *a.s. if $\mathcal{H} \subset \mathcal{G}$.*

(x) $E[ZX|\mathcal{G}] = ZE[X|\mathcal{G}]$ *a.s. if* $X \in L^p(\Omega, \mathcal{F}, P)$, $Z \in L^q(\Omega, \mathcal{G}, P)$, *for* $p = 1$, $q = \infty$ *or* $p > 1$, $\frac{1}{p} + \frac{1}{q} = 1$.

(xi) $E[X|\mathcal{G}] = E[X]$ *a.s. if* X *and* 1_A *are independent for any* $A \in \mathcal{G}$.

Proof. The proof is obtained by integrating over arbitrary sets $A \in \mathcal{G}$ and by using several properties of the integrals. In particular, (v) is immediately from the monotone convergence theorem. For (vi), we apply (v) to $Y_n = \inf_{m \geq n} X_m$, and (vii) follows from (vi). ∎

1.2 Stochastic Processes

1.2.1 General Notations

Definition 1.2.1. *A quadruple* $(\Omega, \mathcal{F}, P, \{\mathcal{F}_t\}_{t\geq 0})$ *is a stochastic basis if the following assertions hold:*

(a) $(\Omega, \mathcal{F}, P)$ *is a complete probability space.*

(b) $\{\mathcal{F}_t\}_{t\geq 0}$ *is a filtration, that is, a nondecreasing family of sub-σ-algebra of* $\mathcal{F}$:

$$\mathcal{F}_s \subset \mathcal{F}_t \subset \mathcal{F} \text{ for } 0 \leq s < t < \infty.$$

(c) *The filtration* $\{\mathcal{F}_t\}_{t\geq 0}$ *satisfies the following ("usual") conditions:*
   (i) $\{\mathcal{F}_t\}_{t\geq 0}$ *is right-continuous:* $\mathcal{F}_t = \mathcal{F}_{t+} := \cap_{\varepsilon>0}\mathcal{F}_{t+\varepsilon}$, *for all* $t \geq 0$.
   (ii) $\mathcal{F}_0$ *contains all* P*-null sets in* $\mathcal{F}$.

Definition 1.2.2. *An N-dimensional stochastic process* $X = \{X_t\}_{t\geq 0}$ *on a complete probability space* $(\Omega, \mathcal{F}, P)$ *is a collection of* $\mathbf{R}^N$*-valued random variables* $X(t, \omega)$, $\omega \in \Omega$. *For fixed* $\omega \in \Omega$, *the set* $\{X(t, \omega) : t \geq 0\}$ *is a path of* X.

Definition 1.2.3. *Let* $(\Omega, \mathcal{F}, P, \{\mathcal{F}_t\}_{t\geq 0})$ *be a stochastic basis, and let* $\mathcal{B}(\mathbf{R})$, $\mathcal{B}(\mathbf{R}^N)$, $\mathcal{B}([0, t])$ *be Borel σ-algebra on* $\mathbf{R}$, $\mathbf{R}^N$, $[0, t]$.

(a) *A stochastic process* $\{X_t\}_{t\geq 0}$ *is* $\{\mathcal{F}_t\}_{t\geq 0}$*-adapted (with respect to the filtration* $\{\mathcal{F}_t\}_{t\geq 0}$*) if* X_t *is* $\mathcal{F}_t$*-measurable for all* $t \geq 0$.
Such a stochastic process will be denoted by $\{(X_t, \mathcal{F}_t)\}_{t\geq 0}$ *or* $(X_t, \mathcal{F}_t)$, $\{(X(t), \mathcal{F}_t)\}$.

(b) *A stochastic process* $\{X_t\}_{t\geq 0}$ *is measurable if, for all* $A \in \mathcal{B}(\mathbf{R}^N)$,

$$\{(t, \omega) : X_t(\omega) \in A\} \in \mathcal{B}([0, \infty)) \otimes \mathcal{F}.$$

(c) *A measurable adapted stochastic process* $\{X_t\}_{t\geq 0}$ *is progressively measurable if, for each* $t \geq 0$ *and* $A \in \mathcal{B}(\mathbf{R}^N)$,

$$\{(s, \omega) : 0 \leq s \leq t,\ X_s(\omega) \in A\} \in \mathcal{B}([0, t]) \otimes \mathcal{F}_t.$$

(d) *A stochastic process* $\{X_t\}_{t\geq 0}$ *is continuous (right-continuous) if there is* $\Omega_0 \in \mathcal{F}$ *with* $P(\Omega_0) = 1$ *such that* $t \to X_t(\omega)$ *is continuous (right-continuous) for every* $\omega \in \Omega_0$.

(e) *Two stochastic processes X and Y are indistinguishable if there is* $\Omega_0 \in \mathcal{F}$ *with* $P(\Omega_0) = 1$ *such that* $X_t(\omega) = Y_t(\omega)$ *for all* $t \geq 0$ *and* $\omega \in \Omega_0$. *The expression* $X = Y$ *will indicate that X and Y are indistinguishable.*
(f) *Y is a modification of X if* $P(X_t = Y_t) = 1$ *for all* $t \geq 0$.

Proposition 1.2.4. *Let* $\{(X_t, \mathcal{F}_t)\}_{t\geq 0}$ *be a stochastic process on a stochastic basis* $(\Omega, \mathcal{F}, P, \{\mathcal{F}_t\})$.

(i) *If* $X = \{X_t\}$ *is a right-continuous modification of* 0, *then* $X = 0$.
(ii) *If* $X = \{X_t\}$ *is right-continuous, then X is progressively measurable.*

Proof. Let $\Omega_0 = \{\omega : t \to X_t(\omega)$ right-continuous$\}$ and $P(\Omega_0) = 1$.

(1) Since $\{X_t\}$ is a modification of 0, $P(X_r \neq 0) = 0$ for each $r \in \mathbf{Q}_+$, and

$$P(\bigcup_{r\in\mathbf{Q}_+} \{\omega : X_r(\omega) \neq 0\}) = 0.$$

Define $\Omega' = (\cap_{r\in\mathbf{Q}_+}\{\omega : X_r(\omega) = 0\}) \cap \Omega_0$. It is obvious that $P(\Omega') = 1$ and $X_t(\omega) = 0$, for all $t \geq 0$ and $\omega \in \Omega'$.
(2) Taking into account the indistinguishable process of $\{X_t\}$, we may consider that $X(t, \omega) = 0$ if $\omega \notin \Omega_0$. Fix $t \geq 0$, and let $\delta = \{t_0 = 0 < t_1 < t_2 < \cdots < t_n = t\}$ be a partition of $[0, t]$, with $t_k = kt/2^n$, $k = 0, 1, \ldots, 2^n$. Define

$$X^\delta(s, \omega) = \begin{cases} X(0, \omega)1_{\{0\}}(s) + \sum_{k=1}^{2^n} X(t_k, \omega)1_{(t_{k-1}, t_k]}(s) & \text{if } \omega \in \Omega_0, \\ 0 & \text{otherwise.} \end{cases}$$

Clearly, the map $(s, \omega) \to X^\delta(s, \omega)$ is $\mathcal{B}([0, t]) \otimes \mathcal{F}_t$-measurable. Letting $n \to \infty$, by right continuity, we see that $X^\delta(s, \omega) \to X(s, \omega)$ for all $s \in [0, t]$ and $\omega \in \Omega$. Thus the map $(s, \omega) \to X(s, \omega)$ is measurable for this σ-algebra, so $\{X_t\}$ is progressively measurable. ■

Remark 1.2.5.

(a) *Let* $\{\mathcal{F}_t\}_{t\geq 0}$ *be a filtration such that* $\mathcal{F}_0$ *contains all* $P-$*null sets. If* $X = Y$ *and X is adapted to* $\{\mathcal{F}_t\}_{t\geq 0}$, *then Y is adapted to* $\{\mathcal{F}_t\}_{t\geq 0}$.
(b) *For any filtration* $\{\mathcal{G}_t\}_{t\geq 0}$ *we can obtain the right-continuous filtration* $\{\mathcal{F}_t\}_{t\geq 0}$ *defined by* $\mathcal{F}_t = \mathcal{G}_{t+} := \bigcap_{\varepsilon>0} \mathcal{G}_{t+\varepsilon}$.
(c) *The filtration in connection with the stochastic process* $\{X_t\}_{t\geq 0}$ *is the* σ*-algebra*

$\sigma(X_s, s \leq t)$ *generated by* $\{X_s, s \leq t\}$, *where*

$$\sigma(X_s, s \leq t) := \sigma(\mathcal{G}_t), \quad \mathcal{G}_t = \{X_s^{-1}(A) : A \in \mathcal{B}(\mathbf{R}^N), s \leq t\}.$$

(d) *Furthermore, a stochastic basis* $(\Omega, \mathcal{F}, P, \{\mathcal{F}_t^X\})$ *is obtained by setting*

$$\mathcal{F}_t^X = \mathcal{H}_{t+}, \quad \mathcal{H}_t = \sigma(X_s, s \leq t) \vee \mathcal{N} := \sigma(\sigma(X_s, s \leq t) \cup \mathcal{N}),$$

where $\mathcal{N}$ *is the collection of P null sets.*

1.2.2 Brownian Motion

Definition 1.2.6.

(a) *The real-valued stochastic process* $\{B_t\}_{t\geq 0}$ *is a (one-dimensional standard) Brownian motion if*
   (i) $\{B_t\}_{t\geq 0}$ *is continuous,* $B_0 = 0$ *a.s.,*
   (ii) $B_t - B_s$ *is independent of* $\mathcal{F}_s^B$,
   (iii) $P(B_t - B_s \in A) = \int_A \frac{1}{\sqrt{2\pi(t-s)}} \exp\{-\frac{x^2}{2(t-s)}\}dx$, *for* $t > s$, $A \in \mathcal{B}(\mathbf{R})$.

(b) *The N-dimemsional stochastic process* $B(t) = (B_1(t), B_2(t), \ldots, B_N(t))$ *is a (standard) N-dimensional Brownian motion if the N-components* $B_i(t)$ *are independent one-dimensional standard Brownian motions.*

Proposition 1.2.7. *A continuous stochastic process* $\{B_t\}_{t\geq 0}$ *with* $B_0 = 0$ *is an N-dimensional Brownian motion if and only if*

$$E[e^{i(\xi, B_t - B_s)}|\mathcal{F}_s^B] = e^{-\frac{|\xi|^2}{2}(t-s)}, \quad t > s, \quad \xi \in \mathbf{R}^N,$$

where $i = \sqrt{-1}$, $(\ ,\ )$ *denotes the inner product of* $\mathbf{R}^N$, *and* $|\xi| = (\xi, \xi)^{1/2}$.

Proof. For $t > s$, the random variable $Y := B_t - B_s$ is normally distributed with mean 0 and covariance $(t - s)I$ (I : identity), that is to say, the characteristic function of Y is given by

$$E[e^{i(\xi, Y)}] = e^{-\frac{|\xi|^2}{2}(t-s)}.$$

Let Y be independent of $\mathcal{F}_s^B$. Then it is easy to see that

$$E[e^{i(\xi, Y)}|\mathcal{F}_s^B] = E[e^{i(\xi, Y)}].$$

Conversely,

$$E[e^{i(\xi, Y)}e^{i(\eta, Z)}] = E[E[e^{i(\xi, Y)}|\mathcal{F}_s^B]e^{i(\eta, Z)}] = E[e^{i(\xi, Y)}]E[e^{i(\eta, Z)}],$$

for any $\eta \in \mathbf{R}^N$ and $\mathcal{F}_s^B$-measurable random variable Z. Thus we conclude that Y and Z are independent. ■

Remark 1.2.8. *The existence of a Brownian motion can be shown by introducing the probability measure P, called a Wiener measure, on the space* $C([0, \infty) : \mathbf{R}^N)$ *of* $\mathbf{R}^N$*-valued contiunuous functions on* $[0, \infty)$. *See I. Karatzas and S. E. Shreve [87] and K. Ito and H. P. Mackean [75] for details. The remarkable properties of the Brownian motion are as follows:*

(a) *The Brownian motion* $\{B_t\}$ *is not differentiable a.s.,*
(b) *The total variation of* $\{B_t\}$ *on* $[0, T]$ *is infinite a.s.*

1.3 Stopping Times

Definition 1.3.1. *Let* $(\Omega, \mathcal{F}, P, \{\mathcal{F}_t\}_{t\geq 0})$ *be a stochastic basis. A map* $\tau : \Omega \to [0, \infty]$ *is a stopping time if*

$$\{\omega : \tau(\omega) \leq t\} \in \mathcal{F}_t, \quad \text{for all } t \geq 0.$$

For any stopping time τ*, the* σ*-algebra* $\mathcal{F}_\tau$ *is defined by*

$$\mathcal{F}_\tau = \{A \in \mathcal{F} : A \cap \{\tau \leq t\} \in \mathcal{F}_t, \text{ for all } t \geq 0\}.$$

Proposition 1.3.2. *Let* $(\Omega, \mathcal{F}, P, \{\mathcal{F}_t\}_{t\geq 0})$ *be a stochastic basis. Let* $\{X_t\}_{t\geq 0}$ *be a continuous,* $\{\mathcal{F}_t\}_{t\geq 0}$*-adapted,* $\mathbf{R}^N$*-valued process and* A *an open or closed subset of* $\mathbf{R}^N$*. Then*

$$\tau(\omega) = \inf\{t \geq 0 : X_t(\omega) \in A\}, \quad \text{with understanding } \inf\{\phi\} = \infty,$$

is a stopping time.

Proof. Let $\Omega_0 = \{t \to X_t(\omega) \text{ continuous}\}$ and $P(\Omega_0) = 1$.

(1) Let A be open. If $\tau(\omega) < t$ and $\omega \in \Omega_0$, then there is $s < t$ such that $X_s(\omega) \in A$. Taking $r_n \in \mathbf{Q}_+$ such that $r_n \searrow s$, we have $X_{r_n}(\omega) \in A$, for some $r_n < t$. Hence,

$$\{\tau < t\} \cap \Omega_0 = \bigcup_{r<t,\ r\in\mathbf{Q}_+} (\{X_r \in A\} \cap \Omega_0) \in \mathcal{F}_t.$$

Thus, by the completeness of $\mathcal{F}_t$,

$$\{\tau \leq t\} = \bigcap_{n\in\mathbf{N}} \{\tau < t + 1/n\} \in \mathcal{F}_{t+} = \mathcal{F}_t, \quad \text{for all } t \geq 0.$$

(2) Let A be closed. For each $n \in \mathbf{N}$, the set $A_n := \{x : d(x, A) < 1/n\}$ is open, where $d(x, A) = \inf\{|x - y| : y \in A\}$. Let $X_{r_m}(\omega) \in A_n$, $\omega \in \Omega_0$, and $r_m \to s \leq t$. Then

$$d(X_{r_m}(\omega), A) \leq d(X_{r_m}(\omega), A_n) + \frac{1}{n}.$$

Passing to the limit, we get $d(X_s(\omega), A) = 0$. This implies $X_s(\omega) \in A$, and then $\tau(\omega) \leq s$. Hence

$$\{\tau \leq t\} \cap \Omega_0 = \bigcap_{n\in\mathbf{N}} \bigcup_{r\in([0,t)\cap\mathbf{Q})\cup\{t\}} (\{X_r \in A_n\} \cap \Omega_0) \in \mathcal{F}_t.$$

By the same argument as above, we have $\{\tau \leq t\} \in \mathcal{F}_t$, for all $t \geq 0$. ∎

Proposition 1.3.3. *Let* $\tau, \sigma,$ *and* $\tau_n, n = 1, 2, \ldots,$ *be stopping times. Then the following assertions hold:*

(i) $\tau \wedge \sigma, \tau \vee \sigma,$ *and* $\tau + a$ $(a \in \mathbf{R}_+)$ *are stopping times.*
(ii) τ *is* $\mathcal{F}_\tau$*-measurable.*
(iii) *If* $\tau \leq \sigma$*, then* $\mathcal{F}_\tau \subset \mathcal{F}_\sigma$.
(iv) *If* $A \in \mathcal{F}_\tau$*, then* $A \cap \{\tau \leq \sigma\} \in \mathcal{F}_\sigma$.
(v) $\mathcal{F}_\tau \cap \mathcal{F}_\sigma = \mathcal{F}_{\tau\wedge\sigma}$.

(vi) $\{\tau < \sigma\}, \{\tau = \sigma\}, \{\tau > \sigma\} \in \mathcal{F}_{\tau\wedge\sigma}$.
(vii) $\sup_n \tau_n$ *and* $\inf_n \tau_n$ *are stopping times.*

For the proof, see I. Karatzas and S. E. Shreve [87, section 1.1.2, pp. 6–11].

Definition 1.3.4. *Let $X = \{X_t\}_{t\geq 0}$ be a measurable process on a stochastic basis $(\Omega, \mathcal{F}, P, \{\mathcal{F}_t\}_{t\geq 0})$, and let τ be a stopping time. We define the random variable X_τ by $X_\tau(\omega) = X(\tau(\omega), \omega)$ on $\{\tau < \infty\}$, and $X_\tau(\omega) = X_\infty(\omega)$ on $\{\tau = \infty\}$ if $X_\infty(\omega)$ is defined for all $\omega \in \Omega$.*

Proposition 1.3.5. *Let $X = \{X_t\}_{t\geq 0}$ be an adapted process on a stochastic basis $(\Omega, \mathcal{F}, P, \{\mathcal{F}_t\}_{t\geq 0})$, and let τ be a stopping time. If X is progressively measurable, then X_τ is $\mathcal{F}_\tau$-measurable and the stopped process $X^\tau = \{X_{t\wedge\tau}\}_{t\geq 0}$ at τ is progressively measurable.*

Proof.

(1) For fixed $t \geq 0$, it is easy to see that $\tau \wedge t$ is an $\mathcal{F}_t$-measurable random variable. Hence the map $(s, \omega) \in [0, t] \times \Omega \to H(s, \omega) := (\tau(\omega) \wedge s, \omega) \in [0, t] \times \Omega$ is $\mathcal{B}([0, t]) \otimes \mathcal{F}_t$ measurable. Thus the map $(s, \omega) \to X(\tau(\omega) \wedge s, \omega) = X \circ H(s, \omega)$ is $\mathcal{B}([0, t]) \otimes \mathcal{F}_t$ measurable, so $\{X_{\tau\wedge t}\}$ is progressively measurable.
(2) Let $B \in \mathcal{B}(\mathbf{R}^N)$. Taking $s = t$ in (1), the map $\omega \in \Omega \to X \circ H(t, \omega)$ is $\mathcal{F}_t$-measurable. Hence

$$\{X_\tau \in B\} \cap \{\tau \leq t\} = \{X_{\tau\wedge t} \in B\} \cap \{\tau \leq t\} \in \mathcal{F}_t, \qquad \text{for all } t \geq 0.$$

This implies that $\{X_\tau \in B\} \in \mathcal{F}_\tau$, that is, X_τ is $\mathcal{F}_\tau$ measurable. ∎

1.4 Martingales

Let $(\Omega, \mathcal{F}, P, \{\mathcal{F}_t\}_{t\geq 0})$ be a stochastic basis.

Definition 1.4.1. *A stochastic process $X = \{(X_t, \mathcal{F}_t)\}_{t\geq 0}$ is a martingale (supermartingale submartingale) if $E[|X_t|] < \infty$, for all $t \geq 0$, and*

$$E[X_t|\mathcal{F}_s] = X_s, \quad P\text{-a.s.}, \quad \text{for any } s \leq t,$$

(if $E[X_t|\mathcal{F}_s] \leq X_s$ or $E[X_t|\mathcal{F}_s] \geq X_s$).

Proposition 1.4.2. *Let $\{B_t\}$ is a one-dimensional Brownian motion. Then*

(i) $(B_t, \mathcal{F}_t^B)$ *is a martingale,*
(ii) $(B_t^2 - t, \mathcal{F}_t^B)$ *is a martingale.*

Proof. Let $t \geq s$.

(1) $E[B_t - B_s|\mathcal{F}_s^B] = E[B_t - B_s] = 0, \quad$ a.s.

(2) By Proposition 1.1.19 (x), we have

$$\begin{aligned} E[B_t^2 - B_s^2|\mathcal{F}_s^B] &= E[\{(B_t - B_s) + B_s\}^2 - B_s^2|\mathcal{F}_s] \\ &= E[(B_t - B_s)^2 + 2(B_t - B_s)B_s|\mathcal{F}_s^B] \\ &= E[(B_t - B_s)^2] = t - s, \quad \text{a.s.} \end{aligned}$$

■

Theorem 1.4.3 (Doob's maximal inequality). *Let $\{(X_t, \mathcal{F}_t)\}_{t\geq 0}$ be a right-continuous stochastic process and $0 < T < \infty$, $\lambda > 0$. Then we have*

(i) *If $\{X_t\}$ is a nonnegative submartingale, then*

$$P(\sup_{0\leq t\leq T} X_t \geq \lambda) \leq \frac{1}{\lambda} E[X_T].$$

(ii) *If $\{X_t\}$ is a nonnegative submartingale or a martingale, then*

$$E[\sup_{0\leq t\leq T} |X_t|^p] \leq (\frac{p}{p-1})^p E[|X_T|^p], \quad \text{for } 1 < p < \infty.$$

Proof. Let D be a countable dense subset of $[0, T]$ defined by $[0, T) \cap \mathbf{Q}$, and let $D_n = \{t_i \in D : t_0 = 0 \leq t_1 \leq \cdots \leq t_n, i = 0, 1, \ldots, n\}$ be a sequence of subsets of D such that $D_n \nearrow \bigcup_n D_n = D$ as $n \to \infty$. By right continuity, we observe that

$$\sup_{0\leq t\leq T} X_t = \sup_{t\in D} X_t = \lim_{n\to\infty} \max_{t\in D_n} X_t, \quad \text{a.s.},$$

and it is $\mathcal{F}_T$-measurable. Furthermore,

$$\begin{aligned} P(\sup_{t\in D} X_t \geq \lambda) &= \lim_{m\to\infty} P(\sup_{t\in D} X_t > \lambda - \frac{1}{m}) \\ &= \lim_{m\to\infty} \lim_{n\to\infty} P(\max_{t\in D_n} X_s > \lambda - \frac{1}{m}). \end{aligned}$$

(1) We claim that

$$P(\max_{t\in D_n} X_t > \lambda) \leq \frac{1}{\lambda} E[X_T].$$

Put $Y_i = X_{t_i}$ and $\mathcal{G}_i = \mathcal{F}_{t_i}$, for $i = 0, 1, \ldots, n$. Let us define

$$\tau = \min\{i \leq n : Y_i > \lambda\}, \ = n \quad \text{if } \{\,\} = \phi.$$

Then τ is a $\mathcal{G}_i$-stopping time such that $\tau \leq n$ a.s. Since $\{Y_i\}$ is a $\mathcal{G}_i$-submartingale,

$$\begin{aligned} E[Y_\tau] &= \sum_{i=0}^n E[Y_i 1_{\{\tau=i\}}] \leq \sum_{i=0}^n E[E[Y_n|\mathcal{G}_i]1_{\{\tau=i\}}] \\ &= \sum_{i=0}^n E[Y_n 1_{\{\tau=i\}}] = E[Y_n] \leq E[X_T]. \end{aligned}$$

We set $Y_n^* = \max_{i \le n} Y_i = \max_{t \in D_n} X_t$ and note that $Y_\tau > \lambda$ on $\{Y_n^* > \lambda\}$ and $\{\tau = n\}$ on $\{Y_n^* \le \lambda\}$. Therefore,

$$\begin{aligned} E[Y_\tau] &= E[Y_\tau 1_{\{Y_n^* > \lambda\}}] + E[Y_\tau 1_{\{Y_n^* \le \lambda\}}] \\ &\ge \lambda P(Y_n^* > \lambda) + E[Y_n 1_{\{Y_n^* \le \lambda\}}], \end{aligned}$$

which implies

$$\lambda P(\max_{t \in D_n} X_t > \lambda) \le E[Y_n] - E[Y_n 1_{\{Y_n^* \le \lambda\}}] = E[Y_n 1_{\{Y_n^* > \lambda\}}] \le E[X_T].$$

(2) Let $E[|X_T|^p] < \infty$. For any $k > 0$,

$$E[(Y_n^* \wedge k)^p] = \int_0^\infty x^p dP(Y_n^* \wedge k \le x) = p \int_0^\infty x^{p-1} P(Y_n^* \wedge k > x) dx.$$

Letting $k \to \infty$, by (1), Fubini's theorem and Hölder's inequality, we have

$$\begin{aligned} E[(Y_n^*)^p] &= p \int_0^\infty x^{p-1} P(Y_n^* > x) dx \le p \int_0^\infty x^{p-2} E[Y_n 1_{\{Y_n^* > x\}}] dx \\ &= pE[Y_n \int_0^{Y_n^*} x^{p-2} dx] = \frac{p}{p-1} E[Y_n (Y_n^*)^{p-1}] \\ &= qE[Y_n (Y_n^*)^{p-1}] \le qE[Y_n^p]^{1/p} E[(Y_n^*)^{(p-1)q}]^{1/q} \\ &= qE[Y_n^p]^{1/p} E[(Y_n^*)^p]^{1/q}, \end{aligned}$$

where $1/p + 1/q = 1$. Put $y = E[(Y_n^*)^p]$. Then y satisfies $y^q \le ay$ with $q > 1$, for some $a > 0$, and thus y is finite. Therefore,

$$E[(Y_n^*)^p]^{1/p} \le qE[Y_n^p]^{1/p}.$$

By Fatou's lemma and Proposition 1.1.19 (viii), we deduce

$$E[\sup_{t \in D} (X_t)^p]^{1/p} \le \liminf_{n \to \infty} E[(Y_n^*)^p]^{1/p} \le \liminf_{n \to \infty} qE[Y_n^p]^{1/p} \le qE[X_T^p]^{1/p}.$$

When $\{X_t\}$ is a martingale, we apply this inequality to a nonnegative submartingale $\{|X_t|\}$ to obtain the result. ∎

Theorem 1.4.4 (Optional sampling theorem). *Let $(X_t, \mathcal{F}_t)$ be a right-continuous martingale (or submartingale) and let τ, σ be two bounded stopping times. Then*

$$E[X_\tau | \mathcal{F}_\sigma] = X_{\tau \wedge \sigma}, \quad P\text{-}a.s. \quad (\text{or } E[X_\tau | \mathcal{F}_\sigma] \ge X_{\tau \wedge \sigma}, \quad P\text{-}a.s.)$$

Proof. We will show only the case when $\{X_t\}$ is a submartingale.

(1) Suppose that $\sigma \le \tau \le T$ a.s. and σ, τ take values in a finite subset $\delta_n := \{t_i : t_0 = 0 < t_1 < \cdots < t_n, i = 0, 1, \ldots, n\}$ of $[0, T]$. Put $Y_i = X_{t_i}$ for each i. Let $A \in \mathcal{F}_\sigma$ and $A_i = A \cap \{\sigma = t_i\}$. For $j \ge i$, $A_i \cap \{\tau \ge t_{j+1}\} \in \mathcal{F}_{t_j}$. Hence, by the submartingale property,

$$\begin{aligned} E[Y_j 1_{A_i \cap \{\tau \ge t_j\}}] &= E[Y_j 1_{A_i \cap \{\tau = t_j\}}] + E[Y_j 1_{A_i \cap \{\tau \ge t_{j+1}\}}] \\ &\le E[X_\tau 1_{A_i \cap \{\tau = t_j\}}] + E[Y_{j+1} 1_{A_i \cap \{\tau \ge t_{j+1}\}}]. \end{aligned}$$

Taking the summation over j with $t_{n+1} > T$, we have

$$E[Y_i 1_{A_i \cap \{\tau \geq t_i\}}] \leq \sum_{j=i}^{n} E[X_\tau 1_{A_i \cap \{\tau = t_j\}}] = E[X_\tau 1_{A_i \cap \{\tau \geq t_i\}}],$$

and then

$$E[X_\sigma 1_{A_i}] = E[X_\sigma 1_{A_i \cap \{\tau \geq \sigma\}}] \leq E[X_\tau 1_{A_i \cap \{\tau \geq \sigma\}}] = E[X_\tau 1_{A_i}].$$

Taking the summation over i, we deduce $E[X_\sigma 1_A] \leq E[X_\tau 1_A]$, that is, $X_\sigma \leq E[X_\tau | \mathcal{F}_\sigma]$ a.s.

(2) Suppose $\sigma \leq \tau \leq T$ a.s. We take a finite set δ_n for each n such that $t_i = iT/2^n, i = 0, 1, \ldots, 2^n$. Define

$$\rho_n = \sum_{i=1}^{2^n} t_i 1_{\{t_{i-1} \leq \rho < t_i\}} + T 1_{\{\rho = T\}} \quad \text{for} \quad \rho = \tau \text{ or } \sigma.$$

Then ρ_n is a stopping time taking values in δ_n, $\rho_n \searrow \rho$, and $\tau_n \geq \sigma_n$. By right continuity, $\lim_{n\to\infty} X_{\tau_n} = X_\tau$ and $\lim_{n\to\infty} X_{\sigma_n} = X_\sigma$ hold a.s. We apply (1) to obtain

$$E[X_{\tau_n} | \mathcal{F}_{\sigma_n}] \geq X_{\sigma_n}, \quad \text{a.s.},$$

and

$$E[X_{\tau_n} 1_A] \geq E[X_{\sigma_n} 1_A] \quad \text{for } A \in \mathcal{F}_\sigma.$$

By (3) below, Proposition 1.1.15 is applicable and so $\lim_{n\to\infty} X_{\rho_n} = X_\rho \in L^1(\Omega)$. Passing to the limit, we deduce $E[X_\tau 1_A] \geq E[X_\sigma 1_A]$.

(3) We claim that $\{X_{\rho_n}\}$ is uniformly integrable. By Proposition 1.1.19 (viii), $\{|X_t|\}$ is a nonnegative submartingale. By Doob's maximal inequality,

$$P(\sup_n |X_{\rho_n}| \geq c) \leq P(\sup_{0 \leq t \leq T} |X_t| \geq c) \leq \frac{1}{c} E[|X_T|] \to 0 \quad \text{as } c \to \infty.$$

Thus, by (1)

$$\begin{aligned}\sup_n E[|X_{\rho_n}| 1_{\{|X_{\rho_n}| \geq c\}}] &\leq \sup_n E[|X_T| 1_{\{|X_{\rho_n}| \geq c\}}] \\ &\leq E[|X_T| 1_{\{\sup_n |X_{\rho_n}| \geq c\}}] \to 0 \quad \text{as } c \to \infty.\end{aligned}$$

(4) In the general case, by (2) and Proposition 1.3.3 (vi), we have

$$\begin{aligned}E[X_\tau | \mathcal{F}_\sigma] 1_{\{\tau \geq \sigma\}} &= E[X_{\tau \vee \sigma} 1_{\{\tau \geq \sigma\}} | \mathcal{F}_\sigma] = E[X_{\tau \vee \sigma} | \mathcal{F}_\sigma] 1_{\{\tau \geq \sigma\}} \\ &\geq X_\sigma 1_{\{\tau \geq \sigma\}} = X_{\tau \wedge \sigma} 1_{\{\tau \geq \sigma\}}.\end{aligned}$$

Furthermore,

$$E[X_\tau | \mathcal{F}_\sigma] 1_{\{\tau < \sigma\}} = E[X_{\tau \wedge \sigma} 1_{\{\tau < \sigma\}} | \mathcal{F}_\sigma] = X_{\tau \wedge \sigma} 1_{\{\tau < \sigma\}}.$$

By addition, we obtain the desired inequality. ■

Remark 1.4.5. *Let $\{X_t\}$ be a right-continuous submartingale such that $E[\sup_{t\geq 0}|X_t|] < \infty$. Then the optional sampling theorem holds by the uniform integrability of $\{X_{\rho\wedge t}\}$ for $\rho = \tau$ or σ if τ, σ are two stopping times such that $P(\tau, \sigma < \infty) = 1$. See I. Karatzas and S. E. Shreve [87] for weaker conditions and the submartingale convergence theorem.*

1.5 Stochastic Integrals

1.5.1 Definition of the Stochastic Integral

Let $B = \{(B_t, \mathcal{F}_t)\}_{t\geq 0}$ be a one-dimensional Brownian motion on a stochastic basis $(\Omega, \mathcal{F}, P, \{\mathcal{F}_t\}_{t\geq 0})$, that is to say, $\{(B_t, \mathcal{F}_t)\}_{t\geq 0}$ is a continuous martingale with $B_0 = 0$ such that

$$\{(B_t^2 - t, \mathcal{F}_t)\}_{t\geq 0} \text{ is a martingale.}$$

We note by Theorem 1.6.2 below, due to Lévy, that $\{B_t\}_{t\geq 0}$ is a Brownian motion. We will construct the stochastic integral, denoted by $I(f) = \int_0^T f_t dB_t$, for appropriate stochastic processes $\{(f_t, \mathcal{F}_t)\}_{t\geq 0}$ and $T > 0$.

Definition 1.5.1.

(a) *For $p \geq 1$, let $\mathcal{M}_p(0, T)$ be the space of progressively measurable processes $f = \{(f_t, \mathcal{F}_t)\}$ such that*

$$E[\int_0^T |f_t|^p dt] < \infty, \quad \textit{for fixed } T > 0.$$

(b) *Let $\mathcal{M}_0(0, T)$ be the space of step processes $\phi \in \mathcal{M}_2(0, T)$, that is, there is a partition $\{t_0 = 0 < t_1 < t_2 < \cdots < t_n = T\}$ such that*

$$\phi_t = \phi_{(0)} 1_{\{0\}}(t) + \sum_{i=1}^n \phi_{(i)} 1_{(t_{i-1}, t_i]}(t), \quad \textit{for some } \phi_{(i)} \in L^\infty(\Omega, \mathcal{F}_{t_{i-1}}, P).$$

(c) *We define the stochastic integral $I(\phi)$ for $\phi \in \mathcal{M}_0(0, T)$ by*

$$I(\phi) = I_T(\phi) = \sum_{i=1}^n \phi_{(i)}(B_{t_i} - B_{t_{i-1}}).$$

Lemma 1.5.2. *The space $\mathcal{M}_0(0, T)$ is dense in $\mathcal{M}_2(0, T)$.*

Proof. Let $f = \{f_t\} \in \mathcal{M}_2(0, T)$, and we will show that there exists a sequence $\phi^{(n)} \in \mathcal{M}_0(0, T)$ such that

$$E[\int_0^T |f_t - \phi_t^{(n)}|^2 dt] \to 0 \quad \text{as } n \to \infty.$$

(1) Suppose that $\{f_t\}$ is continuous and bounded. Define

$$\phi_t^{(n)} = f_{t_0} 1_{\{t=0\}} + \sum_{k=1}^{n} f_{t_{k-1}} 1_{(t_{k-1},t_k]}(t), \qquad \text{for } t_k = kT/n,\ k = 0, 1, \ldots, n.$$

Then $\{\phi^{(n)}\} \in \mathcal{M}_0(0,T)$ and $\phi_t^{(n)} \to f_t$ for each t a.s. By the dominated convergence theorem, we have $\lim_{n\to\infty} E[\int_0^T |f_t - \phi_t^{(n)}|^2 dt] = 0$.

(2) Let $\{f_t\}$ be progressively measurable and bounded. Put $h(t,\omega) = \int_0^t f(s,\omega)ds$. By progressive measurability, $\{h_t\}$ is a measurable process, and the random variable $h(t,\omega)$ is $\mathcal{F}_t$-measurable for each $t \in [0,T]$. Define

$$h^\varepsilon(t,\omega) = \frac{1}{\varepsilon}\int_{(t-\varepsilon)^+}^{t} f(s,\omega)ds = \frac{h(t,\omega) - h((t-\varepsilon)^+,\omega)}{\varepsilon}, \qquad \text{for } \varepsilon > 0.$$

By Lebesgue's Theorem, there exists a derivative $h'(t,\omega)$ for almost all $t \in [0,T]$ a.s., and

$$\lim_{\varepsilon\to 0} |h'(t,\omega) - h^\varepsilon(t,\omega)| = \lim_{\varepsilon\to 0} |f(t,\omega) - h^\varepsilon(t,\omega)| = 0, \quad \text{a.s.}$$

Hence

$$\lim_{\varepsilon\to 0} \int_0^T |h^\varepsilon(t,\omega) - f(t,\omega)|^2 dt = 0, \quad \text{a.s.},$$

and

$$\lim_{\varepsilon\to 0} E[\int_0^T |h^\varepsilon(t,\omega) - f(t,\omega)|^2 dt] = 0.$$

(3) In the general case, we define the bounded progressively measurable process $f^{(m)} \in \mathcal{M}_2(0,T)$ by

$$f_t^{(m)} = f_t 1_{\{|f_t|\le m\}}, \qquad \text{for each } m \in \mathbf{N}.$$

By the dominated convergence theorem, we have

$$E[\int_0^T |f_t - f_t^{(m)}|^2 dt] \to 0 \quad \text{as } m \to \infty.$$

Applying (2) to obtain a sequence of continuous bounded processes $h^{(m,\varepsilon)}$ approximating $f^{(m)}$, and then (1) to $h^{(m,\varepsilon)}$, we obtain the assertion. ■

Remark 1.5.3. *Lebesgue's Theorem states that if $f \in L^1[a,b]$, then*

$$\lim_{h\to 0} \frac{1}{h}\int_x^{x+h} f(t)dt = f(x) \quad \textit{for almost all } x \in [a,b].$$

For the proof, see H. L. Royden [139, chapter 5, Theorems 2 and 9].

Theorem 1.5.4. *Let $f \in \mathcal{M}_2(0,T)$ and let $\phi^{(m)} \in \mathcal{M}_0(0,T)$, $m = 1, 2, \ldots,$ be such that*

$$E[\int_0^T |f_t - \phi_t^{(m)}|^2 dt] \to 0 \quad \textit{as } m \to \infty.$$

Then $I(\phi^{(m)})$ converges to some $I(f)$ in $L^2(\Omega, \mathcal{F}_T, P)$, independent of a choice of $\phi^{(m)}$.

Proof.

(1) We claim that for $\phi \in \mathcal{M}_0(0, T)$,

$$E[|I(\phi)|^2] = E[\int_0^T |\phi_t|^2 dt].$$

For brevity, we set $\Delta_i B = B_{t_i} - B_{t_{i-1}}$ and $\Delta_i t = t_i - t_{i-1}$. Then

$$|I(\phi)|^2 = \sum_{i=0}^{n} \phi_{(i)}^2 (\Delta_i B)^2 + 2 \sum_{i<j} \phi_{(i)}\phi_{(j)}(\Delta_i B)(\Delta_j B).$$

In view of Proposition 1.4.2

$$E[\phi_{(i)}^2(\Delta_i B)^2|\mathcal{F}_{t_{i-1}}] = \phi_{(i)}^2 E[(\Delta_i B)^2|\mathcal{F}_{t_{i-1}}] = \phi_{(i)}^2 \, \Delta_i t, \quad \text{a.s.},$$

and

$$E[\phi_{(i)}\phi_{(j)}(\Delta_i B)(\Delta_j B)|\mathcal{F}_{t_{j-1}}] = \phi_{(i)}\phi_{(j)}(\Delta_i B)E[\Delta_j B|\mathcal{F}_{t_{j-1}}] = 0, \quad \text{a.s.}, \quad i < j.$$

Therefore,

$$E[|I(\phi)|^2] = \sum_{i=1}^{n} E[\phi_{(i)}^2] \, \Delta_j t.$$

On the other hand,

$$|\phi_t|^2 = \phi_{(0)}^2 1_{\{0\}}(t) + \sum_{i=1}^{n} \phi_{(i)}^2 1_{(t_{i-1}, t_i]}(t).$$

This implies that

$$E[\int_0^T |\phi_t|^2 dt] = E[\sum_{i=1}^{n} \phi_{(i)}^2 \, \Delta_i t] = E[|I(\phi)|^2],$$

as required.

(2) Let $f \in \mathcal{M}_2(0, T)$. It is clear that

$$E[\int_0^T |\phi_t^{(m)} - \phi_t^{(n)}|^2 dt]^{1/2} \leq E[\int_0^T |\phi_t^{(m)} - f_t|^2 dt]^{1/2} + E[\int_0^T |f_t - \phi_t^{(n)}|^2 dt]^{1/2} \to 0 \quad \text{as } n, m \to \infty.$$

Hence, by (1)

$$E[|I(\phi^{(m)}) - I(\phi^{(n)})|^2] = E[\int_0^T |\phi_t^{(m)} - \phi_t^{(n)}|^2 dt] \to 0 \quad \text{as } n, m \to \infty.$$

By the completeness of $L^2(\Omega)$, there exists $I(f) \in L^2(\Omega)$ such that

$$E[|I(\phi^{(n)}) - I(f)|^2] \to 0 \quad \text{as } n \to \infty.$$

(3) Let $\psi^{(m)} \in \mathcal{M}_0(0,T)$ be another sequence approximating f. Then the sequence $\{\phi^{(1)}, \psi^{(1)}, \phi^{(2)}, \psi^{(2)}, \ldots\}$ approximates f too, so the sequence $\{I(\phi^{(1)}), I(\psi^{(1)}), I(\phi^{(2)}), I(\psi^{(2)}), \ldots\}$ has a limit $\hat{I}(f) \in L^2(\Omega)$. Thus two subsequences $\{I(\phi^{(1)}), I(\phi^{(2)}), \ldots\}$ and $\{I(\psi^{(1)}), I(\psi^{(2)}), \ldots\}$ have the same limit $\hat{I}(f)$, and $I(f) = \hat{I}(f)$. ■

Definition 1.5.5. *Let $(f_t, \mathcal{F}_t) \in \mathcal{M}_2(0,T)$ and $0 \le a \le b \le T$. We define the stochastic integral:*

$$\int_0^T f_t dB_t := I_T(f) = I(f), \quad \text{and} \quad \int_a^b f_t dB_t = \int_0^T 1_{[a,b]}(t) f_t dB_t.$$

1.5.2 Basic Properties

Theorem 1.5.6. *Let $\{f_t\}, \{g_t\} \in \mathcal{M}_2(0,T)$, $0 \le a \le b \le T$. Then*

(i) $E[\int_a^b f_t dB_t | \mathcal{F}_a] = 0$ *a.s.*,

(ii) $E[(\int_a^b f_t dB_t)^2 | \mathcal{F}_a] = E[(\int_0^b f_t dB_t)^2 - (\int_0^a f_t dB_t)^2 | \mathcal{F}_a]$
$= E[\int_a^b |f_t|^2 dt | \mathcal{F}_a]$ *a.s.*,

(iii) $\int_a^b (\alpha f_t + \beta g_t) dB_t = \alpha \int_a^b f_t dB_t + \beta \int_a^b g_t dB_t$ *a.s.* $(\alpha, \beta \in \mathbf{R})$,

(iv) $\int_a^b f_t dB_t = \int_a^c f_t dB_t + \int_c^b f_t dB_t$ *a.s.* $(a < c < b)$.

Proof. By Definition 1.5.1(c), we observe these properties of stochastic integrals if $\{f_t\}, \{g_t\} \in \mathcal{M}_0(0,T)$. In particular, (i) and (ii) are immediate from Proposition 1.4.2. In the general case, we take the step processes in $\mathcal{M}_0(0,T)$ approximating $\{f_t\}, \{g_t\}$. Passing to the limit, by Theorem 1.5.4, we obtain the assertion. ■

Theorem 1.5.7. *Let $\{f_t\} \in \mathcal{M}_2(0,T)$. Then there exists a continuous $\mathcal{F}_t$-adapted modification $\{\zeta_t\}$ of $\{I_t(f)\}$ uniquely.*

Proof. Let $\phi \in \mathcal{M}_0(0,T)$. By the continuity of $\{B_t\}$, it is easy to see that $I_t(\phi)$ is continuous for $0 \le t \le T$ a.s. Choose the sequence $\phi^{(n)} \in \mathcal{M}_0(0,T)$ converging to $f \in \mathcal{M}_2(0,T)$, so that

$$E[\int_0^T |f_s - \phi_s^{(n)}|^2 ds] \to 0 \text{ as } n \to \infty,$$

and

$$E[\int_0^T |\phi_s^{(n+1)} - \phi_s^{(n)}|^2 ds] \le \frac{1}{8^n}.$$

We set $A_n = \{\sup_{0\le t\le T} |I_t(\phi^{(n+1)}) - I_t(\phi^{(n)})| > 1/2^n\}$. Then, by Doob's maximal inequality,

$$P(A_n) \le (2^n)^2 E[|I_T(\phi^{(n+1)} - \phi^{(n)})|^2] = 4^n E[\int_0^T |\phi_s^{(n+1)} - \phi_s^{(n)}|^2 ds].$$

Summing up, we have

$$\sum_{n=1}^{\infty} P(A_n) \le \sum_{n=1}^{\infty} \frac{1}{2^n} \le 1.$$

By the Borel–Cantelli lemma, $P(\limsup_{n\to\infty} A_n) = 0$. Let $\omega \notin \limsup_{n\to\infty} A_n := \cap_{i=1}^{\infty} \cup_{n=i}^{\infty} A_n$. Then, there is i such that

$$\sum_{n=i}^{\infty} \sup_{0\le t\le T} |I_t(\phi^{(n+1)}) - I_t(\phi^{(n)})| \le \sum_{n=i}^{\infty} \frac{1}{2^n} \le 1.$$

Thus

$$\lim_{n\to\infty} I_t(\phi^{(n)}) = I_t(\phi^{(1)}) + \sum_{n=1}^{\infty} \{I_t(\phi^{(n+1)}) - I_t(\phi^{(n)})\}$$

converges uniformly in $t \in [0, T]$ a.s. Define $\{\zeta_t\}$ by $\zeta_t := \lim_{n\to\infty} I_t(\phi^{(n)})$ if $\omega \notin \limsup_{n\to\infty} A_n$. Clearly $\{\zeta_t\}$ is continuous, $\mathcal{F}_t$-adapted and $\zeta_t \in L^2(\Omega)$. Also, by Fatou's lemma,

$$E[|\zeta_t - I_t(f)|^2] \le \liminf_{n\to\infty} E[|I_t(\phi^{(n)}) - I_t(f)|^2] = 0.$$

This yields that $P(\zeta_t = I_t(f)) = 1$ for each $t \in [0, T]$. The uniqueness is obvious. ■

Notation 1.5.8. *Since the continuous modification $\{\zeta_t\}$ has the properties of Theorem 1.5.6, we may identify $I_t(f)$ with ζ_t. While considering the integral $I_t(f)$, $f \in \mathcal{M}_2(0, T)$, we also denote ζ_t by $I_t(f)$ or $\int_0^t f_s dB_s$.*

Definition 1.5.9.

(a) *Let $\mathcal{M}_2$ be the space of all $(f_t, \mathcal{F}_t) \in \mathcal{M}_2(0, T)$ for every $T > 0$.*
(b) *Let $\mathcal{M}_2^{N\times M}$ be the space of all $\{h(t)\} = \{h_{ij}(t)\}$ such that $(h_{ij}(t), \mathcal{F}_t) \in \mathcal{M}_2$ for every $i = 1, 2, \ldots, N$, $j = 1, 2, \ldots, M$.*
(c) *Let $(B(t), \mathcal{F}_t^B)$ be an M-dimensional Brownian motion and $(h(t), \mathcal{F}_t) \in \mathcal{M}_2^{N\times M}$. We define the stochastic integral:*

$$\int_0^t h(s)dB(s)$$
$$= (\sum_{j=1}^{M} \int_0^t h_{1j}(s)dB_j(s), \sum_{j=1}^{M} \int_0^t h_{2j}(s)dB_j(s), \ldots, \sum_{j=1}^{M} \int_0^t h_{Nj}(s)dB_j(s)).$$

Theorem 1.5.10. *Let* $(h_t, \mathcal{F}_t) \in \mathcal{M}_2^{N\times M}$. *Then the following stochastic processes are continuous martingales:*

(i) *Each component of* $(\int_0^t h(s)dB(s), \mathcal{F}_t^B)$,
(ii) $(|\int_0^t h(s)dB(s)|^2 - \int_0^t |h(s)|^2 ds, \mathcal{F}_t^B)$,
(iii) $(\int_0^t h_{ki}(s)dB_i(s) \int_0^t h_{lj}(s)dB_j(s), \mathcal{F}_t^B)$, $i \neq j$.

Proof. By Theorem 1.5.7, the continuity of these processes is immediate.

(1) We recall Proposition 1.2.7. By differentiation

$$\frac{\partial}{\partial \xi_j} E[e^{i(\xi, B(t)-B(s))}|\mathcal{F}_s^B] = iE[\{B_j(t) - B_j(s)\}e^{i(\xi, B(t)-B(s))}|\mathcal{F}_s^B]$$
$$= -e^{-\frac{|\xi|^2}{2}(t-s)}\xi_j(t-s), \quad \text{a.s.}$$

This implies

$$E[B_j(t) - B_j(s)|\mathcal{F}_s^B] = 0, \quad a.s., \quad t \geq s, \quad 1 \leq j \leq N.$$

Thus we get the martingale property (i).

(2) By Theorem 1.5.6 (i) and (ii), we have (ii) if (3) below holds.

(3) By differentiation, we have

$$\frac{\partial^2}{\partial \xi_k \partial \xi_j} E[e^{i(\xi, B(t)-B(s))}|\mathcal{F}_s^B]$$
$$= -E[\{B_k(t) - B_k(s)\}\{B_j(t) - B_j(s)\}e^{i(\xi, B(t)-B(s))}|\mathcal{F}_s^B]$$
$$= \begin{cases} e^{-\frac{|\xi|^2}{2}(t-s)}\{\xi_j^2(t-s)^2 - (t-s)\} & \text{if } j = k, \\ e^{-\frac{|\xi|^2}{2}(t-s)}\xi_k\xi_j(t-s)^2 & \text{if } j \neq k, \quad \text{a.s.} \end{cases}$$

Hence, for $t \geq s$,

$$E[\{B_j(t) - B_j(s)\}^2|\mathcal{F}_s^B] = E[B_j(t)^2 - B_j(s)^2|\mathcal{F}_s^B] = t - s, \quad \text{a.s.},$$
$$E[\{B_k(t) - B_k(s)\}\{B_j(t) - B_j(s)\}|\mathcal{F}_s^B]$$
$$= E[B_i(t)B_j(t) - B_i(s)B_j(s)|\mathcal{F}_s^B] = 0, \quad \text{a.s.}, \quad j \neq k.$$

Furthermore,

$$E[\{B_k(t') - B_k(s')\}\{B_j(t) - B_j(s)\}|\mathcal{F}_s^B] = 0, \quad \text{a.s.},$$
$$t' \geq s' \geq t \geq s, \ j \neq k.$$

Thus, for $\{f_t\}, \{g_t\} \in \mathcal{M}_0(0, T)$, we can see that

$$E[\int_0^t f_r dB_k(r) \int_0^t g_r dB_j(r) - \int_0^s f_r dB_k(r) \int_0^s g_r dB_j(r)|\mathcal{F}_s^B]$$
$$= E[\int_s^t f_r dB_k(r) \int_s^t g_r dB_j(r)|\mathcal{F}_s^B] = 0, \quad \text{a.s.}$$

Consequently, by Theorem 1.5.4, we get (iii). ■

1.5.3 Extension of the Stochastic Integral

Let $\{(B_t, \mathcal{F}_t)\}_{t\geq 0}$ be a one-dimensional Brownian motion on a stochastic basis $(\Omega, \mathcal{F}, P; \{\mathcal{F}_t\})$.

Lemma 1.5.11. *Let $\{f_t\} \in \mathcal{M}_2(0, T)$, $a, \varepsilon > 0$, and let σ be a stopping time. Then*

(i) $\int_0^{\sigma\wedge T} f_t\, dB_t = \int_0^T f_t 1_{\{t\leq\sigma\}}\, dB_t, \quad a.s.$

(ii) $P(\sup_{0\leq t\leq T} |I_t(f)| > \varepsilon) \leq \frac{a}{\varepsilon^2} + P(\int_0^T |f_s|^2 ds > a)$.

Proof.

(1) In case of $\{f_t\} \in \mathcal{M}_0(0, T)$, let $\{\sigma_n\}$ be a sequence of stopping times such that σ_n is a step function and $\sigma_n \searrow \sigma$ if $n \to \infty$ as in Theorem 1.4.4 (2). From Definition 1.5.1 (c) it follows that

$$\int_0^{r\wedge T} f_t\, dB_t = \int_0^T f_t 1_{\{t\leq r\}}\, dB_t, \quad \text{a.s.}, \quad \forall r \geq 0,$$

and hence

$$\int_0^{\sigma_n\wedge T} f_t\, dB_t = \int_0^T f_t 1_{\{t\leq\sigma_n\}}\, dB_t, \quad \text{a.s.}$$

In the general case, let $\{\phi_t^{(n)}\} \in \mathcal{M}_0(0, T)$ be such that

$$\lim_{n\to\infty} E[\int_0^T |\phi_t^{(n)} - f_t|^2 dt] = 0.$$

Then

$$\int_0^{\sigma_n\wedge T} \phi_t^{(n)}\, dB_t = \int_0^T \phi_t^{(n)} 1_{\{t\leq\sigma_n\}}\, dB_t, \quad \text{a.s.}$$

By Theorem 1.5.10 (ii) and the optional sampling theorem, we have

$$E[|\int_0^{\sigma_n\wedge T} (\phi_t^{(n)} - f_t) dB_t|^2] = E[\int_0^{\sigma_n\wedge T} |\phi_t^{(n)} - f_t|^2 dt] \to 0.$$

Also, by the continuity of $\{I_t(f)\}$,

$$\int_0^{\sigma_n\wedge T} f_t\, dB_t \to \int_0^{\sigma\wedge T} f_t\, dB_t, \quad \text{a.s.}$$

Thus

$$\lim_{n\to\infty} E[|\int_0^{\sigma_n\wedge T} \phi_t^{(n)}\, dB_t - \int_0^{\sigma\wedge T} f_t\, dB_t|^2] = 0.$$

On the other hand,

$$E[\int_0^T |\phi_t^{(n)} 1_{\{t\leq\sigma_n\}} - f_t 1_{\{t\leq\sigma\}}|^2 dt] \leq 2\{E[\int_0^T |\phi_t^{(n)} - f_t|^2 1_{\{t\leq\sigma_n\}} dt]$$
$$+ E[\int_0^T |f_t|^2 (1_{\{t\leq\sigma_n\}} - 1_{\{t\leq\sigma\}})^2 dt]\} \to 0.$$

By Theorem 1.5.4

$$\lim_{n\to\infty} E[|\int_0^T \phi_t^{(n)} 1_{\{t\le\sigma_n\}}\, dB_t - \int_0^T f_t 1_{\{t\le\sigma\}}\, dB_t|^2] = 0.$$

Therefore, we obtain (i).

(2) Let $\rho_a = \inf\{t \in [0,T] : \int_0^t |f_s|^2 ds > a\}, = T$ if $\{\ \} = \phi$ and $f_t^{(a)} = f_t 1_{\{t\le\rho_a\}}$. Clearly, ρ_a is a stopping time. By (1), it is easy to see that

$$\{\int_0^T |f_s|^2 ds \le a\} = \{\rho_a = T\} \subset \{|\int_0^{t\wedge\rho_a} f_s^{(a)} dB_s - \int_0^t f_s dB_s| = 0,\ \forall t \le T\}$$
$$= \{\sup_{0\le t\le T} |\int_0^t (f_s^{(a)} - f_s) dB_s| = 0\}.$$

Thus, by Doob's maximal inequality

$$P(\sup_{0\le t\le T} |I_t(f)| > \varepsilon) \le P(\sup_{0\le t\le T} |I_t(f^{(a)})| > \varepsilon) + P(\sup_{0\le t\le T} |I_t(f^{(a)} - f)| > 0)$$
$$\le \frac{1}{\varepsilon^2} E[\{I_T(f^{(a)})\}^2] + P(\int_0^T |f_s|^2 ds > a)$$
$$\le \frac{a}{\varepsilon^2} + P(\int_0^T |f_s|^2 ds > a),$$

which implies (ii). ■

Definition 1.5.12.

(a) *For $p \ge 1$, let $\mathcal{L}_p(0,T)$ be the space of progressively measurable processes $(f_t, \mathcal{F}_t)$ such that*

$$P(\int_0^T |f_t|^p dt < \infty) = 1, \quad \textit{for fixed } T > 0.$$

(b) *Let $\rho_n = \inf\{t \in [0,T] : \int_0^t |f_s|^2 ds > n\}, = T$ if $\{\ \} = \phi$, and let*

$$f_t^{(n)} := f_t 1_{\{t\le\rho_n\}},\ n = 0, 1, 2, \ldots.$$

Theorem 1.5.13. *Let $(f_t, \mathcal{F}_t) \in \mathcal{L}_2(0,T)$. Then $\{I_t(f^{(n)})\}$ converges uniformly in $t \in [0,T]$ to some random variable $I_t(f)$ in probability as $n \to \infty$.*

Proof. Note that $\{f_t^{(n)}\}$ belongs to $\mathcal{M}_2(0,T)$ and $\lim_{n\to\infty} \int_0^T |f_s^{(n)} - f_s|^2 ds = 0$ in probability. Hence

$$P(\int_0^T |f_s^{(n)} - f_s^{(m)}|^2 ds > a) \to 0 \quad \text{as } n, m \to \infty, \quad a > 0.$$

Applying Lemma 1.5.11 (ii), we get

$$P(\sup_{0\le t\le T} |I_t(f^{(n)} - f^{(m)})| > \varepsilon) \le \frac{a}{\varepsilon^2} + P(\int_0^T |f_s^{(n)} - f_s^{(m)}|^2 ds > a), \quad \varepsilon > 0.$$

Letting $n, m \to \infty$ and then $a \to 0$, we deduce

$$\lim_{n,m\to\infty} P(\sup_{0\le t\le T} |I_t(f^{(n)}) - I_t(f^{(m)})| > \varepsilon) = 0.$$

By extracting a subsequence, also denoted by $\{f_t^{(k)}\}$, we have

$$P(\sup_{0\le t\le T} |I_t(f^{(k+1)}) - I_t(f^{(k)})| > \frac{1}{2^k}) < \frac{1}{2^k}.$$

By the Borel–Cantelli lemma,

$$P(\bigcap_{i=1}^{\infty} \bigcup_{k\ge i} \{\sup_{0\le t\le T} |I_t(f^{(k+1)}) - I_t(f^{(k)})| > \frac{1}{2^k}\}) = 0.$$

Hence, there exists $i \in \mathbf{N}$ such that

$$\sup_{0\le t\le T} |I_t(f^{(k+1)}) - I_t(f^{(k)})| \le \frac{1}{2^k} \quad \text{for} \quad k \ge i, \quad \text{a.s.}$$

Thus

$$\lim_{k\to\infty} I_t(f^{(k)}) = I_t(f^{(0)}) + \sum_k \{I_t(f^{(k+1)}) - I_t(f^{(k)})\}$$

converges uniformly in $t \in [0, T]$ to $I_t(f)$ a.s. Hence, its limit $\{I_t(f)\}$ is continuous. Furthermore, letting $n, k \to \infty$, we deduce

$$\begin{aligned} P(\sup_{0\le t\le T} |I_t(f^{(n)}) - I_t(f)| > \varepsilon) &\le P(\sup_{0\le t\le T} |I_t(f^{(n)}) - I_t(f^{(k)})| > \frac{\varepsilon}{2}) \\ &\quad + P(\sup_{0\le t\le T} |I_t(f^{(k)}) - I_t(f)| > \frac{\varepsilon}{2}) \to 0. \end{aligned}$$

By the similar argument to Theorem 1.5.4 (3), we remark that the limit $I_t(f)$ is independent of a sequence approximating $\{f_t\}$. ■

Definition 1.5.14. *Let $\{(B_t, \mathcal{F}_t)\}_{t\ge 0}$ be a one-dimensional Brownian motion on a stochastic basis $(\Omega, \mathcal{F}, P, \{\mathcal{F}_t\})$.*

(a) *For $(f_t, \mathcal{F}_t) \in \mathcal{L}_2(0, T)$, we define the stochastic integral:*
$I_t(f) = \lim_{n\to\infty} I_t(f^{(n)})$ *uniformly in probability.*
(b) *Let $\mathcal{L}_p$, $p \ge 1$, be the space of all $(f_t, \mathcal{F}_t) \in \mathcal{L}_p(0, T)$, for every $T > 0$.*
(c) *For $(f_t, \mathcal{F}_t) \in \mathcal{L}_2$, we define the stochastic integral:*
$\int_0^t f_s dB_s := I_t(f)$ *if* $0 \le t \le T$.

Definition 1.5.15. *A stochastic process $\{(X_t, \mathcal{F}_t)\}_{t\ge 0}$ on a stochastic basis $(\Omega, \mathcal{F}, P, \{\mathcal{F}_t\})$ is a local martingale if there exists a nondecreasing sequence $\{\tau_n\}$ of stopping times with $\tau_n \nearrow \infty$ a.s as $n \to \infty$ such that the stopped process $X_t^{(n)} := X_{t\wedge\tau_n}$ is a martingale for each $n \in \mathbf{N}$.*

Proposition 1.5.16.

(i) *Every right-continuous martingale is a local martingale.*
(ii) *Every right-continuous local matingale* $\{(X_t, \mathcal{F}_t)\}$ *is a martingale if*

$$E[\sup_{t\geq 0} |X_t|] < \infty.$$

(iii) *Every nonnegative right-continuous local martingale is a supermartingale.*

Proof. Let $t \geq s$.

(1) Let $\{X_t\}$ be a right-continuous martingale. Take $\tau_n = n \nearrow \infty$. By the optional sampling theorem,

$$E[X_{t\wedge n}|\mathcal{F}_s] = X_{t\wedge s\wedge n} = X_{s\wedge n}, \quad \text{a.s.}$$

(2) Clearly $X_{\tau_n\wedge r} \to X_r$ a.s., for $r = t$ or s. Since

$$\sup_n E[|X_{\tau_n\wedge r}| 1_{\{|X_{\tau_n\wedge r}|\geq c\}}] \leq E[\sup_r |X_r| 1_{\{\sup_r |X_r| \geq c\}}] \to 0 \quad \text{as } c \to \infty,$$

$\{X_{\tau_n\wedge r}\}$ is uniformly integrable. So, we can apply Proposition 1.1.15 to the relation

$$E[X_{\tau_n\wedge t} 1_A] = E[X_{\tau_n\wedge s} 1_A], \qquad A \in \mathcal{F}_s.$$

Letting $n \to \infty$, we get $E[X_t 1_A] = E[X_s 1_A]$, that is, $\{X_t\}$ is a martingale.
(3) By Fatou's lemma as in Proposition 1.1.19 (vi)

$$\begin{aligned} E[X_t|\mathcal{F}_s] &= E[\lim_{n\to\infty} X_{\tau_n\wedge t}|\mathcal{F}_s] \leq \liminf_{n\to\infty} E[X_{\tau_n\wedge t}|\mathcal{F}_s] \\ &= \liminf_{n\to\infty} X_{\tau_n\wedge s} = X_s, \quad \text{a.s.} \end{aligned}$$

■

Lemma 1.5.17. *Let* $\{f_t\} \in \mathcal{L}_2(0, T)$, $a, \varepsilon > 0$, *and let* σ *be a stopping time. Then*

(i) $\int_0^{\sigma\wedge T} f_t \, dB_t = \int_0^T f_t 1_{\{t\leq\sigma\}} \, dB_t$, *a.s.*
(ii) $P(\sup_{0\leq t\leq T} |I_t(f)| > \varepsilon) \leq \frac{a}{\varepsilon^2} + P(\int_0^T |f_s|^2 ds > a)$.

Proof.

(1) We recall that

$$\int_0^{\sigma_n\wedge T} f_t^{(n)} \, dB_t = \int_0^T f_t^{(n)} 1_{\{t\leq\sigma_n\}} \, dB_t, \quad \text{a.s.},$$

where σ_n is as in the proof of Lemma 1.5.11 (i) and $\{f_t^{(n)}\} \in \mathcal{M}_2(0, T)$ is a sequence such that

$$\lim_{n\to\infty} \int_0^T |f_t^{(n)} - f_t|^2 dt = 0 \quad \text{in probability.}$$

By Theorem 1.5.13, we have

$$P(|\int_0^{\sigma_n \wedge T}(f_t^{(n)} - f_t)\,dB_t| > \varepsilon) \le P(\sup_{0\le r\le T}|\int_0^r (f_t^{(n)} - f_t)\,dB_t| > \varepsilon) \to 0.$$

Also, by the continuity of $\{I_t(f)\}$,

$$\int_0^{\sigma_n \wedge T} f_t\,dB_t \;\to\; \int_0^{\sigma \wedge T} f_t\,dB_t \;\; a.s.$$

Thus

$$\lim_{n\to\infty}\int_0^{\sigma_n \wedge T} f_t^{(n)}\,dB_t = \int_0^{\sigma\wedge T} f_t\,dB_t \quad \text{in probability.}$$

On the other hand,

$$\int_0^T |f_t^{(n)}1_{\{t\le\sigma_n\}} - f_t 1_{\{t\le\sigma\}}|^2 dt \le 2\{\int_0^T |f_t^{(n)} - f_t|^2 1_{\{t\le\sigma_n\}}dt$$
$$+ \int_0^T |f_t|^2(1_{\{t\le\sigma_n\}} - 1_{\{t\le\sigma\}})^2 dt\} \;\to\; 0 \quad \text{in probability.}$$

By Theorem 1.5.13

$$\lim_{n\to\infty}\int_0^T f_t^{(n)}1_{\{t\le\sigma_n\}}\,dB_t = \int_0^T f_t 1_{\{t\le\sigma\}}\,dB_t \quad \text{in probability.}$$

Therefore, we obtain (i).

(2) By the same line as Lemma 1.5.11 (ii), (ii) follows from (i). ■

Theorem 1.5.18. *Let $(f_t, \mathcal{F}_t) \in \mathcal{L}_2$. Then the following stochastic processes are continuous local martingales:*

(i) $(\int_0^t f_s dB_s, \mathcal{F}_t)$,
(ii) $((\int_0^t f_s dB_s)^2 - \int_0^t |f_s|^2 ds, \mathcal{F}_t)$.

Proof. The continuity of these processes is immediate from Theorem 1.5.13. Define $\tau_n = \inf\{t \ge 0 : \int_0^t |f_s|^2 ds > n\}$, $= \infty$ if $\{\ \} = \phi$. Clearly, $\{\tau_n\}$ is a sequence of stopping times such that $\tau_n \nearrow \infty$ as $n \to \infty$. Since $\{f_t 1_{\{t\le\tau_n\}}\}$ belongs to $\mathcal{M}_2$ for each $n \in \mathbf{N}$, we have by Lemma 1.5.17,

$$E[\int_0^{\tau_n\wedge t} f_r dB_r|\mathcal{F}_s] = E[\int_0^t f_r 1_{\{r\le\tau_n\}}dB_r|\mathcal{F}_s] = \int_0^s f_r 1_{\{r\le\tau_n\}}dB_r$$
$$= \int_0^{\tau_n\wedge s} f_r dB_r, \quad \text{a.s.,} \quad t \ge s,$$

which implies (i). Similarly, the martingale property of (ii) is immediate. ■

1.6 Itô's Formula

Let $(B_t, \mathcal{F}_t)$ be a one-dimensional Brownian motion on a stochastic basis $(\Omega, \mathcal{F}, P, \{\mathcal{F}_t\})$. Let $\{a_t\} \in \mathcal{L}_1$ and $\{b_t\} \in \mathcal{L}_2$, that is, $(a_t, \mathcal{F}_t), (b_t, \mathcal{F}_t)$ are progressively measurable processes such that

$$\int_0^T |a_t|dt < \infty, \quad \int_0^T |b_t|^2 dt < \infty, \quad \text{a.s., for every } T > 0.$$

We define

$$X_t = X_0 + \int_0^t a_s ds + \int_0^t b_s dB_s,$$

which is called an *Itô process*, where X_0 is a $\mathcal{F}_0$-measurable random variable.

Theorem 1.6.1 (Itô's formula). *Let* $\varphi \in C^2(\mathbf{R})$. *Then*

$$\varphi(X_t) = \varphi(X_0) + \int_0^t \varphi'(X_s)a_s ds + \int_0^t \varphi'(X_s)b_s dB_s + \frac{1}{2}\int_0^t \varphi''(X_s)b_s^2 ds, \quad a.s.$$

Proof.

(1) By localization, we may assume that $\{|X_t|\}, \{\int_0^t |a_s|ds\}, \{\int_0^t |b_s|^2 ds\}$ are bounded by a constant $C > 0$ and $\varphi, \varphi', \varphi''$ are bounded and uniformly continuous. Indeed, we define

$$\tau_n = \inf\{t \geq 0 : |X_t| + \int_0^t |a_s|ds + \int_0^t |b_s|^2 ds > n\}, \quad = \infty \text{ if } \{\} = \phi.$$

Then $\{\tau_n\}$ is a sequence of stopping times such that $\tau_n \nearrow \infty$ as $n \to \infty$. Thus each stochastic process stopped at τ_n takes its values in a compact set. Suppose that Itô's formula holds for $X_t^{(n)} := X_{t\wedge\tau_n}$. Then

$$X_t^{(n)} = X_0^{(n)} + \int_0^t a_s^{(n)} ds + \int_0^t b_s^{(n)} dB_s, \quad \text{a.s.,}$$

and

$$\begin{aligned}
\varphi(X_t^{(n)}) &= \varphi(X_0^{(n)}) + \int_0^t \varphi'(X_s^{(n)})a_s^{(n)} ds + \int_0^t \varphi'(X_s^{(n)})b_s^{(n)} dB_s \\
&\quad + \frac{1}{2}\int_0^t \varphi''(X_s^{(n)})(b_s^{(n)})^2 ds \\
&= \varphi(X_0^{(n)}) + \int_0^{t\wedge\tau_n} \varphi'(X_s)a_s ds + \int_0^{t\wedge\tau_n} \varphi'(X_s)b_s dB_s \\
&\quad + \frac{1}{2}\int_0^{t\wedge\tau_n} \varphi''(X_s)b_s^2 ds, \quad \text{a.s.,}
\end{aligned}$$

where $d_s^{(n)} = d_s 1_{\{s\leq\tau_n\}}$ for $d_s = a_s$ or b_s. Letting $n \to \infty$, we obtain Itô's formula for X_t.

(2) Let $\delta = \{t_0 = 0 < t_1 < t_2 < \cdots < t_N = t\}$, with $t_i = it/N$, $i = 0, 1, \ldots, N$, be a partition of $[0, t]$. By Taylor's formula,

$$\varphi(X_{t_i}) - \varphi(X_{t_{i-1}}) = \varphi'(X_{t_{i-1}})(X_{t_i} - X_{t_{i-1}}) + \frac{1}{2}\varphi''(\tilde{X}_{t_i})(X_{t_i} - X_{t_{i-1}})^2,$$

where $\tilde{X}_{t_i} = X_{t_{i-1}} + \theta_i(X_{t_i} - X_{t_{i-1}})$, for some $\theta_i \in (0, 1)$. Hence

$$\begin{aligned}
\varphi(X_t) - \varphi(X_0) &= \sum_{i=1}^{N}\{\varphi(X_{t_i}) - \varphi(X_{t_{i-1}})\} \\
&= \sum_{i=1}^{N}\varphi'(X_{t_{i-1}})\int_{t_{i-1}}^{t_i} a_s ds + \sum_{i=1}^{N}\varphi'(X_{t_{i-1}})\int_{t_{i-1}}^{t_i} b_s dB_s \\
&\quad + \frac{1}{2}\sum_{i=1}^{N}\varphi''(\tilde{X}_{t_i})(\int_{t_{i-1}}^{t_i} a_s ds)^2 \\
&\quad + \sum_{i=1}^{N}\varphi''(\tilde{X}_{t_i})(\int_{t_{i-1}}^{t_i} a_s ds)(\int_{t_{i-1}}^{t_i} b_s dB_s) \\
&\quad + \frac{1}{2}\sum_{i=1}^{N}\{\varphi''(\tilde{X}_{t_i}) - \varphi''(X_{t_{i-1}})\}(\int_{t_{i-1}}^{t_i} b_s dB_s)^2 \\
&\quad + \frac{1}{2}\sum_{i=1}^{N}\varphi''(X_{t_{i-1}})(\int_{t_{i-1}}^{t_i} b_s dB_s)^2 \\
&\equiv I_1 + I_2 + I_3 + I_4 + I_5 + I_6.
\end{aligned}$$

(3) Let $\{M_t\}$ be the stopped process of $\{\int_0^t b_s dB_s\}$. Clearly, $\{M_t\}$ is a continuous martingale such that $|M_t| \le C$ for all $t \ge 0$ a.s. Put $\Delta_i M = M_{t_i} - M_{t_{i-1}}$ and $A_t = \sum_{i=1}^{N}(\Delta_i M)^2$. Then

$$A_t^2 = \sum_{i=1}^{N}(\Delta_i M)^4 + 2\sum_{i=1}^{N-1}\sum_{j=i+1}^{N}(\Delta_i M)^2(\Delta_j M)^2.$$

Since

$$\begin{aligned}
E[\sum_{j=i+1}^{N}(\Delta_j M)^2|\mathcal{F}_{t_i}] &= \sum_{j=i+1}^{N} E[M_{t_j}^2 - 2E[M_{t_j}|\mathcal{F}_{t_{j-1}}]M_{t_{j-1}} + M_{t_{j-1}}^2|\mathcal{F}_{t_i}] \\
&= E[\sum_{j=i+1}^{N}(M_{t_j}^2 - M_{t_{j-1}}^2)|\mathcal{F}_{t_i}] \\
&= E[M_{t_N}^2 - M_{t_i}^2|\mathcal{F}_{t_i}] \le C^2, \quad \text{a.s.},
\end{aligned}$$

we have

$$E[\sum_{i=1}^{N-1}\sum_{j=i+1}^{N}(\Delta_i M)^2(\Delta_j M)^2] \le C^2 E[A_t] = C^2 E[M_t^2 - M_0^2] \le C^4.$$

Also,

$$E[\sum_{i=1}^{N}(\Delta_i M)^4] \le 2E[\sum_{i=1}^{N}(\Delta_i M)^2(M_{t_i}^2 + M_{t_{i-1}}^2)] \le 4C^2 E[A_t] \le 4C^4.$$

Thus we get

$$E[\{\sum_{i=1}^{N}(\Delta_i M)^2\}^2] \le 6C^4,$$

Furthermore, by the dominated convergence theorem,

$$\begin{aligned}E[\sum_{i=1}^{N}(\Delta_i M)^4] &\le E[\max_i(\Delta_i M)^2 \sum_{i=1}^{N}(\Delta_i M)^2] \\ &\le E[(\max_i(\Delta_i M)^2)^2]^{1/2} E[\{\sum_{i=1}^{N}(\Delta_i M)^2\}^2]^{1/2} \\ &\le E[\max_i(\Delta_i M)^4]^{1/2}\sqrt{6}C^2 \ \to\ 0 \quad \text{as} \quad N \to \infty.\end{aligned}$$

(4) Put $\psi_s^{(N)} = \varphi'(X_0)1_{\{0\}}(s) + \sum_{i=1}^{N} \varphi'(X_{t_{i-1}})1_{(t_{i-1},t_i]}(s)$ for $0 \le s \le t$. By the uniform continuity of φ',

$$\begin{aligned}|\psi_s^{(N)} - \varphi'(X_s)| &\le \sum_{i=1}^{N} |\varphi'(X_{t_{i-1}}) - \varphi'(X_s)| 1_{(t_{i-1},t_i]}(s) \\ &\le \sup\{|\varphi'(X_r) - \varphi'(X_{\bar{r}})| : |r - \bar{r}| \\ &< 2\frac{t}{N},\ 0 \le r, \bar{r} \le t\} \to 0, \quad \text{a.s.}\end{aligned}$$

Hence, we can show that

$$\begin{aligned}&|I_1 - \int_0^t \varphi'(X_s)a_s ds| \le \int_0^t |\psi_s^{(N)} - \varphi'(X_s)||a_s| ds \ \to\ 0, \quad \text{a.s.}, \\ &E[|I_2 - \int_0^t \varphi'(X_s)b_s dB_s|^2] = E[\int_0^t |\psi_s^{(N)} - \varphi'(X_s)|^2 |b_s|^2 ds] \ \to\ 0, \\ &|I_3| \ \to\ 0, \quad \text{a.s.}, \\ &|I_4| \ \to\ 0, \quad \text{a.s.},\end{aligned}$$

and by (3),

$$\begin{aligned}&E[|I_5|] \le E[\max_i |\varphi''(\tilde{X}_{t_i}) - \varphi''(X_{t_{i-1}})| \sum_{i=1}^{N}(\Delta_i M)^2] \ \to\ 0, \\ &\tilde{I}_6 := \frac{1}{2}\sum_{i=1}^{N} \varphi''(X_{t_{i-1}}) \int_{t_{i-1}}^{t_i} b_s^2 ds \ \to\ \frac{1}{2}\int_0^t \varphi''(X_s) b_s^2 ds, \quad \text{a.s.}, \\ &E[|I_6 - \tilde{I}_6|^2] = \frac{1}{4}E[\{\sum_{i=1}^{N} \varphi''(X_{t_{i-1}})((\Delta_i M)^2 - \int_{t_{i-1}}^{t_i} b_s^2 ds)\}^2] \ \to\ 0.\end{aligned}$$

Consequently, $\sum_{k=1}^{6} I_k$ converges to the right-hand side of Itô's formula in probability as $N \to \infty$ for each $t \ge 0$. By the continuity of stochastic processes of both sides, Itô's formula holds for all $t \ge 0$ a.s. ■

Theorem 1.6.2 (Lévy). *Let* $(\Omega, \mathcal{F}, P, \{\mathcal{F}_t\})$ *be a stochastic basis. Let* $\{(B_t, \mathcal{F}_t)\}_{t\geq 0}$ *be a continuous martingale with* $B_0 = 0$ *such that* $\{B_t^2 - t\}$ *is a martingale. Then* $\{B_t\}_{t\geq 0}$ *is a Brownian motion.*

Proof. We will show that

$$E[e^{i\xi(B_t - B_s)}|\mathcal{F}_s^B] = e^{-\frac{\xi^2}{2}(t-s)}, \quad \text{a.s.,} \quad t \geq s, \ \xi \in \mathbf{R}.$$

We remark that Itô's formula holds for a complex valued function $\varphi(x) = e^{i\xi x}$ by linearity. So,

$$\varphi(B_t) = 1 + \int_0^t i\xi\varphi(B_s)\,dB_s + \frac{1}{2}\int_0^t i^2\xi^2\varphi(B_s)ds, \quad \text{a.s.}$$

Hence

$$E[\varphi(B_t) - \varphi(B_s)|\mathcal{F}_s] = -\frac{1}{2}\xi^2 E[\int_s^t \varphi(B_r)dr|\mathcal{F}_s], \quad \text{a.s.}$$

Put $\psi(t) = E[\varphi(B_t)1_A]$ for fixed $A \in \mathcal{F}_s^B \subset \mathcal{F}_s$. Then, $\psi(t)$ is continuous, and by Fubini's Theorem,

$$\psi(t) = \psi(s) - \frac{1}{2}\xi^2\int_s^t \psi(r)dr.$$

Differentiating both sides, we have

$$\frac{d}{dt}\psi(t) = -\frac{1}{2}\xi^2\psi(t), \quad t \geq s.$$

This equation has a solution

$$\psi(t) = \psi(s)e^{-\frac{\xi^2}{2}(t-s)},$$

which implies the desired result. ∎

Notation 1.6.3. *We adopt the following notations:*

(a) $tr\{\ \}$ *is the trace of a matrix,*
(b) $\cdot^{\top}$ *is the transpose of a vector,*
(c) $D\varphi = (\frac{\partial\varphi}{\partial x_i}; i = 1, 2, \ldots, N)$,
$D^2\varphi = (\frac{\partial^2\varphi}{\partial x_i \partial x_j}; i = 1, 2, \ldots, N, j = 1, 2, \ldots, N)$.

Theorem 1.6.4 (Itô). *Let* $(B(t), \mathcal{F}_t)$ *be an* M*-dimensional Brownian motion on a stochastic basis* $(\Omega, \mathcal{F}, P, \{\mathcal{F}_t\})$. *Let* $(a(t), \mathcal{F}_t) \in \mathcal{L}_1^{N\times 1}$ *and* $(b(t), \mathcal{F}_t) \in \mathcal{L}_2^{N\times M}$. *Let* X_0 *be an* $\mathbf{R}^N$*-valued,* $\mathcal{F}_0$*-measurable random variable, and let* $X_t = (X_1(t), \ldots, X_N(t))$ *be the Itô process:*

$$X_t = X_0 + \int_0^t a(s)ds + \int_0^t b(s)dB(s).$$

Let $\varphi(t,x)$ *be in* $C^{1,2}([0,\infty)\times\mathbf{R}^N)$. *Then*

$$\begin{aligned}\varphi(t,X_t) = \varphi(0,X_0) &+ \int_0^t \{\frac{\partial\varphi}{\partial t}(s,X_s) + (D\varphi(s,X_s),a(s))\}\,ds \\ &+ \int_0^t (D\varphi(s,X_s), b(s)\,dB(s)) \\ &+ \frac{1}{2}\int_0^t tr\{b(s)b(s)^\top D^2\varphi(s,X_s)\}ds, \quad a.s.\end{aligned}$$

Proof. $(B(t),\mathcal{F}_t)$, is the Brownian motion – $\{\mathcal{F}_t\}$ replaced $\{\mathcal{F}_t^B\}$. In view of Theorem 1.5.10, each component $B_k(t)$, $k=1,\ldots,M$, is a continuous martingale with $B_k(0)=0$ such that $\{B_k(t)^2 - t\}$ is a martingale, and so is $\{B_k(t)B_l(t)\}$ if $k\neq l$. Furthermore, by virtue of Theorem 1.6.1 (2), we may use the formal calculation:

$$\begin{cases} dB_k(t)dB_l(t) = dt & \text{if } k=l,\\ dB_k(t)dB_l(t) = 0 & \text{otherwise},\\ dB_k(t)dt = 0. \end{cases}$$

We note that

$$\int_0^t (D\varphi(s,X_s), b(s)\,dB(s)) = \sum_{i=1}^N \int_0^t \sum_{j=1}^M \frac{\partial\varphi}{\partial x_i}(s,X_s)b_{ij}(s)dB_j(s),$$

$$\int_0^t tr\{b(s)b(s)^\top D^2\varphi(s,X_s)\}ds = \sum_{i,j=1}^N \int_0^t \{\sum_{k=1}^M b_{ik}(s)b_{jk}(s)\}\frac{\partial^2\varphi}{\partial x_i\partial x_j}(s,X_s)ds.$$

Applying Taylor's formula, we can rewrite the second-order term of $\varphi(t,X_t)$ as follows:

$$\begin{aligned}&\frac{1}{2}\sum_{i,j=1}^N \frac{\partial^2\varphi}{\partial x_i\partial x_j}\left(\sum_{k=1}^M b_{ik}(t)dB_k(t)\right)\left(\sum_{l=1}^M b_{jl}(t)dB_l(t)\right)\\ &= \frac{1}{2}\sum_{i,j=1}^N \frac{\partial^2\varphi}{\partial x_i\partial x_j}\left(\sum_{k,l=1}^M b_{ik}(t)b_{jl}(t)dB_k(t)dB_l(t)\right)\\ &= \frac{1}{2}\sum_{i,j=1}^N \frac{\partial^2\varphi}{\partial x_i\partial x_j}\left(\sum_{k=1}^M b_{ik}(t)b_{jk}(t)dt\right).\end{aligned}$$

Thus the assertion follows from the same line as shown in Theorem 1.6.1. ■

1.7 Stochastic Differential Equations: Strong Formulation

Let $b(t,x) = \{b_i(t,x); i = 1,2,\ldots,N\}$, $\sigma(t,x) = \{\sigma_{ij}(t,x); i = 1,2,\ldots,N, j = 1,2,\ldots,M\}$, and

$$b_i(t,x) : [0,\infty) \times \mathbf{R}^N \to \mathbf{R},$$
$$\sigma_{ij}(t,x) : [0,\infty) \times \mathbf{R}^N \to \mathbf{R}$$

be Borel measurable functions. Let $(B_t, \mathcal{F}_t)$ be an M-dimensional Brownian motion on a stochastic basis $(\Omega, \mathcal{F}, P, \{\mathcal{F}_t\})$, and $\xi = (\xi_1, \xi_2, \ldots, \xi_N)$ is $\mathcal{F}_0$ measurable.

Definition 1.7.1. *A stochastic process* $\{(X_t, \mathcal{F}_t)\}_{t\geq 0}$ *on a stochastic basis* $(\Omega, \mathcal{F}, P, \{\mathcal{F}_t\})$ *is a strong solution of a stochastic differential equation*

$$\begin{cases} dX_t = b(t,X_t)dt + \sigma(t,X_t)dB_t, & t > 0 \\ X_0 = \xi, \end{cases} \tag{1.1}$$

that is,

$$\begin{cases} dX_i(t) = b_i(t,X_t)dt + \sum_{j=1}^{M} \sigma_{ij}(t,X_t)dB_j(t), & t > 0, \\ X_i(0) = \xi_i, \quad i = 1,2,\ldots,N, \end{cases}$$

if the following assertions hold:

(a) $X_0 = \xi, \quad P\text{-}a.s.$
(b) $P(\int_0^T |b_i(t,X_t)| + |\sigma_{ij}(t,X_t)|^2 dt < \infty \quad \textit{for } \forall T > 0,\ i,\ j) = 1,$
(c) $X_t = \xi + \int_0^t b(s,X_s)ds + \int_0^t \sigma(s,X_s)dB_s, \quad t \geq 0,\ P\text{-}a.s.,$
that is,

$$X_i(t) = \xi_i + \int_0^t b_i(s,X_s)ds + \sum_{j=1}^{M} \int_0^t \sigma_{ij}(s,X_s)dB_j(s),$$
$$t \geq 0, i = 1,2,\ldots,N,\ P\text{-}a.s.$$

Theorem 1.7.2. *We assume that there exists* $K > 0$ *such that*

(i) $|b(t,x) - b(t,y)| + \|\sigma(t,x) - \sigma(t,y)\| \leq K|x-y|$
$\forall x,y \in \mathbf{R}^N,\ \forall t \geq 0,$
(ii) $|b(t,x)|^2 + \|\sigma(t,x)\|^2 \leq K(1+|x|^2)$, *and*
(iii) $\xi \in L^2(\Omega, \mathcal{F}_0, P)$.
Then there exists a unique solution $\{(X_t, \mathcal{F}_t)\}_{t\geq 0}$ *to* (1.1) *in* $\mathcal{L}_2$ *such that*
(iv) $E[\sup_{0\leq t\leq T} |X_t|^2] \leq C(1 + E[|\xi|^2])$,
for some $C > 0$*, depending only on* K *and* T*, where* $\|\sigma\|^2 := \sum_{i,j} \sigma_{i,j}^2$.

Proof. Fix $T > 0$ and we will show the existence of a solution $\{X_t\}$ to (1.1) on $[0, T]$. Let $\mathcal{H}$ be the space of $\mathbf{R}^N$-valued continuous stochastic process $\{z_t\} = \{(z_t, \mathcal{F}_t)\}_{0\le t\le T}$ satisfying

$$\|z\| := (E[\sup_{0\le t\le T} |z_t|^2])^{1/2} < \infty.$$

(1) $\mathcal{H}$ is a Banach space with norm $\|\cdot\|$. Indeed, it is easy to see that $\mathcal{H}$ is a normed linear space. Let $\{z_n(t)\}$, $n = 1, 2, \ldots,$ be a Cauchy sequence in $\mathcal{H}$. For any $k \in \mathbf{N}$, there is $n_k \in \mathbf{N}$ such that $\|z_n - z_m\| < 1/4^k$ if $n, m \ge n_k$. Take $n_1 < n_2 < \ldots, \nearrow \infty$ such that

$$\|z_{n_{k+1}} - z_{n_k}\| < 1/4^k.$$

We set $A_k = \{\sup_{0\le t\le T} |z_{n_{k+1}}(t) - z_{n_k}(t)| > 1/2^k\}$. Then

$$\begin{aligned}\sum_{k=1}^{\infty} P(A_k) &= \sum_{k=1}^{\infty} P(\sup_{0\le t\le T} |z_{n_{k+1}}(t) - z_{n_k}(t)| > 1/2^k)\\ &\le \sum_{k=1}^{\infty} 2^{2k} E[\sup_{0\le t\le T} |z_{n_{k+1}}(t) - z_{n_k}(t)|^2]\\ &\le \sum_{k=1}^{\infty} \frac{2^{2k}}{4^{2k}} < \infty.\end{aligned}$$

By the Borel–Cantelli lemma, $P(\limsup_{k\to\infty} A_k) = 0$. Let $\omega \notin \limsup_{k\to\infty} A_k := \cap_{i=1}^{\infty} \cup_{k=i}^{\infty} A_k$. Then, there is i such that

$$\sum_{k=i}^{\infty} \sup_{0\le t\le T} |z_{n_{k+1}}(t) - z_{n_k}(t)| \le \sum_{k=i}^{\infty} \frac{1}{2^k} \le 1.$$

Thus

$$\lim_{k\to\infty} z_{n_k}(t) = z_{n_1}(t) + \sum_{i=1}^{\infty} \{z_{n_{i+1}}(t) - z_{n_i}(t)\}$$

converges uniformly in $t \in [0, T]$. Define $\{z(t)\}$ by $z(t) := \lim_{k\to\infty} z_{n_k}(t)$ if $\omega \notin \limsup_{k\to\infty} A_k$. Clearly, $\{z(t)\}$ is continuous and adapted. We need to show that $z := \{z(t)\}$ belongs to $\mathcal{H}$ and $\|z_m - z\| \to 0$ as $m \to \infty$. Note that $\sup_{0\le t\le T} |z_{n_k}(t) - z(t)| \to 0$ a.s., and hence

$$\lim_{k\to\infty} \sup_{0\le t\le T} |z_{n_k}(t) - z_m(t)| = \sup_{0\le t\le T} |z(t) - z_m(t)|, \quad \text{a.s.}$$

Given $\varepsilon > 0$, choose $N_0 \in \mathbf{N}$ such that $\|z_n - z_m\| < \varepsilon$ for $m, n \ge N_0$. Using Fatou's lemma, we have

$$E[\sup_{0\le t\le T} |z(t) - z_m(t)|^2] \le \liminf_{k\to\infty} E[\sup_{0\le t\le T} |z_{n_k}(t) - z_m(t)|^2] \le \varepsilon^2.$$

Hence $z - z_m \in \mathcal{H}$ and so $z = z_m + (z - z_m) \in \mathcal{H}$, and we have $\|z - z_m\| < \varepsilon$ for all $m \geq N_0$. Thus $z_m \to z$ in $\mathcal{H}$ as required.

(2) Define the stochastic process $\{(Sz_t, \mathcal{F}_t)\}$ by

$$Sz_t = \xi + \int_0^t b(s, z_s)ds + \int_0^t \sigma(s, z_s)dB_s, \qquad \text{for each} \quad z \in \mathcal{H}.$$

We claim that S maps $\mathcal{H}$ into $\mathcal{H}$. Clearly $\{Sz_t\}$ is continuous and adapted. It follows from (ii) and Doob's maximal inequality that

$$\begin{aligned} E[\sup_{0\leq t\leq T} |Sz_t|^2] &\leq 3\{E[|\xi|^2] + E[\sup_{0\leq t\leq T} |\int_0^t b(s, z_s)ds|^2] \\ &\quad + E[\sup_{0\leq t\leq T} |\int_0^t \sigma(s, z_s)dB_s|^2]\} \\ &\leq 3\{E[|\xi|^2] + TE[\int_0^T |b(s, z_s)|^2 ds] \\ &\quad + 4E[\int_0^T \|\sigma(s, z_s)\|^2 ds]\} \\ &\leq C_1(E[|\xi|^2] + E[\int_0^T (1 + |z_s|^2)ds]) < \infty, \end{aligned}$$

for some constant $C_1 > 0$. Therefore, $\{Sz_t\}$ belongs to $\mathcal{H}$.

(3) There exists $k \in \mathbf{N}$ such that S^k is a contraction mapping, where $S^2x := S(S(x)), \ldots, S^kx := S(S^{k-1}(x))$ for $x \in \mathcal{H}$. Indeed, let $z, y \in \mathcal{H}$. By (i), we have

$$\begin{aligned} E[\sup_{0\leq r\leq t} |Sz_r - Sy_r|^2] &\leq C_2\{E[\int_0^t |b(s, z_s) - b(s, y_s)|^2 ds] \\ &\quad + E[\int_0^t \|\sigma(s, z_s) - \sigma(s, y_s)\|^2 ds]\} \\ &\leq C_3 E[\int_0^t |z_s - y_s|^2 ds] \\ &\leq C_3\, tE[\sup_{0\leq r\leq t} |z_r - y_r|^2], \end{aligned}$$

for some constants $C_2, C_3 > 0$. Hence

$$\begin{aligned} E[\sup_{0\leq r\leq t} |S^2 z_r - S^2 y_r|^2] &\leq C_3 E[\int_0^t |Sz_s - Sy_s|^2 ds] \\ &\leq C_3^2 \int_0^t sE[\sup_{0\leq r\leq s} |z_r - y_r|^2]ds \\ &\leq C_3^2 \frac{t^2}{2!} E[\sup_{0\leq r\leq t} |z_r - y_r|^2]. \end{aligned}$$

Repeating this argument, we get

$$E[\sup_{0\le r\le t}|S^n z_r - S^n y_r|^2] \le C_3^n \frac{t^n}{n!} E[\sup_{0\le r\le t}|z_r - y_r|^2], \quad n = 1, 2, \ldots.$$

Therefore, taking k sufficiently large as $\alpha := \sqrt{C_3^k \frac{T^k}{k!}} < 1$, we have

$$\|S^k z - S^k y\| \le \alpha \|y - z\|.$$

(4) By the contraction mapping theorem (cf. A. Friedman [69]), S^k has a fixed point $X \in \mathcal{H}$; $S^k(X) = X$. Clearly, $S^{k+1}(X) = S(X)$, which implies that $S(X)$ is also a fixed point. By uniqueness, $S(X) = X$. Therefore, $X = \{X_t\}_{t\ge 0}$ solves

$$X_t = \xi + \int_0^t b(s, X_s)ds + \int_0^t \sigma(s, X_s)dB_s, \quad 0 \le t \le T, \quad \text{a.s.}$$

Furthermore, by (ii)

$$\psi(t) \le C(1 + E[|\xi|^2]) + C\int_0^t \psi(s)ds, \quad 0 \le t \le T,$$

where $\psi(t) = E[\sup_{0\le s\le t}|X_s|^2]$. By monotonicity, we see that $\psi(t)$ is continuous. Thus (iv) follows from Gronwall's inequality (cf. L. C. Evans [59, pp. 624–625]).

(5) Let $X, \hat{X}$ be two solutions to (1.1) on $[0, T]$ satisfying (iv), and let $\eta(t) = E[|X_t - \hat{X}_t|^2]$. We note by the dominated convergence theorem that $\eta(t)$ is continuous on $[0, T]$. By the same calculation as (3), under (i) and (ii), we have

$$\begin{aligned}
\eta(t) &\le C_4\{E[\int_0^t |b(s, X_s) - b(s, \hat{X}_s)|^2 ds] \\
&\quad + E[\int_0^t \|\sigma(s, X_s) - \sigma(s, \hat{X}_s)\|^2 ds]\} \\
&\le C_5 \int_0^t E[|X_s - \hat{X}_s|^2]ds \\
&= C_5 \int_0^t \eta(s)ds, \qquad 0 \le t \le T,
\end{aligned}$$

for some constants $C_4, C_5 > 0$. Applying Gronwall's inequality to $\eta(t)$, we obtain $\eta(t) = 0$, and thus

$$E[|X_t - \hat{X}_t|^2] = 0 \quad \text{for all } t \in [0, T].$$

Hence

$$\Omega_T := \bigcap_{t\in\mathbf{Q}\cap[0,T]} \{\omega : |X_t(\omega) - \hat{X}_t(\omega)| = 0\} \in \mathcal{F}_T \quad \text{and} \quad P(\Omega_T) = 1.$$

By the continuity of $t \to |X_t - \hat{X}_t|$, we have

$$\Omega_{c,T} := \{\omega : t \in [0,T] \to |X_t - \hat{X}_t| \text{ is continuous}\} \in \mathcal{F} \quad \text{and} \quad P(\Omega_{c,T}) = 1,$$

and thus

$$|X_t(\omega) - \hat{X}_t(\omega)| = 0 \quad \text{for all } t \in [0,T] \quad \text{if} \quad \omega \in \Omega_T \cap \Omega_{c,T}.$$

This implies that $P(X_t = \hat{X}_t \text{ for all } t \in [0,T]) = 1$.

(6) We have a unique solution to (1.1) on $[0,T]$, denoted by $\{X_t^T\}$, and extend it on $[T, 2T]$ starting from X_T^T at $T > 0$. Then the extension constructed on $[0, 2T]$ coincides with the solution on $[0, 2T]$ a.s. Therefore we can define the solution $\{X_t\}$ to (1.1) on $[0, \infty)$ by $X_t = X_t^T$ if $t \in [0,T]$. (iv) is immediate.

(7) Let $X, \hat{X}$ be two solutions to (1) satisfying (iv) and let $T > 0$ be arbitrary. By (5), we have

$$|X_t(\omega) - \hat{X}_t(\omega)| = 0 \quad \text{for all } t \in [0,\infty) \quad \text{if} \quad \omega \in \Omega_0,$$

where $\Omega_0 := \bigcap_{T \in \mathbf{N}} (\Omega_T \cap \Omega_{c,T}) \in \mathcal{F}$ and $P(\Omega_0) = 1$. This implies that $X = \hat{X}$. ■

Example 1.7.3. *Let A be an $N \times N$ matrix, and we consider the N-dimensional stochastic differential equation,*

$$\begin{cases} dX_t = AX_t dt + dB_t, & t \geq 0, \\ X_0 = x \in \mathbf{R}^N. \end{cases}$$

Then, by Itô's formula, this equation admits a unique solution $(X_t, \mathcal{F}_t)$ of the form:

$$X_t = e^{At}\{x + \int_0^t e^{-As} dB_s\}, \quad a.s.$$

Example 1.7.4. *Let b, $\sigma \neq 0$ be constants, and we consider the one-dimensional stochastic differential equation,*

$$\begin{cases} dX_t = bX_t dt + \sigma X_t dB_t, & t \geq 0, \\ X_0 = x \in \mathbf{R}. \end{cases}$$

Then, by Itô's formula, this equation admits a unique solution $(X_t, \mathcal{F}_t)$ of the form:

$$X_t = x \exp(bt + \sigma B_t - \frac{1}{2}\sigma^2 t), \quad a.s.$$

1.8 Martingale Moment Inequalities

Let $\{B_t, \mathcal{F}_t\}$ be a one-dimensional Brownian motion on a stochastic basis $(\Omega, \mathcal{F}, P, \{\mathcal{F}_t\})$ and $\{(f_t, \mathcal{F}_t)\} \in \mathcal{L}_2$.

Theorem 1.8.1 (Burkholder-Davis-Gundy). *Let $p > 0$. Then there exists a constant $C > 0$, depending only on p, such that*

$$E[\sup_{0\le t\le T} |\int_0^t f_s dB_s|^{2p}] \le CE[(\int_0^T (f_s)^2 ds)^p].$$

Proof. Let $M_t = \int_0^t f_s dB_s$, $M_t^* = \sup_{0\le s\le t} |M_s|$, and $A_t = \int_0^t (f_s)^2 ds$.

(1) Let $p \ge 1$ and let $\{|M_t| + A_t\}$ be bounded, that is, $\sup_{0\le t\le T}(|M_t| + A_t) \in L^\infty$. By Itô's formula,

$$|M_T|^{2p} = \int_0^T 2p|M_s|^{2p-1}\mathrm{sgn}(M_s) f_s dB_s + p(2p-1) \times \int_0^T |M_s|^{2p-2} dA_s, \quad \text{a.s.}$$

Since the first term of the right-hand side is a martingale, we have, by Hölder's inequality,

$$\begin{aligned} E[|M_T|^{2p}] &= p(2p-1)E[\int_0^T |M_s|^{2p-2} dA_s] \\ &\le p(2p-1)E[(M_T^*)^{2p-2} A_T] \\ &\le p(2p-1)E[(M_T^*)^{(2p-2)q}]^{1/q}\ E[A_T^p]^{1/p}, \end{aligned}$$

where $1/p + 1/q = 1$. By Doob's maximal inequality

$$E[(M_T^*)^{2p}] \le (\frac{2p}{2p-1})^{2p} E[|M_T|^{2p}].$$

Thus

$$E[(M_T^*)^{2p}] \le (\frac{2p}{2p-1})^{2p} p(2p-1)\ E[(M_T^*)^{2p}]^{1/q}\ E[A_T^p]^{1/p},$$

and hence

$$\begin{aligned} E[(M_T^*)^{2p}] &\le \{(\frac{2p}{2p-1})^{2p} p(2p-1)E[A_T^p]^{1/p}\}^p \\ &= \{(\frac{2p}{2p-1})^{2p} p(2p-1)\}^p E[A_T^p]. \end{aligned}$$

(2) Let $0 < p < 1$ and let $\{|M_t| + A_t\}$ be bounded. We set $N_t = \int_0^t A_s^{(p-1)/2} f_s dB_s$, which is a martingale. By Itô's formula,

$$N_t^2 = \int_0^t 2N_s A_s^{(p-1)/2} f_s dB_s + \int_0^t A_s^{p-1} dA_s, \quad \text{a.s.},$$

and then

$$pE[N_t^2] = E[p\int_0^t A_s^{p-1} dA_s] = E[\int_0^t d(A_s^p)] = E[A_t^p].$$

By Itô's formula

$$N_t A_t^{(1-p)/2} = \int_0^t A_s^{(1-p)/2} A_s^{(p-1)/2} f_s dB_s + \int_0^t N_s dA_s^{(1-p)/2}$$
$$= M_t + \int_0^t N_s dA_s^{(1-p)/2}, \quad \text{a.s.}$$

Hence

$$M_T^* \le \{N_T^* A_T^{(1-p)/2} + \sup_{0\le t\le T} \int_0^t |N_s| dA_s^{(1-p)/2}\} \le 2N_T^* A_T^{(1-p)/2}, \quad \text{a.s.}$$

By Hölder's inequality, we have

$$\begin{aligned}
E[(M_T^*)^{2p}] &\le 2^{2p} E[(N_T^*)^{2p} A_T^{p(1-p)}] \\
&\le 2^{2p}\ E[\{(N_T^*)^{2p}\}^{\frac{1}{p}}]^p\ E[\{A_T^{p(1-p)}\}^{1/(1-p)}]^{1-p} \\
&\le 2^{2p}\ (4E[N_T^2])^p\ E[A_T^p]^{1-p} \\
&= 2^{2p} 4^p\ (E[A_T^p]/p)^p\ E[A_T^p]^{1-p} \\
&= (\frac{16}{p})^p E[A_T^p].
\end{aligned}$$

(3) Let $\{|M_t| + A_t\}$ be unbounded. Let $\{\tau_n\}$ be a sequence of stopping times $\tau_n \nearrow \infty$ defined by

$$\tau_n = \inf\{t : |M_t| + A_t \ge n\}, \ = \infty \ \text{ if } \{\ \} = \phi.$$

Then the stopped processes M^{τ_n} and A^{τ_n} defined by $M_t^{\tau_n} = M_{t\wedge\tau_n}$, $A_t^{\tau_n} = A_{t\wedge\tau_n}$ are bounded. By (1) and (2), we have

$$E[\sup_{0\le t\le T} |M_t^{\tau_n}|^{2p}] \le CE[A_{T\wedge\tau_n}^p] \le CE[A_T^p].$$

Note that

$$\sup_{0\le t\le T} |M_t^{\tau_n}|^{2p} = \sup_{0\le s\le T\wedge\tau_n} |M_s|^{2p} \to \sup_{0\le s\le T} |M_s|^{2p}, \quad \text{a.s.}$$

Letting $n \to \infty$ and using Fatou's lemma, we obtain the assertion, in which the right-hand side is possibly infinite. ■

Remark 1.8.2. *See N. Ikeda and S. Watanabe [75] for the reverse inequality,*

$$E[\sup_t\{|\int_0^t f_s dB_s|^p\}] \ge C_0 E[\int_0^\infty (f_s)^2 ds)^{p/2}], \quad (C_0 > 0).$$

Proposition 1.8.3. *Let $\{X_t\}$ be a unique solution to* (1.1) *for $b(t,x)$, $\sigma(t,x)$ satisfying the conditions of Theorem 1.7.2. If $\xi \in L^{2p}(\Omega, \mathcal{F}_0, P)$ for $p \ge 1$, then*

(i) $E[|X_t|^{2p}] \le C(1 + e^{Ct}), \quad 0 \le t \le T,$
(ii) $E[\sup_{0\le r\le t} |X_r - \xi|^{2p}] \le Ct^p, \quad 0 \le t \le T,$

for some constant $C > 0$, depending on p and $T > 0$.

Proof.

(1) By the localizing argument as in Theorem 1.8.1 (3), we may consider that $\{X_t\}$ is bounded. By the Burkholder–Davis–Gundy inequality and Theorem 1.7.2 (ii)

$$
\begin{aligned}
E[\sup_{0\le r\le t} |\int_0^r \sigma(s, X_s)dB_s|^{2p}] &\le C_0 E[(\int_0^t \|\sigma(s, X_s)\|^2 ds)^p] \\
&\le C_0 t^{p-1} E[\int_0^t \|\sigma(s, X_s)\|^{2p} ds] \\
&\le C_1 t^{p-1} E[\int_0^t (1 + |X_s|^{2p}) ds],
\end{aligned}
$$

for some $C_0, C_1 > 0$. Hence

$$
\begin{aligned}
E[|X_t|^{2p}] &\le 3^{2p} E[|\xi|^{2p} + |\int_0^t b(s, X_s)ds|^{2p} + |\int_0^t \sigma(s, X_s)dB_s|^{2p}] \\
&\le 3^{2p}\{E[|\xi|^{2p} + C_1 t^{2p-1} E[\int_0^t (1 + |X_s|^{2p})ds] \\
&\quad + C_1 t^{p-1} E[\int_0^t (1 + |X_s|^{2p})ds]\} \\
&\le C(1 + E[\int_0^t |X_s|^{2p} ds]).
\end{aligned}
$$

By Gronwall's inequality, we obtain (i).

(2) By the same calculation as above, we have

$$
\begin{aligned}
E[\sup_{0\le r\le t} |X_r - \xi|^{2p}] &\le 2^{2p-1} E[\sup_{0\le r\le t} |\int_0^r b(s, X_s)ds|^{2p} \\
&\quad + \sup_{0\le r\le t} |\int_0^r \sigma(s, X_s)dB_s|^{2p}] \\
&\le 2^{2p-1} C_1 t^{p-1} E[\int_0^t (1 + |X_s|^{2p})ds].
\end{aligned}
$$

This together with (1) yields (ii). ■

1.9 Existence and Uniqueness: Locally Lipschitz Case

We improve the existence result on a unique solution of

$$
dX(t) = b(X(t))dt + \sigma(X(t))dB(t), \quad X(0) = x \in \mathbf{R}^N, \tag{1.2}
$$

on a stochastic basis $(\Omega, \mathcal{F}, P, \{\mathcal{F}_t\})$ carrying an M-dimensional standard Brownian motion $\{B(t)\}$, where

$$
b : \mathbf{R}^N \to \mathbf{R}^N, \quad \sigma : \mathbf{R}^N \to \mathbf{R}^N \times \mathbf{R}^M, \quad \text{continuous.}
$$

Furthermore, we make the following assumptions on b, σ:

There exist $K > 0$ and $K_R > 0$, depending on each $R > 0$, such that

$$|b(x)| + \|\sigma(x)\| \leq K(1 + |x|), \quad x \in \mathbf{R}^N, \tag{1.3}$$

$$|b(x) - b(\bar{x})| + \|\sigma(x) - \sigma(\bar{x})\| \leq K_R|x - \bar{x}| \quad \text{if } |x|, |\bar{x}| \leq R. \tag{1.4}$$

Theorem 1.9.1. *We assume* (1.3), (1.4). *Then there exists a unique solution* $\{X(t)\}$ *of* (1.2) *such that*

$$E[\sup_{0 \leq t \leq T} |X(t)|^2] \leq C_T(1 + |x|^2), \tag{1.5}$$

for each $T > 0$, *where the constant* $C_T > 0$ *depends only on* T, K.

Proof.

(1) For every $n \in \mathbf{N}$, let

$$\xi_n(x) = \begin{cases} x & \text{if } |x| \leq n, \\ \dfrac{x}{|x|}n & \text{otherwise.} \end{cases}$$

It is easy to see that

$$|\xi_n(x) - \xi_n(y)| \leq |x - y|, \quad x, y \in \mathbf{R}^N,$$
$$|\xi_n(x)| \leq \min(|x|, n).$$

By Theorem 1.7.2, there exists a unique solution $\{X_n(t)\}$ of

$$dX_n(t) = b \circ \xi_n(X_n(t))dt + \sigma \circ \xi_n(X_n(t))dB(t), \quad X_n(0) = x.$$

(2) We claim that if $n < m$,

$$X_n(t \wedge \tau_n) = X_m(t \wedge \tau_n), \quad \text{for all } t \geq 0, \text{ a.s.},$$

where $\tau_n = \inf\{t \geq 0 : |X_n(t)| > n\}$, $= \infty$ if $\{\,\} = \phi$. Let $Z(t) = X_n(t) - X_m(t)$. It is obvious that

$$h \circ \xi_n(x) = h \circ \xi_m(x) \quad \text{if} \quad |x| \leq n, \ n < m.$$

Then, we see, by (1.4), that if $t \leq \tau_n$,

$$\begin{aligned} |h \circ \xi_n(X_n(t)) - h \circ \xi_m(X_m(t)))| &= |h \circ \xi_m(X_n(t)) - h \circ \xi_m(X_m(t))| \\ &\leq K_m|Z_t| \end{aligned}$$

for $h = b$ or σ. Furthermore, as in the proof of Theorem 1.7.2 (5), we have

$$\begin{aligned} E[|Z_{t\wedge\tau_n}|^2] &\le K_m E[\int_0^{t\wedge\tau_n} |Z_s|^2 ds] \\ &= K_m E[\int_0^t |Z_s|^2 1_{\{s\le\tau_n\}} ds] \\ &\le K_m E[\int_0^t |Z_{s\wedge\tau_n}|^2 ds]. \end{aligned}$$

By Gronwall's inequality and the continuity of $\{Z_t\}$, we get $Z_{t\wedge\tau_n} = 0$ for all $t \ge 0$ a.s.

(3) From (2) it follows that $|X_m(\tau_n)| < m$ for $n < m$, and τ_n is nondecreasing. Moreover, there exists a null set N such that

$$X_n(t,\omega) = X_m(t,\omega), \qquad \text{for } t \le \tau_n(\omega) \quad \text{if } \omega \notin N.$$

Define $\tau = \lim_{n\to\infty} \tau_n$. Then, for $\omega \notin N$ and $t < \tau(\omega)$, $\{X_n(t,\omega)\}$ is a Cauchy sequence, that is,

$$X(t,\omega) := \lim_{n\to\infty} X_n(t,\omega) \text{ exists.}$$

Clearly, $X(t,\omega)$ is continuous for $\omega \notin N,\ t < \tau(\omega)$, and $\{X(t)\}$ is adapted since $X(t) = X_n(t)$ if $t \le \tau_n$. Also, $\lim_{n\to\infty} |X(\tau_n(\omega),\omega)| = \infty$ if $\omega \notin N,\ \tau(\omega) < \infty$. By (1), we see that

$$\begin{aligned} X(t\wedge\tau_n) &= x + \int_0^{t\wedge\tau_n} b(X_n(s))ds + \int_0^{t\wedge\tau_n} \sigma(X_n(s))dB(s) \\ &= x + \int_0^{t\wedge\tau_n} b(X(s))ds + \int_0^{t\wedge\tau_n} \sigma(X(s))dB(s), \quad t \ge 0, \text{ a.s.} \end{aligned}$$

Letting $n \to \infty$, we deduce that

$$X(t) = x + \int_0^t b(X(s))ds + \int_0^t \sigma(X(s))dB(s) \quad \text{on } \{t < \tau\}, \text{ a.s.}$$

(4) By (1.3) and Proposition 1.8.3, we have

$$E[\sup_{0\le t\le T} |X_n(t)|^2] \le C_T(1 + E[|X(0)|^2]),$$

for some constant $C_T > 0$, independent of n. Hence,

$$\begin{aligned} P(\tau_n \le T) &= P(\sup_{0\le t\le T} |X_n(t)| \ge n) \\ &\le \frac{1}{n^2} E[\sup_{0\le t\le T} |X_n(t)|^2] \\ &\le \frac{C_T}{n^2}(1 + E[|X(0)|^2]) \to 0. \end{aligned}$$

Letting $n \to \infty$ and $T \to \infty$, we obtain $P(\tau < \infty) = 0$. This implies that $\{X(t)\}$ satisfies (1.2) for all $t \geq 0$ a.s. Moreover, by Fatou's lemma

$$
\begin{aligned}
E[\sup_{0\le t\le T} |X(t)|^2] &= E[\sup_{0\le t\le T} \lim_{m\to\infty} |X_m(t)|^2] \\
&\le E[\liminf_{m\to\infty} \sup_{0\le t\le T} |X_m(t)|^2] \\
&\le \liminf_{m\to\infty} E[\sup_{0\le t\le T} |X_m(t)|^2] \\
&\le C_T(1 + E[|X(0)|^2]),
\end{aligned}
$$

which implies (1.5).

(5) For uniqueness, let $X^i(t), i = 1, 2$, be two solutions of (1.2). Let $\tau_n^i = \inf\{t \geq 0 : |X^i(t)| > n\}$, $= \infty$ if $\{\,\} = \phi$ and $\bar{\tau} = \tau_n^1 \wedge \tau_n^2$. Then, by (1.4),

$$
\begin{aligned}
E[|X^1(t\wedge\bar{\tau}) - X^2(t\wedge\bar{\tau})|^2] &\le 2E[|\int_0^{t\wedge\bar{\tau}} \{b(X^1(s)) - b(X^2(s))\}ds|^2] \\
&\quad + 2E[|\int_0^{t\wedge\bar{\tau}} \{\sigma(X^1(s)) - \sigma(X^2(s))\}dB(s)|^2] \\
&\le K_n E[\int_0^t |X^1(s\wedge\bar{\tau}) - X^2(s\wedge\bar{\tau})|^2 ds],
\end{aligned}
$$

for some constant $K_n > 0$. Using Gronwall's inequality, we get

$$E[|X^1(t\wedge\bar{\tau}) - X^2(t\wedge\bar{\tau})|^2] = 0,$$

and then

$$P(X^1(t) \neq X^2(t),\ t \le \tau_n^1 \wedge \tau_n^2) = 0.$$

Letting $n \to \infty$, we have

$$P(X^1(t) \neq X^2(t),\ t \le \tau^1 \wedge \tau^2) = 0, \qquad \text{for all } t \ge 0,$$

where $\tau^i = \lim_{n\to\infty} \tau_n^i$. By the continuity of $X^1(t)$, $X^2(t)$, we deduce that $\tau^1 = \tau^2$ and $X^1(t) = X^2(t)$ for $t < \tau^1$ a.s. Under (1.3), we have $\tau^i = \infty$ a.s. ∎

Definition 1.9.2. *Let b, σ be locally Lipschitz functions, that is,* (1.4) *holds. A stopping time* e *is called an explosion time of a stochastic process* $\{(X_t, \mathcal{F}_t^B)\}$ *if there exist* e *and* $\{X_t\}$ *(possibly* ∞*) uniquely on a stochastic basis* $(\Omega, \mathcal{F}, P, \{\mathcal{F}_t^B\})$ *such that*

$$
\begin{aligned}
&X_t \text{ is continuous on } [0, \mathrm{e}) \quad a.s., \\
&X_t = x + \int_0^t b(X_s)ds + \int_0^t \sigma(X_s)dB_s, \qquad \text{on } \{t < \mathrm{e}\} \quad a.s., \\
&\limsup_{t\nearrow \mathrm{e}} |X_t| = \infty \quad a.s. \text{ on } \{\mathrm{e} < \infty\}.
\end{aligned}
$$

Remark 1.9.3. *Theorem 1.9.1 shows that the explosion time of the solution* $\{X_t\}$ *is given by* τ *if* (1.3) *is dropped.*

Example 1.9.4. *We take the locally Lipschitz functions to be* $b(x) = x^2, \sigma(x) = 0$. *The equation,*

$$\frac{dx_t}{dt} = x_t^2, \quad x_0 = x > 0,$$

has a positive solution

$$x_t = \frac{x}{1 - tx}, \quad \text{for } t < 1/x.$$

Therefore, the explosion time e *of* $\{x_t\}$ *is* $1/x$.

Example 1.9.5. *We take the locally Lipschitz functions to be* $b(x) = x^{1/2}, \sigma(x) = 0$. *The equation,*

$$\frac{dx_t}{dt} = \theta(x_t)^{1/2}, \quad x_0 = 0,$$

has two nonnegative solutions:

$$x_t = \frac{t^2}{4}; \quad x_t = 0 \quad \text{if } \theta = 1.$$

On the other hand, if $\theta = -1$, *then* $x_t = 0$ *is a unique solution since* x_t *is non-increasing.*

1.10 Comparison Results

Let $\{X_t\}$ and $\{Y_t\}$ be two continuous adapted processes on a stochastic basis $(\Omega, \mathcal{F}, P, \{\mathcal{F}_t\})$ carrying a one-dimensional standard Brownian motion $\{B_t\}$. Let $\{X_t\}$ and $\{Y_t\}$ satisfiy one-dimensional stochastic differential equations,

$$dX_t = b(X_t)dt + \sigma(X_t)dB_t, \quad X_0 = x \in \mathbf{R}, \text{ and} \tag{1.6}$$

$$dY_t = c_t dt + \sigma(Y_t)dB_t, \quad Y_0 = x, \tag{1.7}$$

where $\{c_t\}$ is progressively measurable, $P(\int_0^t |c_s|ds < \infty) = 1$, for all $t \geq 0$,

$$b : \text{ locally Lipschitz on } \mathbf{R}, \tag{1.8}$$

and σ fulfills the following condition:

There exists a nondecreasing continuous function $\rho : [0, \infty) \to [0, \infty)$ such that

$$\begin{cases} \rho(0) = 0, \quad \rho(x) > 0 & \text{for } x > 0, \\ \int_{(0,\varepsilon)} \rho(x)^{-2} dx = \infty, & \forall \varepsilon > 0, \\ |\sigma(x) - \sigma(y)| \leq \rho(|x - y|), & x, y \in \mathbf{R}. \end{cases} \tag{1.9}$$

We state the comparison result in the following theorem.

Theorem 1.10.1. *We assume* (1.8)*, and* (1.9) *and that*

$$P(b(Y_t) \leq c_t \ \textit{for all}\ t \geq 0) = 1. \tag{1.10}$$

Then $P(X_t \leq Y_t \textit{ for all } t \geq 0) = 1$.

Proof. We may consider that, for $m > 0$,

$$Z_t := |X_t| + |Y_t| + \int_0^t |c_s| ds \leq m,$$

taking stopped processes $X_{t\wedge\tau_m}$, $Y_{t\wedge\tau_m}$, $\int_0^{t\wedge\tau_m} |c_s|ds$ at $\tau_m = \inf\{t \geq 0 : Z_t > m\}, = \infty$ if $\{\ \} = \phi$. By (1.9) we choose a sequence $\{a_n\}$ such that

$$\int_{a_n}^{a_{n-1}} \rho(x)^{-2} dx = n \quad \text{and} \quad a_0 = 1 > a_1 > \cdots > a_n \cdots, \searrow 0.$$

Then there exists $\eta_n \in C[0,\infty)$ such that $\eta_n(x) = 0$ for $x \notin (a_n, a_{n-1})$, $0 \leq \eta_n(x) \leq \rho(x)^{-2}/n$, and $1/2 \leq c_n := \int_{a_n}^{a_{n-1}} \eta_n(x)dx$. Setting $\psi_n(x) = \eta_n(x)/c_n$, we see that $\psi_n \in C[0,\infty)$ satisfies

$$\psi_n(x) = \begin{cases} 0, & 0 \leq x \leq a_n, \\ \text{between 0 and } \dfrac{2}{n}\rho(x)^{-2}, & a_n < x < a_{n-1}, \\ 0, & a_{n-1} \leq x, \end{cases}$$

and

$$\int_{a_n}^{a_{n-1}} \psi_n(x) dx = 1.$$

Define

$$\varphi_n(x) = \int_0^{x^+} dy \int_0^y \psi_n(z) dz, \quad n = 1, 2, \ldots.$$

Clearly, $\varphi_n \in C^2(\mathbf{R})$, $\varphi_n(x) \nearrow x^+$ as $n \to \infty$, and

$$\varphi_n'(x) = \begin{cases} 0, & x \leq a_n, \\ \text{between 0 and 1}, & a_n < x < a_{n-1}, \\ 1, & a_{n-1} \leq x. \end{cases}$$

Since

$$d(X_t - Y_t) = (b(X_t) - c_t)dt + (\sigma(X_t) - \sigma(Y_t))dB_t,$$

Itô's formula gives

$$\begin{aligned}
\varphi_n(X_t - Y_t) &= \int_0^t \varphi_n'(X_s - Y_s)\{b(X_s) - c_s\}ds \\
&\quad + \int_0^t \varphi_n'(X_s - Y_s)\{\sigma(X_s) - \sigma(Y_s)\}dB_s \\
&\quad + \frac{1}{2}\int_0^t \varphi_n''(X_s - Y_s)\{\sigma(X_s) - \sigma(Y_s)\}^2 ds \\
&=: I_1 + I_2 + I_3, \quad \text{say.}
\end{aligned}$$

By boundedness, we have $E[I_2] = 0$. Also,

$$E[I_1] \to E[\int_0^t 1_{\{X_s - Y_s > 0\}}\{b(X_s) - c_s\}ds],$$

and

$$\begin{aligned}
E[I_3] &\le \frac{1}{2}E[\int_0^t \psi_n(X_s - Y_s)\rho(|X_s - Y_s|)^2 ds] \\
&\le \frac{t}{n} \to 0 \quad \text{as} \quad n \to \infty.
\end{aligned}$$

Thus, by (1.10)

$$\begin{aligned}
E[(X_t - Y_t)^+] &= E[\int_0^t 1_{\{X_s - Y_s > 0\}}\{b(X_s) - c_s\}ds] \\
&\le E[\int_0^t 1_{\{X_s - Y_s \ge 0\}}\{b(X_s) - b(Y_s)\}ds].
\end{aligned}$$

Note by (1.8) that for $x \ge y$ and $|x|, |y| \le m$,

$$b(x) - b(y) \le K_m|x - y| = K_m(x - y)^+,$$

for some constant $K_m > 0$. Hence

$$E[(X_t - Y_t)^+] \le K_m E[\int_0^t (X_s - Y_s)^+ ds].$$

By Gronwall's inequality

$$E[(X_t - Y_t)^+] = 0, \quad \text{for all } t \ge 0.$$

Therefore, we deduce by the continuity of X_t, Y_t that $X_t \le Y_t$, for all $t \ge 0$ a.s. ■

Corollary 1.10.2. *Under the assumptions of Theorem 1.10.1, the pathwise uniqueness for* (1.6) *holds.*

Proof. We take $c_t = b(Y_t)$ to obtain $X_t \le Y_t$, for all $t \ge 0$. Changing the role of X_t and Y_t, we obtain $\{X_t\} = \{Y_t\}$.

1.11 The Martingale Representation Theorem

Theorem 1.11.1. *Let $(B_t, \mathcal{F}_t^B)$ be a one-dimensional Brownian motion and let $(X_t, \mathcal{F}_t^B)$ be a right-continuous martingale with $E[|X_t|^2] < \infty$, for all $t \geq 0$. Then there exists $(f_t, \mathcal{F}_t^B) \in \mathcal{M}_2$ with respect to $\{\mathcal{F}_t^B\}$ such that*

$$X_t = X_0 + \int_0^t f_s dB_s, \quad \textit{for all } t \geq 0, \quad a.s.$$

Proof. For fixed $T > 0$, let $\mathbf{H}_0$ be the space of all $\{\mathcal{F}_t^B\}$-adapted, right-continuous stochastic processes $z = \{z_t\}_{t\in[0,T]}$ such that

$$\|z\| = (E[\sup_{0\leq t\leq T} |z_t|^2])^{1/2} < \infty.$$

Define

$$\mathbf{H} = \{z = \{z_t\}_{t\in[0,T]} \in \mathbf{H}_0 : \{z_t\} \text{ is a martingale, } z_0 = 0\},$$
$$\mathbf{M} = \{I(f) = \{I_t(f)\}_{t\in[0,T]} : f \in \mathcal{M}_2(0,T) \text{ with respect to } \{\mathcal{F}_t^B\}\}.$$

(1) By analogy with Theorem 1.7.2 (1), we note that $\mathbf{H}_0$ is a Banach space with norm $\|\cdot\|$. Then it is easy to see that $\mathbf{H}$ is a closed subspace of $\mathbf{H}_0$. By Doob's maximal inequality,

$$E[\sup_{0\leq t\leq T} |z_t|^2] \leq 4E[z_T^2] \leq 4E[\sup_{0\leq t\leq T} |z_t|^2], \quad \text{for} \quad z \in \mathbf{H}.$$

Thus $\mathbf{H}$ is a Hilbert space with inner product

$$(z, \hat{z}) = E[z_T \cdot \hat{z}_T].$$

By analogy with Proposition 1.1.10, $\mathcal{M}_2(0,T)$ is complete. Since

$$E[|I_T(f) - I_T(\bar{f})|^2] = E[\int_0^T |f_t - \bar{f}_t|^2 dt],$$

we observe that $\mathbf{M}$ is a closed subspace of $\mathbf{H}$. By the projection theorem, we have

$$\mathbf{H} = \mathbf{M} \oplus \mathbf{M}^\perp,$$

where $\mathbf{M}^\perp$ denotes the orthogonal complement of $\mathbf{M}$. Therefore, for the proof, it is enough to show that $\mathbf{M}^\perp = \{0\}$.

(2) Let $m = \{m_t\} \in \mathbf{M}^\perp$ and $E[m_T z_T] = 0$ for every $z \in \mathbf{M}$. Put $\hat{z}_t = z_{\tau\wedge t}$ for any stopping time $\tau \leq T$. Since

$$\hat{z}_t = \int_0^t f_s 1_{\{s\leq\tau\}} dB_s, \quad f \in \mathcal{M}_2(0,T),$$

we get $\{\hat{z}_t\} \in \mathbf{M}$ and $\hat{z}_T = z_\tau$. Hence, by the optional sampling theorem,

$$E[m_\tau z_\tau] = E[E[m_T|\mathcal{F}_\tau]z_\tau] = E[m_T z_\tau] = E[m_T \hat{z}_T] = 0.$$

Let $Y_t = m_t z_t$ and set $\bar{\tau} = s1_{A^c} + t1_A$ for $A \in \mathcal{F}_s$, $s \le t \le T$. It is clear that $\bar{\tau}$ is a stopping time. Then

$$0 = E[Y_{\bar{\tau}}] = E[Y_s 1_{A^c} + Y_t 1_A],$$

and

$$0 = E[Y_s] = E[Y_s 1_{A^c} + Y_s 1_A].$$

Thus $E[Y_t 1_A] = E[Y_s 1_A]$. This implies that $\{Y_t\}$ is a martingale.

(3) Let $Z_t = \exp\{i\xi B_t + \frac{1}{2}\xi^2 t\} - 1$ for $\xi \in \mathbf{R}$. By Itô's formula, $\{Z_t\}$ fulfills

$$Z_t = \int_0^t i\xi(Z_s + 1)dB_s.$$

Hence the real and imaginary parts of $\{Z_t\}$ belong to $\mathbf{M}$. Thus, by (2), $\{m_t Z_t\}$ is a complex-valued martingale. This yields that

$$E[m_t \exp\{i\xi(B_t - B_s)\}|\mathcal{F}_s^B] = m_s \exp\{-\frac{1}{2}\xi^2(t-s)\}, \quad \text{a.s.,} \quad t \ge s.$$

(4) Let J denote the collection of all finite partitions of $[0, T]$, and let $\delta = \{t_0 = 0 \le t_1 \le \cdots \le t_n = T\} \in J$. Then we have

$$\sigma(\underset{\delta\in J}{\cup} \mathcal{G}_\delta) = \mathcal{F}_T^B,$$

where

$$\mathcal{G}_\delta = \sigma(B_{t_0}, B_{t_1}, \ldots, B_{t_n}) = \sigma(B_{t_0}, B_{t_1} - B_{t_0}, \ldots, B_{t_n} - B_{t_{n-1}}).$$

(5) From (3) it follows that for $(\xi_1, \ldots, \xi_n) \in \mathbf{R}^n$,

$$\begin{aligned} E[\exp\{i\sum_{j=1}^n \xi_j(B_{t_j} - B_{t_{j-1}})\}m_T] &= E[\exp\{i\sum_{j=1}^{n-1} \xi_j(B_{t_j} - B_{t_{j-1}})\}m_{t_{n-1}}] \\ &\quad \times \exp\{-\frac{1}{2}\xi_n^2(t_n - t_{n-1})\} \\ &= E[m_0]\exp\{-\sum_{j=1}^n \frac{1}{2}\xi_j^2(t_j - t_{j-1})\} = 0. \end{aligned}$$

Clearly,

$$E[m_T] = E[m_T^+] - E[m_T^-] = 0.$$

Suppose that $\mu := E[m_T^+] = E[m_T^-] > 0$. Then we define two probability measures P_1, P_2 by

$$dP_1 = \frac{m_T^+}{\mu}dP, \quad dP_2 = \frac{m_T^-}{\mu}dP.$$

Hence

$$\int \exp\{i\sum_{j=1}^n \xi_j(B_{t_j} - B_{t_{j-1}})\}dP_1 = \int \exp\{i\sum_{j=1}^n \xi_j(B_{t_j} - B_{t_{j-1}})\}dP_2.$$

Since the characteristic function is uniquely determined by the distribution function, we get

$$P_1(B_{t_1} - B_{t_0} \in C_1, \ldots, B_{t_n} - B_{t_{n-1}} \in C_n) \\ = P_2(B_{t_1} - B_{t_0} \in C_1, \ldots, B_{t_n} - B_{t_{n-1}} \in C_n),$$

for any Borel sets $C_1, \ldots, C_n$. By (4), we see that

$$P_1 = P_2 \quad \text{on} \quad \mathcal{F}_T^B.$$

Therefore, we conclude that $m_T^+ = m_T^-$ a.s., that is, $m = 0$.

(6) Let $0 < T_1 < T_2$. By (5), we have

$$\begin{cases} X_t = X_0 + \int_0^t f_s^{(T_1)} dB_s & \text{if } t \le T_1, \\ X_t = X_0 + \int_0^t f_s^{(T_2)} dB_s & \text{if } t \le T_2, \end{cases}$$

for some $f^{(T_i)} \in \mathcal{M}_2(0, T_i),\ i = 1, 2$. Hence

$$E[\int_0^{T_1} |f_s^{(T_1)} - f_s^{(T_2)}|^2 ds] = 0,$$

and then

$$f^{(T_1)}(s, \omega) = f^{(T_2)}(s, \omega) \quad \text{on } [0, T_1] \times \Omega, \quad ds \times dP\text{-a.e.}$$

Thus we can define $f(s, \omega)$ for almost all $(s, \omega) \in [0, \infty) \times \Omega$ by

$$f(s, \omega) = f^{(T_1)}(s, \omega) \quad \text{if } s \le T_1.$$

Consequently, we get

$$X_t = X_0 + \int_0^t f_s^{(t)} dB_s = X_0 + \int_0^t f_s dB_s, \quad \text{for all } t \ge 0, \quad a.s. \quad \blacksquare$$

Example 1.11.2. *Following D. Duffie [48], we consider a financial asset that pays the cumulative dividend D_t up to time $t \in [0, T]$. The price process $\{S_t\}$ of the corresponding asset represents the market value of a claim to all future dividends paid by the asset. We assume that $\{D_t\}$ and $\{S_t\}$ evolve according to*

$$dD_t = bD_t dt + \sigma D_t dB_t, \quad 0 \le t \le T, \quad D_0 = x > 0, \\ S_t = E[D_T - D_t | \mathcal{F}_t^B],$$

for some positive constants b, σ. Then the gain process of the asset $\{G_t\}$ is defined by

$$G_t = S_t + D_t = E[D_T | \mathcal{F}_t^B].$$

By Example 1.7.4, we get

$$D_t = x \exp(bt + \sigma B_t - \frac{1}{2}\sigma^2 t) > 0.$$

By Proposition 1.2.7, we recall that

$$E[\exp\{\sigma(B_T - B_t)\}|\mathcal{F}_t^B] = \exp\{\frac{1}{2}\sigma^2(T-t)\}, \quad a.s.$$

Hence

$$\begin{aligned} G_t &= x\exp(bT)E[\exp(\sigma B_T - \frac{1}{2}\sigma^2 T)|\mathcal{F}_t^B] \\ &= x\exp(bT)\exp(\sigma B_t - \frac{1}{2}\sigma^2 t), \quad a.s. \end{aligned}$$

Then G_t satisfies

$$dG_t = \sigma G_t dB_t = \sigma e^{b(T-t)} D_t dB_t.$$

Let $(X_t, \mathcal{F}_t^B)$ be any right-continuous martingale with $E[|X_t|^2] < \infty$. By the martingale representation theorem, there is a progressively measurable process $\{f_t\}$ such that

$$\begin{aligned} X_t &= X_0 + \int_0^t f_s dB_s, \quad 0 \le t \le T, \quad a.s., \\ E[&\int_0^T f_t^2 dt] < \infty. \end{aligned}$$

Taking $g_t = f_t/\sigma e^{b(T-t)} D_t$, we observe that $\{X_t\}$ admits a representation:

$$X_t = X_0 + \int_0^t g_s dG_s, \quad 0 \le t \le T, \quad a.s.$$

Remark 1.11.3. *See R. S. Liptser and A. N. Shiryayev [106], N. Ikeda and S. Watanabe [75], and I. Karatzas and S. E. Shreve [87] for more details of the martingale representation theorem.*

2

Stochastic Differential Equations: Weak Formulation

In the linear stochastic differential equation,

$$dX(t) = AX(t)dt + \sigma dB(t), \quad X(0) = x,$$

we let $Y(t) := X(t+r)$ for fixed $r \geq 0$. It is easily seen that

$$dY(t) = AY(t)dt + \sigma dB(t), \quad Y(0) = X(r)$$

fails in the strong sense (see Chapter 1) if $\sigma \neq 0$. But, we have

$$\begin{aligned} Y(t) &= X(r) + \int_r^{t+r} AX(s)ds + \int_r^{t+r} \sigma dB(s) \\ &= X(r) + \int_0^t AY(s)ds + \int_0^t \sigma d\tilde{B}(s), \quad a.s., \end{aligned}$$

where $\tilde{B}(t) = B(t+r) - B(r)$. This shows that $\{Y(t)\}$ solves the equation for another Brownian motion $\{\tilde{B}(t)\}$ on some different probability space. We present the formulation of such $\{Y(t)\}$, called a *weak solution*, on the probability space defined by a regular conditional probability. We also discuss the weak solutions shown by the Girsanov measure transformation and the Markov property of $\{X(t)\}$.

N. Ikeda and S. Watanabe [75], J. Neveu [128], D. Stroock [151], R. F. Bass [9], R. M. Blumenthal and R. K. Getoor [27], and L. C. G. Rogers and D. Williams [138] are the basic references for this chapter.

2.1 Probability Laws

We consider the probability law induced by the solution $\{X_t\}$ to the stochastic differential equation,

$$dX_t = b(X_t)dt + \sigma(X_t)dB_t, \quad X_0 = x \in \mathbf{R}^N, \tag{2.1}$$

for $b(x) = \{b_i(x)\}$ and $\sigma(x) = \{\sigma_{ij}(x)\}$, where b_i, σ_{ij} $(i = 1, 2, \ldots, N; j = 1, 2, \ldots, M)$ are Borel measurable functions on $\mathbf{R}^N$, and $\{B_t\}$ is an M-dimensional Brownian motion.

Definition 2.1.1.

(a) *Let W denote the class defined by*

$$W = W^N = \{w : t \in [0, \infty) \to w(t) \in \mathbf{R}^N, \ \textit{continuous}\}.$$

(b) *Let $d : W \times W \to \mathbf{R}_+$ be the metric for W defined by*

$$d(w_1, w_2) = \sum_{n=1}^{\infty} \frac{1}{2^n} \frac{\|w_1 - w_2\|_n}{1 + \|w_1 - w_2\|_n}, \quad \textit{for} \quad w_1, w_2 \in W,$$

where $\|w\|_n := \sup_{t \in [0,n]} |w(t)|$.

(c) *Let $\mathcal{B}(W)$ be the Borel σ-algebra of W.*

(d) *Let $\pi_t(w) = w(t)$, for all $w \in W$, and we define the σ-algebra $\mathcal{B}_t(W)$ by*

$$\mathcal{B}_t(W) = \sigma\{\pi_s^{-1}(A) : A \in \mathcal{B}(\mathbf{R}^N), s \le t\}, \quad t \ge 0.$$

Proposition 2.1.2. *W is a Polish space, that is, a separable complete metric space with metric d.*

Proof. The space $C([0, n]; \mathbf{R}^N)$ of continuous functions w on $[0, n]$ for fixed n, taking values in $\mathbf{R}^N$, is a Banach space with the norm $\|w\|_n$. Define C_0 as the class of all $w \in C([0, n]; \mathbf{R}^N)$ such that each component of w is a polynomial with rational coefficients. Clearly, C_0 is countable. By Weiestrass' approximation theorem, C_0 is dense in $C([0, n]; \mathbf{R}^N)$. Note that

$$\|w_k\|_n \to 0 \text{ for each } n \quad \text{iff} \quad \sum_{n=1}^{\infty} \frac{1}{2^n} \frac{\|w_k\|_n}{1 + \|w_k\|_n} \to 0 \quad (k \to \infty).$$

Since

$$\frac{|a + b|}{1 + |a + b|} \le \frac{|a|}{1 + |a|} + \frac{|b|}{1 + |b|}, \quad a, b \in \mathbf{R},$$

we see that d is a metric. By an extension of w on $[0, \infty)$, we obtain the countable dense set and the completeness of W. ■

Proposition 2.1.3.

(i) *$\{\mathcal{B}_t(W)\}_{t \ge 0}$ is a nondecreasing family of sub-σ-algebra of $\mathcal{B}(W)$ such that $\sigma(\cup_{t \ge 0} \mathcal{B}_t(W)) = \mathcal{B}(W)$.*

(ii) *Let $\{X_t, \mathcal{F}_t\}$ be an N-dimensional continuous adapted stochastic process, and let $X(\omega) = X_\cdot(\omega) \in W$ for $\omega \in \Omega$. Then $X : (\Omega, \mathcal{F}, P) \to (W, \mathcal{B}(W))$ is a random variable, and X: $(\Omega, \mathcal{F}_t, P) \to (W, \mathcal{B}_t(W))$ is measurable for each $t \ge 0$.*

Proof. It is obvious that $\mathcal{B}_t(W)$ is nondecreasing.

(1) Let $0 \le t \le n$. Let $w_m, w \in W$ and $d(w_m, w) \to 0$ as $m \to \infty$. Then

$$|\pi_t(w_m) - \pi_t(w)| \le \|w_m - w\|_n \to 0.$$

This yields that each π_t is continuous on W. Hence

$$\{w \in W : \pi_t(w) \in A\} \in \mathcal{B}(W), \quad A \subset \mathbf{R}^N \text{ closed.}$$

Thus we get $\sigma(\cup_{t\ge 0} \mathcal{B}_t(W)) \subset \mathcal{B}(W)$.

(2) Let $W_0 = \{e_1, e_2, \dots\}$ be a countable dense subset of W. By continuity,

$$d(w, e_j) = \sum_{n=1}^{\infty} \frac{1}{2^n} \sup_{t\in\mathbf{Q}\cap[0,n]} \frac{|\pi_t(w) - \pi_t(e_j)|}{1 + |\pi_t(w) - \pi_t(e_j)|}.$$

Hence $d(\cdot, e_j)$ is $\sigma(\cup_{t\ge 0} \mathcal{B}_t(W))$-measurable. Let G be open, and $B_\varepsilon(\omega)$ a ball with its radius ε centered at ω such that $w \in B_\varepsilon(w) \subset G$, for some $\varepsilon > 0$. Taking $a \in \mathbf{Q}$, $a > 0$ such that $d(w, e_j) < a \le \varepsilon/2$, we have $w \in B_a(e_j) \subset B_\varepsilon(w) \subset G$. Thus G is a countable union of open balls of the form:

$$B_{a_j}(e_j) := \{w \in W : d(w, e_j) < a_j\}, \quad \text{for } a_j > 0,\ a_j \in \mathbf{Q}.$$

Therefore, $\mathcal{B}(W) \subset \sigma(\bigcup_{t\ge 0} \mathcal{B}_t(W))$, which implies (i).

(3) Let $t_0 = 0 < t_1 < \cdots < t_n = t$ and let A be a cylinder set of the form:

$$A = \{w \in W : (\pi_{t_0}(w), \pi_{t_1}(w), \dots, \pi_{t_n}(w)) \in C\} \in \mathcal{B}_t(W),$$
$$C \in \mathcal{B}(\mathbf{R}^N) \times \cdots \times \mathcal{B}(\mathbf{R}^N).$$

Then

$$\{\omega \in \Omega : X(\omega) \in A\} = \{\omega \in \Omega : (X_{t_0}(\omega), X_{t_1}(\omega), \dots, X_{t_n}(\omega)) \in C\} \in \mathcal{F}_t.$$

Thus (ii) follows. ∎

Theorem 2.1.4. *Let $(W_i, \mathcal{B}(W_i))$, $i = 1, 2$, be two Polish spaces. Let $X : (\Omega, \mathcal{F}, P) \to (W_1, \mathcal{B}(W_1))$ and $Y : (\Omega, \mathcal{F}, P) \to (W_2, \mathcal{B}(W_2))$ be two random variables. Then Y is $\sigma(X)$-measurable if and only if there exists a measurable function $\eta : (W_1, \mathcal{B}(W_1)) \to (W_2, \mathcal{B}(W_2))$ such that*

$$Y(\omega) = \eta(X(\omega)), \quad \omega \in \Omega.$$

Proof. We will show the "only if" part.

(1) We claim that there exists a sequence of simple random variables Y_n converging to Y. Let $W' = \{e_1, e_2 \dots\}$ be a countable dense subset of W_2, and let ρ denote the metric of W_2. It is easy to see that $\rho(Y, e_j)$ is $\sigma(X)$-measurable. Put $\xi_n(\omega) = \min_{1\le j\le n} \rho(Y(\omega), e_j)$ for $\omega \in \Omega$. Since there is a sequence $\{m_n\}$, depending on ω, such that $\rho(Y(\omega), e_{m_n}) \to 0$, we have

$$\xi_n(\omega) \to 0 \quad \text{as} \quad n \to \infty.$$

Let $k_n(\omega)$ be the smallest minimizer of $\xi_n(\omega)$ and $Y_n(\omega) = e_{k_n(\omega)}$. Since

$$\{i < k_n\} = \bigcap_{j \le i} \{\rho(Y, e_j) > \xi_n\}, \quad i = 1, \dots, n,$$

we see that k_n is $\sigma(X)$-measurable and $\rho(Y(\omega), Y_n(\omega)) \to 0$ as $n \to \infty$.

(2) We observe that $Y_n(\omega)$ can be rewritten as

$$Y_n(\omega) = \sum_{j=1}^{n} 1_{F_j}(\omega) e_j, \quad \omega \in \Omega,$$

where $F_j := \{k_n = j\} \in \sigma(X)$. As in the proof of Proposition 1.1.7, we find $G_j \in \mathcal{B}(W_1)$ such that $1_{F_j}(\omega) = 1_{G_j}(X(\omega))$. Define

$$\eta_n(w) = \sum_{j=1}^{n} 1_{G_j}(w) e_j, \quad w \in W_1.$$

Note that for any closed subset C of W_2

$$\{w \in W_1 : \eta_n(w) \in C\} = \bigcap_k \{w \in W_1 : \rho(\eta_n(w), C) < \frac{1}{k}\} \in \mathcal{B}(W_1).$$

Hence $\eta_n : W_1 \to W_2$ is a measurable function such that $Y_n(\omega) = \eta_n(X(\omega))$. Let A denote the set of $w \in W_1$ such that $\lim_n \eta_n(w)$ exists. Define

$$\eta(w) = \begin{cases} \lim_{n\to\infty} \eta_n(w) & \text{if } w \in A, \\ 0 & \text{otherwise.} \end{cases}$$

Then $A \in \mathcal{B}(W_1)$, η is $\mathcal{B}(W_1)$-measurable and, by (1), $X(\omega) \in A$, for every $\omega \in \Omega$. Therefore,

$$Y(\omega) = \lim_n Y_n(\omega) = \lim_n \eta_n(X(\omega)) = \eta(X(\omega)), \quad \omega \in \Omega.$$

■

Definition 2.1.5. *Let $X : (\Omega, \mathcal{F}, P) \to (W, \mathcal{B}(W))$ and $X' : (\Omega', \mathcal{F}', P') \to (W, \mathcal{B}(W))$ be two random variables induced by N-dimensional continuous adapted stochastic processes $\{X_t, \mathcal{F}_t\}$ and $\{X'_t, \mathcal{F}'_t\}$, respectively. We say that X and X' have the same probability law, denoted by $X \overset{\mathcal{L}}{\approx} X'$, if*

$$P(X \in A) = P'(X' \in A), \quad \textit{for all} \quad A \in \mathcal{B}(W),$$

or, equivalently,

$$E[\exp(i \sum_{j=1}^{n} (\xi_j, X_{t_j}))] = E'[\exp(i \sum_{i=1}^{n} (\xi_j, X'_{t_j}))],$$

$$\textit{for any } \xi_1, \dots, \xi_n \in \mathbf{R}^N, \ 0 \le t_1 < \cdots < t_n,$$

where E' denotes the expectation with respect to P'.

Proposition 2.1.6. *Let $\{X_t, B_t\}$ and $\{X'_t, B'_t\}$ be two stochastic processes satisfying* (2.1) *under P and P'. Suppose b and σ are Lipschitz. Then X and X' have the same probability law, that is,*

$$X \overset{\mathcal{L}}{\approx} X'.$$

Proof.

(1) Since $\{X_t\}$ is continuous and $\{\mathcal{F}_t^B\}$-adapted, X is $\sigma(B)$-measurable. By Theorem 2.1.4, there exists a measurable function $\eta : W^M \to W^N$ such that $X = \eta(B)$. Put $Y(\omega) = \eta(B'(\omega))$ and $Y_t(\omega) = \pi_t(Y(\omega))$. Clearly, $(X, B) \overset{\mathcal{L}}{\approx} (Y, B')$ for two Brownian motions $\{B_t\}$ and $\{B'_t\}$. Let $B_{\cdot\wedge t} = \{B_{s\wedge t}\}_{s\geq 0}$ for fixed $t \geq 0$. Since $\mathcal{F}_t^B = \mathcal{F}^{B_{\cdot\wedge t}}$, Theorem 2.1.4 gives

$$X_t = \hat{\eta}(B_{\cdot\wedge t})$$

for some measurable function $\hat{\eta} : (W^M, \mathcal{B}_t(W^M)) \to (\mathbf{R}^N, \mathcal{B}(\mathbf{R}^N))$. Since

$$\begin{aligned} E'[\exp\{i(\xi, \pi_t \circ \eta(B') - \hat{\eta}(B'_{\cdot\wedge t}))\}] &= E[\exp\{i(\xi, \pi_t \circ \eta(B) - \hat{\eta}(B_{\cdot\wedge t}))\}] \\ &= E[\exp\{i(\xi, X_t - \hat{\eta}(B_{\cdot\wedge t}))\}] = 1, \\ &\quad \xi \in \mathbf{R}^N, \end{aligned}$$

we have

$$\pi_t \circ \eta(B') = \hat{\eta}(B'_{\cdot\wedge t}), \qquad P'\text{-}a.s.$$

Therefore $\{Y_t\}$ is $\{\mathcal{F}_t^{B'}\}$-adapted.

(2) We claim that $\{Y_t, B'_t\}$ satisfies (2.1). Let $\delta = \{0 = t_0 < t_1 < t_2 < \cdots < t_n = t\}$, with $t_j = jt/n$, and

$$R_n(X, B) = x + \sum_{j=1}^n b(X_{t_{j-1}})(t_j - t_{j-1}) + \sum_{j=1}^n \sigma(X_{t_{j-1}})(B_{t_j} - B_{t_{j-1}}).$$

Then, by (1)

$$E'[\exp\{i(\xi, Y_t - R_n(Y, B'))\}] = E[\exp\{i(\xi, X_t - R_n(X, B))\}], \quad \xi \in \mathbf{R}^N.$$

Letting $n \to \infty$, we have

$$R_n(X, B) \to x + \int_0^t b(X_s)ds + \int_0^t \sigma(X_s)dB_s \quad \text{in probability,}$$

and, hence,

$$E'[\exp\{i(\xi, Y_t - (x + \int_0^t b(Y_s)ds + \int_0^t \sigma(Y_s)dB'_s))\}] = 1, \quad t \geq 0.$$

By continuity,

$$Y_t = x + \int_0^t b(Y_s)ds + \int_0^t \sigma(Y_s)dB'_s, \quad t \geq 0, \quad P'\text{-a.s.}$$

(3) By uniqueness, $X' = Y$. Thus $X \overset{\mathcal{L}}{\approx} X'$. ∎

2.2 Linear Functionals and Probabilities

We present the Daniell theorem, which extends the Riesz representation theorem of a continuous linear functional on $C(\Omega)$ for a compact metric space Ω.

Theorem 2.2.1 (Daniell). *Let H be a linear space of real-valued functions defined on a set Ω, which contains 1 and is closed under the operations $\wedge$ and $\vee$. Let Λ be a linear functional on H such that*

(i) $\Lambda(1) = 1$,
(ii) Λ *is positive, that is,* $\Lambda(f) \geq 0$ *if* $f \geq 0$,
(iii) *if* $f_n \in H$, $f_1 \geq f_2 \geq \cdots \geq f_n \geq \cdots$, $\lim_{n\to\infty} f_n = 0$, *then* $\lim_{n\to\infty} \Lambda(f_n) = 0$.

Then there exists a unique probability P on the measurable space $(\Omega, \mathcal{F})$ with the σ-algebra $\mathcal{F} := \sigma(f, f \in H) = \sigma(H)$ generated by H such that

(iv) $\Lambda(f) = \int_\Omega f(\omega)P(d\omega), \quad f \in H.$

Proof.

(1) Let $\bar{H}$ denote the collection of functions $f : \Omega \to \bar{\mathbf{R}} := \mathbf{R} \cup \{\infty\}$ of the form $f = \lim_{n\to\infty} \nearrow f_n$ for a nondecreasing sequence $f_n \in H^+ := \{h \in H : h \geq 0\}$. Let $f, f' \in \bar{H}$, $f \leq f'$ and $f_n \nearrow f$, $f'_m \nearrow f'$. Since $f_n \wedge f'_m \nearrow f_n \wedge f' = f_n$ as $m \to \infty$, we have by (ii) and (iii)

$$\Lambda(f_n) = \lim_m \Lambda(f_n \wedge f'_m) \leq \lim_m \Lambda(f'_m).$$

Hence $\lim_n \Lambda(f_n) \leq \lim_m \Lambda(f'_m)$. Therefore, we can define $\Lambda(f)$ for $f \in \bar{H}$ by $\Lambda(f) = \lim_{n\to\infty} \nearrow \Lambda(f_n)$, independent of a choice of f_n.

(2) Λ on $\bar{H}$ has the following properties:

① $0 \leq \Lambda(f) \leq \infty$ for all $f \in \bar{H}$,
② $f, g \in \bar{H},\ f \leq g \Rightarrow \Lambda(f) \leq \Lambda(g)$,
③ $f \in \bar{H},\ 0 \leq a < \infty \Rightarrow af \in \bar{H},\ \Lambda(af) = a\Lambda(f)$,
④ $f, g \in \bar{H} \Rightarrow f + g, f \wedge g, f \vee g \in \bar{H},\ \Lambda(f \wedge g) + \Lambda(f \vee g) = \Lambda(f) + \Lambda(g)$,
⑤ $f_n \in \bar{H},\ f_n \nearrow f \Rightarrow f \in \bar{H},\ \lim_n \Lambda(f_n) = \Lambda(f)$.

①–③ are immediate from (1). Let $f_n,\ g_n \in H^+,\ f_n \nearrow f,\ g_n \nearrow g$. Clearly, $f_n + g_n \nearrow f + g,\ f_n \wedge g_n = f_n - (f_n - g_n)^+ \nearrow f \wedge g,\ f_n \vee g_n \nearrow f \vee g$, and $f_n + g_n = f_n \wedge g_n + f_n \vee g_n$. Hence,

$$\Lambda(f_n \wedge g_n) + \Lambda(f_n \vee g_n) = \Lambda(f_n + g_n) = \Lambda(f_n) + \Lambda(g_n).$$

Letting $n \to \infty$, we obtain ④. Let $f_{m,n} \in H^+$, $f_{m,n} \nearrow f_n$ as $m \to \infty$. Setting $g_m = f_{m,1} \vee f_{m,2} \vee \cdots \vee f_{m,m} \in H^+$, we have

$$f_{m,n} \le g_m \le f_m, \quad g_m \le g_{m+1} \quad \text{for} \quad n \le m.$$

Letting $m \to \infty$, we get

$$f_n = \lim_m \nearrow f_{m,n} \le \lim_m \nearrow g_m \le f.$$

Hence,

$$f = \lim_n f_n = \lim_n \lim_m \nearrow f_{m,n} \le \lim_m \nearrow g_m \le f.$$

Therefore, $f \in \bar{H}$ and $\lim_n \Lambda(f_n) = \lim_m \Lambda(g_m) = \Lambda(f)$, which implies ⑤.

(3) Define $\mu(A) = \Lambda(1_A)$ for $A \in \mathcal{J} := \{A \subset \Omega : 1_A \in \bar{H}\}$. Then, we see by (2) that $(\mu, \mathcal{J})$ satisfies the following:

① $0 \le \mu(A) \le 1$ for all $A \in \mathcal{J}$, $\phi, \Omega \in \mathcal{J}$, $\mu(\Omega) = 1$, $\mu(\phi) = 0$,

② $A_1, A_2 \in \mathcal{J} \Rightarrow A_1 \cup A_2, A_1 \cap A_2 \in \mathcal{J}, \mu(A_1 \cap A_2) + \mu(A_1 \cup A_2) = \mu(A_1) + \mu(A_2)$,

③ $A_1, A_2 \in \mathcal{J}$, $A_1 \subset A_2 \Rightarrow \mu(A_1) \le \mu(A_2)$,

④ $A_n \in \mathcal{J}, n = 1, 2, \ldots$, $A_n \nearrow A \Rightarrow A \in \mathcal{J}$, $\mu(A_n) \nearrow \mu(A)$.

(4) Define $\mu^*(A) = \inf\{\mu(B) : A \subset B, B \in \mathcal{J}\}$ for $A \subset \Omega$. Then μ^* satisfies the following:

① $0 \le \mu^*(A) \le 1$, for all $A \subset \Omega$, $\mu^*(A) = \mu(A)$ for $A \in \mathcal{J}$,

② $A_1, A_2 \subset \Omega \Rightarrow \mu^*(A_1 \cap A_2) + \mu^*(A_1 \cup A_2) \le \mu^*(A_1) + \mu^*(A_2)$,

③ $\mu^*(A) + \mu^*(A^c) \ge 1$,

④ $A_1 \subset A_2 \subset \Omega \Rightarrow \mu^*(A_1) \le \mu^*(A_2)$,

⑤ $A_n \subset \Omega$, $n = 1, 2, \ldots$, $A_n \nearrow A \Rightarrow \mu^*(A_n) \nearrow \mu^*(A)$.

① and ④ are immediate from (3) ①, ③. For ②, we take $B_1, B_2 \in \mathcal{J}$ for any $\varepsilon > 0$ such that $A_i \subset B_i$ and $\mu(B_i) \le \mu^*(A_i) + \varepsilon/2$ $(i = 1, 2)$. Then, by (3) ②,

$$\begin{aligned} \mu^*(A_1 \cap A_2) + \mu^*(A_1 \cup A_2) &\le \mu(B_1 \cap B_2) + \mu(B_1 \cup B_2) \\ &= \mu(B_1) + \mu(B_2) \\ &\le \mu^*(A_1) + \mu^*(A_2) + \varepsilon, \end{aligned}$$

which implies ②. Taking $A_1 = A$, $A_2 = A^c$ in ② and noting $\mu^*(\Omega) = \mu(\Omega) = 1$, we have ③. To see ⑤, we notice that $\mu^*(A) \ge \lim_n \mu^*(A_n)$. We take $B_n \in \mathcal{J}$ for $\varepsilon > 0$ such that $A_n \subset B_n$ and $\mu(B_n) \le \mu^*(A_n) + \varepsilon 2^{-n}$.

Let $B'_n = \cup_{i=1}^n B_i$. Clearly, $A \subset \cup_n B_n = \cup_n B'_n \in \mathcal{J}$. Hence

$$\mu^*(A) \le \mu(\bigcup_n B'_n) = \lim_n \mu(B'_n) \le \lim_n(\mu^*(A_n) + \varepsilon \sum_{i=1}^n 2^{-i})$$
$$\le \lim_n \mu^*(A_n) + \varepsilon.$$

It remains to show that

$$\mu(B'_n) \le \mu^*(A_n) + \varepsilon \sum_{i=1}^n 2^{-i}, \quad n = 1, 2, \ldots.$$

This is true for $n = 1$. Suppose that it holds for n. Since $A_n = A_n \cap A_{n+1} \subset B'_n \cap B_{n+1} \in \mathcal{J}$, we get by (3) ②

$$\mu(B'_{n+1}) = \mu(B'_n) + \mu(B_{n+1}) - \mu(B'_n \cap B_{n+1})$$
$$\le \mu^*(A_n) + \varepsilon \sum_{i=1}^n 2^{-i} + \mu^*(A_{n+1}) + \varepsilon 2^{-(n+1)} - \mu^*(A_n)$$
$$= \mu^*(A_{n+1}) + \varepsilon \sum_{i=1}^{n+1} 2^{-i}.$$

(5) Define $\mathcal{K} = \{A \subset \Omega : \mu^*(A) + \mu^*(A^c) = 1\}$. Then $(\Omega, \mathcal{K}, \mu^*)$ is a probability space. Indeed, it is clear that $\Omega \in \mathcal{K}$ and $\mathcal{K}$ is closed under complementation. Let $A_1, A_2 \in \mathcal{K}$. By (4) ②

$$\mu^*(A_1 \cap A_2) + \mu^*(A_1 \cup A_2) \le \mu^*(A_1) + \mu^*(A_2),$$
$$\mu^*((A_1 \cap A_2)^c) + \mu^*((A_1 \cup A_2)^c) = \mu^*(A_1^c \cup A_2^c) + \mu^*(A_1^c \cap A_2^c)$$
$$\le \mu^*(A_1^c) + \mu^*(A_2^c).$$

Putting together, we have by (4) ③

$$\mu^*(A_1 \cap A_2) + \mu^*(A_1 \cup A_2) + \mu^*((A_1 \cap A_2)^c) + \mu^*((A_1 \cup A_2)^c) = 2.$$

Hence $A_1 \cap A_2, A_1 \cup A_2 \in \mathcal{K}$, which implies $\mathcal{K}$ is an algebra. Furthermore, it is easy to see that the equalities hold for two relations above. Thus μ^* is finitely additive on $\mathcal{K}$. Let $A_n \in \mathcal{K}$, $A_n \nearrow A$. By (4) ⑤, there is $A_m \in \mathcal{K}$ for $\varepsilon > 0$ such that $A_m \subset A$ and $\mu^*(A) \le \mu^*(A_m) + \varepsilon$. Also, by (4) ④, $\mu^*(A^c) \le \mu^*(A_m^c)$. Hence $\mu^*(A) + \mu^*(A^c) \le 1$ and then, by (4) ③, $A \in \mathcal{K}$. This shows that $\mathcal{K}$ is a σ-algebra. By (4) ⑤, μ^* is countably additive.

(6) We claim that if $f \in \bar{H}$, $a \in \mathbf{R}$, then $\{f > a\} \in \mathcal{J}$. Let $f_n \in H^+$, $f_n \nearrow f$ and $k > 0$. Since $k(f_n - a)^+ \in H^+$ and $k(f_n - a)^+ \nearrow k(f - a)^+$, we get $k(f - a)^+ \in \bar{H}$. Clearly, $1 \wedge k(f - a)^+ \nearrow 1_{\{f>a\}}$ as $k \to \infty$. By (2) ⑤, $1_{\{f>a\}} \in \bar{H}$, which implies $\{f > a\} \in \mathcal{J}$.

(7) We have $\sigma(\mathcal{J}) = \sigma(\bar{H}) = \sigma(H)$. If $A \in \mathcal{J}$, then $1_A \in \bar{H}$. Hence $A = \{1_A = 1\} \in \sigma(\bar{H})$. Therefore, $\sigma(\mathcal{J}) \subset \sigma(\bar{H})$. If $f \in \bar{H}$, then (6) gives $\{f > a\} \in \mathcal{J}$. Hence $\sigma(\bar{H}) \subset \sigma(\mathcal{J})$. If $f \in \bar{H}$, $f_n \in H$, $f_n \nearrow f$, then f is $\sigma(H)$-measurable. Thus $\sigma(\bar{H}) \subset \sigma(H)$. If $f \in H$, $f = f^+ - f^-$, then

$f^+, f^- \in H^+ \subset \bar{H}$. Hence f is $\sigma(\bar{H})$-measurable. Therefore, $\sigma(H) \subset \sigma(\bar{H})$.

(8) We claim that $\mu^*(A) = \inf\{\Lambda(f) : 1_A \le f, f \in \bar{H}\}$ for $A \subset \Omega$. By definition,

$$\mu^*(A) = \inf\{\Lambda(1_B) : 1_A \le 1_B, 1_B \in \bar{H}\}$$
$$\ge \inf\{\Lambda(f) : 1_A \le f, f \in \bar{H}\}.$$

To show the reverse inequality, let $f \in \bar{H}, 1_A \le f$ and $0 < a < 1$. Then $A \subset \{f > a\} \in \mathcal{J}$. Hence

$$\mu^*(A) \le \mu(f > a) = \Lambda(1_{\{f>a\}}) \le \Lambda(\frac{f}{a}) = \frac{1}{a}\Lambda(f).$$

Letting $a \to 1$, we deduce $\mu^*(A) \le \Lambda(f)$.

(9) Let $A \in \mathcal{J}$ and take $f_n \in H^+$ such that $f_n \nearrow 1_A$. Then $1_{A^c} \le 1 - f_n \in H^+ \subset \bar{H}$. By (8),

$$\mu^*(A^c) \le \Lambda(1 - f_n) = 1 - \Lambda(f_n) \to 1 - \Lambda(1_A) = 1 - \mu^*(A) \text{ as } n \to \infty.$$

Therefore, $A \in \mathcal{K}$. This yields that $\mathcal{J} \subset \mathcal{K}$ and hence, by (7), $\sigma(H) \subset \mathcal{K}$.

(10) By (5) and (7), we define the probability measure P on $\sigma(H)$ by $P(A) = \mu^*(A)$. To prove (iv), let F denote the collection of $\sigma(H)$-measurable functions f satisfying (iv). Since $\Lambda(1_A) = \mu^*(A) = P(A)$ for $A \in \mathcal{J}$, we get $1_A \in F$. Let $f \in \bar{H}$. By (6), $1_{\{f>a\}} \in F$. Hence $1_{\{a<f\le b\}} = 1_{\{f>a\}} - 1_{\{f>b\}} \in F$ for $a < b$. Thus

$$f_n = \sum_{i=1}^{n2^n} \frac{i-1}{2^n} 1_{\{\frac{i-1}{2^n} < f \le \frac{i}{2^n}\}} + n 1_{\{f>n\}} \in F, \ n = 1, 2, \ldots .$$

Since $f_n \nearrow f$, we have $f \in F$ by the monotone convergence theorem. Therefore, $\bar{H} \subset F$ and then $H \subset F$.

(11) To prove uniqueness, let P' be another such measure. Let $\mathcal{A}$ be the collection of $A \in \sigma(H)$ such that $P(A) = P'(A)$. Since $\int f dP = \int f dP'$, for $f \in H$ and $f \in \bar{H}$, we see that $\mathcal{J} \subset \mathcal{A}$. It is easy to check that $\mathcal{A}$ is a σ-algebra. Thus we conclude $\sigma(H) = \sigma(\mathcal{J}) \subset \mathcal{A}$. ■

Theorem 2.2.2. *Let $(\Omega, \mathcal{F}, P)$ be a probability space. Suppose that Ω is a Polish space with metrics d and $\mathcal{F}$ is the σ-algebra of Borel sets. Then*

$$P(A) = \sup\{P(K) : \ K \subset A, K \text{ compact}\}, \quad A \in \mathcal{F}. \tag{2.2}$$

Proof. Let $\mathcal{F}^*$ be the class of sets $A \in \mathcal{F}$ for which (2.2) is valid, and $\mathcal{F}_0$ the class of sets B of the form:

$$B = \bigcup_{n=1}^{\infty} F_n = \bigcap_{n=1}^{\infty} G_n, \quad F_n \text{ closed}, \quad G_n \text{ open}. \tag{2.3}$$

(1) Claim: $\bar{\mathcal{F}} := \{F : \ \text{closed}\} \subset \mathcal{F}^*$. Let $\{\omega_n : n \in \mathbf{N}\}$ be the countable dense set of Ω. For $n \in \mathbf{N}$, let $\bar{B}_{1/k}(\omega_n)$ denote the sequence of closed

balls of radius $1/k (k \in \mathbf{N})$ centered at ω_n. Then $\cup_n \bar{B}_{1/k}(\omega_n) = \Omega$ and $\cup_n (\bar{B}_{1/k}(\omega_n) \cap F) = F$ for $F \in \bar{\mathcal{F}}$. For each $\varepsilon > 0$ and k, we choose n_k such that

$$P(\bigcup_{n=1}^{n_k} (\bar{B}_{1/k}(\omega_n) \cap F)) \geq P(F) - \varepsilon/2^k.$$

Define

$$\bar{B}_k = \bigcup_{n=1}^{n_k} (\bar{B}_{1/k}(\omega_n) \cap F) \quad \text{and} \quad K_\varepsilon = \bigcap_{k=1}^{\infty} \bar{B}_k.$$

Clearly,

$$P(F \setminus K_\varepsilon) \leq \sum_{k=1}^{\infty} P(F \setminus \bar{B}_k) \leq \varepsilon.$$

Furthermore, K_ε is closed and $K_\varepsilon \subset \cup_{n=1}^{n_k} B_{2/k}(\omega_n)$ for every $k \in \mathbf{N}$. Hence K_ε is both complete and totally bounded, that is, K_ε is compact. This implies $F \in \mathcal{F}^*$.

(2) Claim: $\mathcal{F}_0$ is an algebra and $\bar{\mathcal{F}} \subset \mathcal{F}_0$. By (2.3), it is easy to see that $\mathcal{F}_0$ is closed under the operations $\cdot^c$ and $\cup$. For $F \in \bar{\mathcal{F}}$, we have $F = \cap_n \{\omega : d(\omega, F) < \frac{1}{n}\}$, which fulfills (2.3).

(3) Claim: $\mathcal{F}^*$ is monotone and $\mathcal{F}_0 \subset \mathcal{F}^*$. Let $A_n \in \mathcal{F}^*, n = 1, 2, \ldots,$ be a nondecreasing sequence and $A = \cup_n A_n$. Let K_n be a compact set such that $K_n \subset A_n$ and $P(A_n) - \varepsilon/2^{n+1} < P(K_n)$ for $\varepsilon > 0$. Take $N \in \mathbf{N}$ such that $P(A_N) > P(A) - \varepsilon/2$. Then, setting $K_N = \cup_{n=1}^{N} K_n$, we have

$$P(A \setminus K_N) \leq P(A \setminus A_N) + \sum_{n=1}^{N} P(A_n \setminus K_n) < \varepsilon.$$

Thus $A \in \mathcal{F}^*$. Furthermore, B of (2.3) belongs to $\mathcal{F}^*$. By analogy, we can show that $\mathcal{F}^*$ is closed under the limit operation for a nonincreasing sequence.

(4) A standard result on σ-algebras gives that $\mathcal{F} = \sigma(\bar{\mathcal{F}}) \subset \sigma(\mathcal{F}_0) \subset \mathcal{F}^*$. ■

2.3 Regular Conditional Probabilities

Definition 2.3.1. *Let $(\Omega, \mathcal{F}, P)$ be a probability space and $\mathcal{G}$ a sub-σ-algebra of $\mathcal{F}$. A function $Q : (\omega, A) \in \Omega \times \mathcal{F} \to Q(\omega, A) \in [0, 1]$ is a regular conditional probability for $E[\cdot|\mathcal{G}]$ if*

(a) *$Q(\omega, \cdot)$ is a probability measure on $(\Omega, \mathcal{F})$ for each $\omega \in \Omega$;*
(b) *for each $A \in \mathcal{F}, \omega \in \Omega \to Q(\omega, A)$ is $\mathcal{G}$-measurable;*
(c) *for every $f \in L^\infty(\Omega, \mathcal{F}, P)$, $E[f|\mathcal{G}](\omega) = \int_\Omega f(\omega') Q(\omega, d\omega')$, P-a.s. $\omega \in \Omega$.*

Theorem 2.3.2. *Let $(\Omega, \mathcal{F}, P)$ be a probability space and $\mathcal{G}$ a sub-σ-algebra of $\mathcal{F}$. Suppose that Ω is a Polish space with metric d and that $\mathcal{F} = \mathcal{B}(\Omega)$. Then there exists a regular conditional probability Q for $E[\cdot|\mathcal{G}]$, uniquely.*

Proof. Let $H = U_b(\Omega)$ be the Banach space of bounded, uniformly continuous, $\mathbf{R}$-valued functions h on Ω with norm $\|h\| = \sup_\omega |h(\omega)|$.

(1) We claim that if Ω is a compact metric space, H is separable. According to the theory of topological spaces, Ω is totally bounded, that is, there exists $\Omega_n = \{\omega_i \in \Omega : i = 1, 2, \ldots, q(n)\}$ for every $n \in \mathbf{N}$ such that $\Omega = \bigcup_{i=1}^{q(n)} B_{\frac{1}{n}}(\omega_i)$. Define a countable subset $\Omega' = \bigcup_{n \in \mathbf{N}} \Omega_n$ of Ω. Clearly, the closure $\bar{\Omega}'$ of Ω' is Ω, which implies that Ω is separable.

Put $f_{\hat{\omega},r}(\omega) = d(\omega, \Omega \setminus B_r(\hat{\omega}))$ for each $(\hat{\omega}, r) \in \Omega' \times \mathbf{Q}$, and let D denote the linear space of polynomials of $f_{\hat{\omega},r}$ with rational coefficients. Then D is a countable subalgebra of $H = C(\Omega)$, closed under the multiplication, with $1 \in D$. Let $\omega_1 \neq \omega_2$. There is $B_r(\hat{\omega})$, $(\hat{\omega}, r) \in \Omega' \times \mathbf{Q}$, such that $\omega_1 \in B_r(\hat{\omega})$ and $\omega_2 \notin B_r(\hat{\omega})$, hence, by definition, $f_{\hat{\omega},r}(\omega_1) \neq f_{\hat{\omega},r}(\omega_2) = 0$. This yields that D separates points of Ω. By the Stone–Weierstrass Theorem (Appendix B), D is dense in $C(\Omega)$.

(2) We turn to the case when Ω is a Polish space. We set $\hat{d}(\omega, \tilde{\omega}) := d(\omega, \tilde{\omega})/(1 + d(\omega, \tilde{\omega}))$ for $\omega, \tilde{\omega} \in \Omega$, $[0, 1]^\infty := \prod_{n=1}^\infty I_n$, $I_n = [0, 1]$, and take a countable dense subset $\{\omega_n, n \in \mathbf{N}\}$ of Ω. Define the metric ρ on $[0, 1]^\infty$ by

$$\rho(x, y) := \sum_{n=1}^\infty \frac{1}{2^n}|x_n - y_n| \quad \text{for} \quad x = \{x_n\},\ y = \{y_n\} \in [0, 1]^\infty.$$

By Tychonoff's theorem, we observe that $[0, 1]^\infty$ is compact under ρ since

$$\frac{1}{2^m}|x_m - y_m| \leq \rho(x, y) \leq \sum_{n \leq m} \frac{1}{2^n}|x_n - y_n| + \frac{4}{2^{m+1}}, \qquad \text{for all } m \in \mathbf{N}.$$

Let ξ be the map $\xi : \Omega \to [0, 1]^\infty$ defined by

$$\xi(\omega) = (\hat{d}(\omega, \omega_1), \hat{d}(\omega, \omega_2), \ldots).$$

It is easy to see that

$$\begin{aligned}
\rho(\xi(\omega), \xi(\tilde{\omega})) &\leq \hat{d}(\omega, \tilde{\omega}), \qquad \omega, \tilde{\omega} \in \Omega,\\
\hat{d}(\omega, \tilde{\omega}) &\leq |\hat{d}(\omega, \omega_n) - \hat{d}(\tilde{\omega}, \omega_n)| + 2\hat{d}(\tilde{\omega}, \omega_n)\\
&\leq 2^n \rho(\xi(\omega), \xi(\tilde{\omega})) + 2\hat{d}(\tilde{\omega}, \omega_n),
\end{aligned}$$

for all $n \in \mathbf{N}$. This implies that ξ is a homeomorphism of Ω to $F := \xi(\Omega) \subset [0, 1]^\infty$, that is, ξ is continuous and one-to-one, with a continuous inverse. Hence H can be identified with $U_b(F)$. Since the closure $\bar{F}$ of F is compact in $[0, 1]^\infty$, we have, by (1) that $U_b(\bar{F})$ is separable and so is $U_b(F)$.

It remains to show that every $f \in U_b(F)$ is uniquely extended to $\bar{f} \in U_b(\bar{F})$ satisfying $\|f\| := \sup_{x\in F}|f(x)| = \sup_{x\in\bar{F}}|\bar{f}(x)|$. By the uniform continuity of f, we find $\delta > 0$ and $C_\varepsilon := 2\|f\|/\delta \geq 0$, for any $\varepsilon > 0$, such that

$$|f(x)-f(y)| \leq \{\sup_{\rho(x,y)\geq\delta}\frac{|f(x)-f(y)|}{\delta}\}\rho(x,y) + \sup_{\rho(x,y)<\delta}|f(x)-f(y)|$$
$$\leq C_\varepsilon\rho(x,y)+\varepsilon, \quad x,y\in F.$$

Let $x \in \bar{F}\setminus F$ and $x_n \in F \to x$. Then $\{f(x_n)\}$ is a Cauchy sequence. Hence we can define $\bar{f}(x) = \lim_{n\to\infty} f(x_n)$, independent of a choice of $\{x_n\}$. Consequently, we deduce that $\bar{f}$ has the desired properties, and hence $\bar{f} \in U_b(\bar{F})$.

(3) Let H_1 be the class of all f_n taken from the dense subset H_0 of H such that $f_1 = 1$ and $f_1, \ldots, f_n$ are linearly independent for each $n \in \mathbf{N}$. Then the linear span H_2 of H_1 is dense in H. Note that $f \in H_2$ can be uniquely written as $f = \sum_{i=1}^n t_i f_i,\ t_1, \ldots, t_n \in \mathbf{R}$. Define the linear functional λ_ω on H_2 by

$$\lambda_\omega(f) = E[f|\mathcal{G}](\omega), \quad \text{for} \quad f = \sum_{i=1}^n t_i f_i \in H_2,\ t_1,\ldots,t_n \in \mathbf{R}, \quad \omega\in\Omega.$$

If $f = \sum_{i=1}^n r_i f_i \geq 0$, for $r_1, \ldots, r_n \in \mathbf{Q}$, let

$$N(r_1,\ldots,r_n) = \{\omega : \lambda_\omega(f) < 0\}.$$

Clearly, $P(N(r_1,\ldots,r_n)) = 0$. Let N_0 be the union of all such $N(r_1,\ldots,r_n)$ with $n \in \mathbf{N}, r_i \in \mathbf{Q}$. Then $N_0 \in \mathcal{G}$ and $P(N_0) = 0$. For $\omega \notin N_0$, $\lambda_\omega(1) = 1$.

(4) We claim that λ_ω is positive for $\omega \notin N_0$. If $f = \sum_{i=1}^n t_i f_i \geq 0, t_1, \ldots, t_n \in \mathbf{R}$, then there exists $r_1, \ldots, r_n \in \mathbf{Q}$, for $\varepsilon > 0, \varepsilon \in \mathbf{Q}$, such that $r_i \to t_i$ as $\varepsilon \to 0$ and $\varepsilon f_1 + \sum_{i=1}^n r_i f_i \geq 0$. Since $\omega \notin N_0$, we get $E[\varepsilon f_1 + \sum_{i=1}^n r_i f_i|\mathcal{G}](\omega) \geq 0$. Letting $\varepsilon \to 0$, we have $\lambda_\omega(f) \geq 0$.

(5) We claim that for $\omega \notin N_0$, λ_ω admits a unique extension on H, denoted by Λ_ω. Since $-\|f\|f_1 \leq f \leq \|f\|f_1$, we get

$$|\lambda_\omega(f)| \leq \|f\|\lambda_\omega(f_1) = \|f\|, \quad f \in H_2.$$

Hence we can define Λ_ω, for $\omega \notin N_0$, by

$$\Lambda_\omega(f) = \lim_{n\to\infty}\lambda_\omega(g_n), \quad f\in H,$$

by choosing $g_n \in H_2$ such that $g_n \to f$ as $n \to \infty$. It is easy to check that $\Lambda_\omega(f)$ is independent of a choice of g_n and that $\Lambda_\omega(1) = 1$ for $\omega \notin N_0$. If $f \in H$ and $f \geq 0$, then $\varepsilon f_1 + g_n \geq \varepsilon + f - \|g_n - f\| \geq 0$ for $\varepsilon > 0$ and sufficiently large n. By (4),

$$0 \leq \lambda_\omega(\varepsilon f_1 + g_n) = \varepsilon + \lambda_\omega(g_n) \ \to\ \varepsilon + \Lambda_\omega(f), \quad \omega\notin N_0,$$

which implies $\Lambda_\omega(f) \geq 0$. Thus Λ_ω is a positive continuous linear functional on H, for $\omega \notin N_0$.

(6) We claim that if $\varphi_m \in H$, $\varphi_m \searrow 0$, then $\lim_{m\to\infty} \Lambda_\omega(\varphi_m) = 0$ for P-almost all $\omega \in \Omega$. By Theorem 2.2.2, we choose a compact set K_n for each n such that $P(K_n) \geq 1 - 1/3^n$. By Dini's theorem (Appendix A), φ_m converges uniformly to zero on K_n. Define

$$\zeta_{k,n}(\omega) = \zeta_{k,K_n}(\omega) = \frac{kd(\omega, K_n)}{1 + kd(\omega, K_n)}, \quad \text{for } k, n \in \mathbf{N}.$$

Since $|d(\omega, K_n) - d(\omega', K_n)| \leq d(\omega, \omega')$ for $\omega, \omega' \in \Omega$ and $\zeta_{k,n}(\omega) = 0$ if $\omega \in K_n$, we can easily see that

$$\zeta_{k,n} \in H, \quad \zeta_{k,n}(\omega) \nearrow 1_{K_n^c}(\omega) \text{ as } k \to \infty.$$

Hence

$$\begin{aligned}
|\Lambda_\omega((1 - \zeta_{k,n})\varphi_m)| &\leq \|(1 - \zeta_{k,n})\varphi_m\| \\
&\leq \sup_{d(\omega,K_n)\leq 1} |(1 - \zeta_{k,n})\varphi_m(\omega)| \\
&\quad + \sup_{d(\omega,K_n)>1} |(1 - \zeta_{k,n})\varphi_m(\omega)| \\
&\leq \sup_{\omega\in K_n} \varphi_m(\omega) + \frac{1}{1+k} \sup_\omega \varphi_1(\omega) \\
&\to 0 \quad \text{as } m \to \infty, \text{ and then } \ n, k \to \infty.
\end{aligned}$$

On the other hand, by the monotonicity of $\zeta_{k,n}$ with respect to k,

$$\begin{aligned}
\int_{N_0^c} \sup_k \Lambda_\omega(\zeta_{k,n}) P(d\omega) &= \lim_{k\to\infty} \int_{N_0^c} \Lambda_\omega(\zeta_{k,n}) P(d\omega) \\
&= \lim_{k\to\infty} \int_{N_0^c} E[\zeta_{k,n}|\mathcal{G}](\omega) P(d\omega) \\
&= \lim_{k\to\infty} \int_{N_0^c} \zeta_{k,n}(\omega) P(d\omega) \\
&= E[1_{K_n^c}] \leq 1/3^n,
\end{aligned}$$

from which

$$\sum_{n=1}^{\infty} P(\sup_k |\Lambda_\omega(\zeta_{k,n})| > 1/2^n) \leq \sum_{n=1}^{\infty} 2^n/3^n < \infty.$$

By the Borel–Cantelli lemma, we have $P(N_1) = 0$, where $N_1 := \bigcap_{n_0\in\mathbf{N}} \bigcup_{n\geq n_0} \{\sup_k |\Lambda_\omega(\zeta_{k,n})| > 1/2^n\}$. Clearly, $M(\omega) := \sup_n 2^n \sup_k |\Lambda_\omega(\zeta_{k,n})| < \infty$ for $\omega \notin N_2 := N_0 \cup N_1$. Furthermore,

$$|\Lambda_\omega(\zeta_{k,n}\varphi_m)| \leq \Lambda_\omega(\zeta_{k,n})\|\varphi_m\| \leq \frac{M(\omega)}{2^n}\|\varphi_1\| \quad \to \quad 0 \text{ as } n \to \quad \infty.$$

Consequently, we deduce $\limsup_{m\to\infty} |\Lambda_\omega(\varphi_m)| = 0$, for every $\omega \notin N_2$.

(7) By the Daniell theorem, there exists a unique probability measure $\hat{Q}(\omega, d\omega')$, for each $\omega \notin N_2$, such that

$$\Lambda_\omega(f) = \int_\Omega f(\omega')\hat{Q}(\omega, d\omega'), \quad f \in H.$$

Define the probability measure $Q(\omega, d\omega')$ for $\omega \in \Omega$ by $Q(\omega, d\omega') = \hat{Q}(\omega, d\omega')$ if $\omega \notin N_2$, $Q(\omega, d\omega') = P(d\omega')$ otherwise. Then, the mapping $\omega \to Q(\omega, A)$ is $\mathcal{G}$-measurable, and

$$E[f 1_A] = \int_A \int_\Omega f(\omega')Q(\omega, d\omega')P(d\omega), \quad A \in \mathcal{G},$$

for all $f \in H$. Since there is $\zeta_{k,G} \in H$ for a open set G such that $\zeta_{k,G} \to 1_G$ as $k \to \infty$, this relation holds for all bounded $\mathcal{B}(\Omega)$-measurable functions f.

(8) Let $Q_i(\omega, \cdot)$, $i = 1, 2$, be two regular conditional probabilities. Let $A \in \mathcal{F}$. Then $Q_1(\omega, A) = Q_2(\omega, A)$, for P-almost all $\omega \in \Omega$. This implies uniqueness. ∎

2.4 Weak Solutions

Definition 2.4.1. *A system* $(\Omega, \mathcal{F}, P, \{\mathcal{F}_t\}; X, B)$ *is a weak solution of* (2.1) *for Borel measurable functions* $b(x), \sigma(x)$ *as in Section 2.1 if*

(a) $(\Omega, \mathcal{F}, P)$ *is a complete probability space, and* $\{\mathcal{F}_t\}$ *is a filtration of sub-*σ*-field of* $\mathcal{F}$ *satisfying the usual conditions;*
(b) $(B_t, \mathcal{F}_t)$ *is an* M*-dimensional Brownian motion under* P*; and*
(c) $(X_t, \mathcal{F}_t)$ *is a continuous adapted process defined on it;*

$$X_t = X_0 + \int_0^t b(X_s)ds + \int_0^t \sigma(X_s)dB_s, \quad P\text{-}a.s.,$$

$$\text{where} \quad P(\int_0^t \{|b(X_s)| + \|\sigma(X_s)\|^2\}ds < \infty) = 1, \quad \text{for all } t \ge 0.$$

Definition 2.4.2. *We say that the pathwise uniqueness holds for* (2.1) *if, for two weak solutions* $(\Omega, \mathcal{F}, P, \{\mathcal{F}_t\}; X, B)$ *and* $(\Omega', \mathcal{F}', P', \{\mathcal{F}'_t\}; X', B')$ *such that*

$$(\Omega, \mathcal{F}, P; B) = (\Omega', \mathcal{F}', P'; B'), \quad X_0 = X'_0 = x \in \mathbf{R}^N \ (a.s.),$$

we have

$$P(X_t = X'_t \text{ for all } t \ge 0) = 1.$$

We consider the existence of a weak solution by the Girsanov transformation of measures. Let $X_t = x + B_t$ and $f_t = b(X_t)$ for an N-dimensional Brownian motion $(\Omega, \mathcal{F}, P, \{\mathcal{F}_t\}; B)$. We write

$$L_t = \exp(\int_0^t f_s dB_s - \frac{1}{2}\int_0^t |f_s|^2 ds).$$

Lemma 2.4.3. *Assume that*

$$E[\exp\{\frac{1}{2}(1+\delta)\int_0^T |f_s|^2 ds\}] < \infty, \quad \textit{for some } \delta > 0. \tag{2.4}$$

Then we have $E[L_T] = 1$.

Proof.

(1) By Itô's formula,

$$L_t = 1 + \int_0^t L_s f_s \, dB_s, \quad \text{a.s.}$$

Hence, by Theorem 1.5.18, $\{L_t\}$ is a positive local martingale, that is,

$$E[L_{t\wedge\tau_n}|\mathcal{F}_s] = L_{s\wedge\tau_n}, \quad \text{a.s.}, \quad t \geq s, \tag{2.5}$$

for a localizing sequence of stopping times $\tau_n \nearrow \infty$ of $\{L_t\}$. By Proposition 1.5.16

$$\{L_t\} \text{ is a supermartingale,} \quad E[L_t] \leq 1. \tag{2.6}$$

(2) We will show that $\{L_t\}$ is a martingale, that is, $E[L_T] = 1$. Let $\tau \leq T$ be a stopping time and $p, r > 1$. Then, by Hölder's inequality,

$$\begin{aligned}
E[L_\tau^p] &= E[\exp\{\int_0^\tau pf_s \, dB_s - \frac{1}{2}\int_0^\tau p|f_s|^2 ds\}] \\
&= E[\exp\{\int_0^\tau pf_s \, dB_s - \frac{1}{2}\int_0^\tau p^2 r|f_s|^2 ds\} \\
&\quad \times \exp\{\frac{1}{2}\int_0^\tau (p^2 r - p)|f_s|^2 ds\}] \\
&\leq E[(\hat{L}_\tau)^r]^{1/r} E[(\exp\{\frac{1}{2}\int_0^\tau (p^2 r - p)|f_s|^2 ds\})^{r'}]^{1/r'},
\end{aligned}$$

where $\hat{L}_t = \exp\{\int_0^t pf_s \, dB_s - \frac{1}{2}\int_0^t p^2 r|f_s|^2 ds\}$ and $1/r + 1/r' = 1$. By virtue of (2.6), we have $E[\hat{L}_\tau^r] \leq 1$. Hence

$$\begin{aligned}
E[L_\tau^p] &\leq E[(\exp\{\frac{1}{2}(p^2 r - p)\int_0^\tau |f_s|^2 ds\})^{r'}]^{1/r'} \\
&= E[\exp\{\frac{1}{2}(p^2 r - p)r' \int_0^T |f_s|^2 ds\}]^{1/r'}.
\end{aligned}$$

Since

$$\lim_{p\to 1}(p^2 r - p)r' = \lim_{p\to 1}(p^2 r - p)\frac{r}{r-1} = r,$$

we can take $r, p > 1$ such that $r < 1 + \delta$ and $(p^2 r - p)r' < 1 + \delta$. Thus, by (2.4)

$$E[L_\tau^p] \leq E[\exp\{\frac{1}{2}(1+\delta)\int_0^T |f_s|^2 ds\}]^{1/r'} < \infty,$$

and hence

$$\sup_n E[L^p_{t\wedge\tau_n}] < \infty.$$

It is easy to see that

$$Y_n := L_{t\wedge\tau_n} - L_t \quad \to \quad 0, \quad \text{a.s.}$$

Furthermore, $\{Y_n\}$ is uniformly integrable because

$$\sup_n E[|Y_n| 1_{\{|Y_n|\geq c\}}] \leq \frac{1}{c^{p-1}} \sup_n E[|Y_n|^p] \to 0 \quad \text{as } c \to \infty.$$

Also,

$$\begin{aligned} E[|Y_n|] &= E[|Y_n| 1_{\{|Y_n|\leq c\}}] + E[|Y_n| 1_{\{|Y_n|\geq c\}}] \\ &\leq (1+c)E[\frac{|Y_n|}{1+|Y_n|}] + \sup_n E[|Y_n| 1_{\{|Y_n|\geq c\}}]. \end{aligned}$$

Letting $n \to \infty$ and then $c \to \infty$, we get $\limsup_{n\to\infty} E[|Y_n|] = 0$. Integrating (2.5) over $A \in \mathcal{F}_s$ and passing to the limit, we obtain $E[L_t|\mathcal{F}_s] = L_s$ a.s. ∎

Theorem 2.4.4 (Girsanov). *We assume that*

$$\sigma = I \ (I : \textit{identity}), \quad b : \mathbf{R}^N \to \mathbf{R}^N \ \textit{bounded measurable}. \tag{2.7}$$

Then (2.1) *admits a weak solution* $(\Omega, \mathcal{F}_T, P', \{\mathcal{F}_t\}_{0\leq t\leq T}; X, B')$ *for* $0 \leq t \leq T$, *defined by*

$$P'(A) = \int_A L_T \, dP, \quad \textit{for } A \in \mathcal{F}_T,$$

and

$$B'_t = X_t - x - \int_0^t b(X_s)ds.$$

Proof. By Lemma 2.4.3, we notice that the probability measure P' is defined on $\mathcal{F}_T$.

(1) We claim that

$$P'(L_t > 0 \quad \text{for } 0 \leq t \leq T) = 1.$$

Let $\tau = \inf\{t > 0 : L_t = 0\}, = \infty$ if $\{\ \} = \phi$, and $\Omega_0 = \{L_t > 0 \quad \text{for } 0 \leq t \leq T\}$. Since $L_\tau = 0$ on $\{\tau \leq T\} \in \mathcal{F}_\tau$, we get, by the optional sampling theorem,

$$\begin{aligned} P'(\Omega_0^c) = P'(\tau \le T) &= \int_{\{\tau \le T\}} L_T dP \\ &= \int_{\{\tau \le T\}} E[L_T|\mathcal{F}_\tau]dP \\ &= \int_{\{\tau \le T\}} L_{\tau\wedge T} dP = 0. \end{aligned}$$

(2) We show that for $\zeta \in L^1(\Omega, \mathcal{F}_t, P')$ and $t \ge s$,

$$E^{P'}[\zeta|\mathcal{F}_s] = E[\frac{\zeta L_t}{L_s}|\mathcal{F}_s], \quad P'\text{-a.s.},$$

where $E^{P'}$ denotes the expectation with respect to P'. Let $A \in \mathcal{F}_s$. By the definition of the conditional expectation, we have

$$\begin{aligned} \int_A E^{P'}[\zeta|\mathcal{F}_s]dP' = \int_A \zeta dP' = \int_A \zeta L_T dP &= \int_A L_s \frac{\zeta L_t}{L_s} dP \\ &= \int_A E[L_T|\mathcal{F}_s]E[\frac{\zeta L_t}{L_s}|\mathcal{F}_s]dP \\ &= \int_A E[\frac{\zeta L_t}{L_s}|\mathcal{F}_s]L_T dP \\ &= \int_A E[\frac{\zeta L_t}{L_s}|\mathcal{F}_s]dP'. \end{aligned}$$

(3) To complete the proof, it suffices to show that

$$B_t' = X_t - x - \int_0^t b(X_s)ds \quad \text{is an } \mathcal{F}_t\text{-Brownian motion under } P',$$

or, equivalently,

$$E^{P'}[e^{i(\xi, B_t' - B_s')}|\mathcal{F}_s] = e^{-\frac{|\xi|^2}{2}(t-s)}, \quad P'\text{-}a.s., \quad \xi \in \mathbf{R}^N,\ t > s. \tag{2.8}$$

We note that (2.4) is satisfied when $i\xi + f_s$ replaces f_s. Hence, by virtue of Lemma (2.4.3), we see that

$$\tilde{L}_t := \exp\{\int_0^t (i\xi + f_s)\, dB_s - \frac{1}{2}\int_0^t (i\xi + f_s, i\xi + f_s)ds\}$$

is a P-martingale.

By (2), we have

$$
\begin{aligned}
&E^{P'}[e^{i(\xi, B'_t - B'_s)}|\mathcal{F}_s] \\
&= E[\exp\{i(\xi, B_t - B_s) - i\int_s^t (f_r, \xi)\, dr\}L_t L_s{}^{-1}|\mathcal{F}_s] \\
&= E[\exp\{\int_s^t (i\xi + f_r)\, dB_r - i\int_s^t (f_r, \xi)\, dr - \frac{1}{2}\int_s^t |f_r|^2 dr\}|\mathcal{F}_s] \\
&= E[\exp\{\int_s^t (i\xi + f_r)\, dB_r - \frac{1}{2}\int_s^t (i\xi + f_r, i\xi + f_r) dr|\mathcal{F}_s] \\
&\quad \times \exp\{-\frac{1}{2}|\xi|^2(t-s)\} \\
&= E[\tilde{L}_t \tilde{L}_s^{-1}|\mathcal{F}_s]\exp\{-\frac{1}{2}|\xi|^2(t-s)\},
\end{aligned}
$$

which implies (2.8). ■

Definition 2.4.5. *Let $\{\tilde{X}_t\}$ be a continuous adapted stochastic process on a stochastic basis $(\tilde{\Omega}, \tilde{\mathcal{F}}, \tilde{P}, \{\tilde{\mathcal{F}}_t\})$. The system $(\Omega, \mathcal{F}, P, \{\mathcal{F}_t\}; X)$ is the canonical representation of $\{\tilde{X}_t\}$ if*

(a) $(\Omega, \mathcal{F}, \{\mathcal{F}_t\}) = (W, \mathcal{B}(W), \{\mathcal{B}_t(W)\})$ *with the usual conditions,*
(b) $X_t(w) = \pi_t(w)$, *for* $w \in W$, *and*
(c) $P(A) = \tilde{P}(\tilde{X} \in A)$, *for* $A \in \mathcal{B}(W)$.

Theorem 2.4.6. *Let $(\Omega, \mathcal{F}, P, \{\mathcal{F}_t\}; B)$ be the canonical representation of a Brownian motion. We assume* (2.7). *Then* (2.1) *admits a weak solution*

$$(\Omega, \mathcal{F}, \tilde{P}, \{\mathcal{F}_t\}; X, B'), \quad \textit{for } 0 \le t < \infty.$$

Proof. Let P_T denote P' on $\mathcal{F}_T$ in Theorem 2.4.4 for each $T \ge 0$.

(1) By Theorem 2.4.4, we observe that $\{P_T : T \ge 0\}$ is a consistent family of absolutely continuous probability measures with respect to P, that is,

$$P_T|_{\mathcal{F}_S} = P_S \quad \text{for } T > S \ge 0.$$

Define $\tilde{P}$ on $\mathcal{G} := \cup_{t\ge 0}\mathcal{F}_t$ by

$$\tilde{P}(A) = P_T(A) \quad \text{if} \quad A \in \mathcal{F}_T,$$

for each $T \ge 0$.

(2) We claim that $\tilde{P}$ admits an extension, also denoted by $\tilde{P}$, on $\mathcal{F} = \vee_{t\ge 0}\mathcal{F}_t$, where $\vee_{t\ge 0}\mathcal{F}_t$ denotes the σ-algebra $\sigma(\mathcal{G})$ generated by $\mathcal{G}$. By Carathéodory's Theorem, it is sufficient to check that

$$
\begin{aligned}
&\text{if } A_n \in \mathcal{G},\ A_1 \supset A_2 \supset \cdots \supset A_n \supset \ldots,\ \bigcap_n A_n = \phi, \\
&\quad \text{then } \tilde{P}(A_n) \downarrow 0.
\end{aligned} \tag{2.9}
$$

Let Ω_T be the class of all $w \in \Omega = W$ such that $w(t) = w(t \wedge T)$ for every $t \geq 0$. Since Ω_T is closed in Ω, we note that Ω_T is a Polish space and the σ-algebra of Borel sets of Ω_T coincides with $\mathcal{F}_T$. By Theorem 2.2.2, we obtain

$$P_T(A) = \sup\{P_T(K) : K \subset A, K \text{ compact}\}, \quad A \in \mathcal{F}_T.$$

Hence

$$\tilde{P}(A) = \sup\{\tilde{P}(K) : K \subset A, K \text{ compact}\}, \quad A \in \mathcal{G}.$$

Let $\varepsilon > 0$. Then we find a compact set K_n, for each n, such that

$$K_n \subset A_n, \quad \tilde{P}(A_n) \leq \tilde{P}(K_n) + \varepsilon 2^{-n}.$$

Since $\cap_n K_n \subset \cap_n A_n = \phi$, we have, by compactness, that $\cap_{n=1}^{\nu} K_n = \phi$ for some $\nu \in \mathbf{N}$. Hence $A_\nu = \cap_{n=1}^{\nu} A_n \subset \cup_{n=1}^{\nu}(A_n \setminus K_n)$. Thus, for $n \geq \nu$,

$$\tilde{P}(A_n) \leq \tilde{P}(A_\nu) \leq \sum_{n=1}^{\nu} \tilde{P}(A_n \setminus K_n) \leq \sum_n (\tilde{P}(A_n) - \tilde{P}(K_n)) \leq \varepsilon,$$

which implies (2.9).

(3) From (1) and (2) it follows that $\tilde{P}|_{\mathcal{F}_T} = P_T$ and for $T > S$,

$$E^{\tilde{P}}[1_A|\mathcal{F}_S] = E^{P_T}[1_A|\mathcal{F}_S], \quad \tilde{P}\text{-a.s.}, \quad A \in \mathcal{F}_T.$$

Thus, by Theorem 2.4.4, we deduce that $\{X_t - x - \int_0^t b(X_s)ds\}$ is the Brownian motion under $\tilde{P}$. ■

2.5 Uniqueness in Law

Definition 2.5.1. *We say that the uniqueness in law holds for*

$$dX_t = b(X_t)dt + \sigma(X_t)dB_t, \quad X_0 = x_0 \in \mathbf{R}^N \tag{2.10}$$

if, for two weak solutions $(\Omega, \mathcal{F}, P, \{\mathcal{F}_t\}; X, B)$ *and* $(\Omega', \mathcal{F}', P', \{\mathcal{F}'_t\}; X', B')$ *with* $X_0 = x_0$, $X'_0 = x_0$ *(a.s.), we have*

$$X \overset{\mathcal{L}}{\approx} X'.$$

Lemma 2.5.2. *Let* $(\Omega, \mathcal{F}, P, \{\mathcal{F}_t\}; X, B)$ *and* $(\Omega', \mathcal{F}', P', \{\mathcal{F}'_t\}; X', B')$ *be two weak solutions of* (2.10). *Then there exist two weak solutions* (x, β) *and* (x', β) *on the same stochastic basis* $(\Omega^*, \mathcal{F}^*, P^*, \{\mathcal{F}^*_t\})$ *such that*

(i) $(x, \beta) \overset{\mathcal{L}}{\approx} (X, B)$,
(ii) $(x', \beta) \overset{\mathcal{L}}{\approx} (X', B')$, *and*
(iii) $\{\beta_t\}$ *is an* $\{\mathcal{F}^*_t\}$*-Brownian motion.*

Proof. Let P_* denote the Wiener measure defined by $P_*(A) = P(B \in A)$ for $A \in \mathcal{B}(W^M)$, where $\mathcal{B}(W^M)$ is the completion of the Borel σ-algebra of W^M.

(1) Let $(\hat{\Omega}, \hat{\mathcal{F}}, \hat{P}, \{\hat{\mathcal{F}}_t\}; x, \beta)$ and $(\hat{\Omega}, \hat{\mathcal{F}}, \hat{P}', \{\hat{\mathcal{F}}_t\}; x', \beta)$ be the canonical representations of $\{X_t, B_t\}$ and $\{X'_t, B'_t\}$ on

$$(\hat{\Omega}, \hat{\mathcal{F}}, \{\hat{\mathcal{F}}_t\}) = (W^N, \mathcal{B}(W^N), \{\mathcal{B}_t(W^N)\}) \times (W^M, \mathcal{B}(W^M), \{\mathcal{B}_t(W^M)\}).$$

By Theorem 2.3.2, there exists a regular conditional probability $Q(\hat{\omega}, \cdot)$ for the conditional expectation $E^{\hat{P}}[\,\cdot\,|\{W^N, \phi\} \otimes \mathcal{B}(W^M)]$, that is,

$$Q(\hat{\omega}, A) = E^{\hat{P}}[1_A|\{W^N, \phi\} \otimes \mathcal{B}(W^M)](\hat{\omega}), \quad \hat{P} \text{ a.s.},$$

where $\hat{\omega} = (w_1, w_3) \in W^N \times W^M$ and $A \in \mathcal{B}(W^N) \otimes \mathcal{B}(W^M)$. Since the σ-algebra $\{W^N, \phi\}$ is trivial, $Q(\hat{\omega}, \cdot)$ depends only on w_3, also denoted by $Q(w_3, \cdot)$. Clearly, $Q(w_3, \cdot)$ is $\mathcal{B}(W^M)$-measurable. Let $Q'(w_3, \cdot)$ denote $Q(w_3, \cdot)$ for the probability law $\hat{P}'$ of (X', B') replacing (X, B). Define $(\Omega^*, \mathcal{F}^*, P^*, \{\mathcal{F}^*_t\}; x, x', \beta)$ by

$$\begin{aligned}
&\Omega^* = W^N \times W^N \times W^M, \quad \mathcal{F}^* = \mathcal{B}(\Omega^*), \quad \mathcal{F}^*_t = \mathcal{B}_t(\Omega^*),\\
&P^*(dw_1, dw_2, dw_3) = Q(w_3, dw_1)Q'(w_3, dw_2)P_*(dw_3),\\
&(x, x', \beta)(\omega) = (w_1, w_2, w_3), \quad \text{for} \quad \omega = (w_1, w_2, w_3) \in \Omega^*,
\end{aligned}$$

where every σ-algebra denotes its completion under P^*.

(2) By (1), for $A_1 \times A_3 \in \mathcal{B}(W^N) \otimes \mathcal{B}(W^M)$, we have

$$\hat{P}(W^N \times A_3) = P(X \in W^N, B \in A_3) = P_*(A_3).$$

Hence

$$\begin{aligned}
\int_{A_3} Q(w_3, A_1)P_*(dw_3) &= \int_{A_3} E^{\hat{P}}[1_{A_1}|\{W^N, \phi\} \otimes \mathcal{B}(W^M)](w_3)P_*(dw_3)\\
&= \int_{W^N \times A_3} E^{\hat{P}}[1_{A_1 \times W^M}|\{W^N, \phi\}\\
&\qquad \otimes \mathcal{B}(W^M)](w_1, w_3)\hat{P}(dw_1, dw_3)\\
&= \hat{P}(A_1 \times A_3).
\end{aligned}$$

Thus

$$P^*(A_1 \times W^N \times A_3) = \int_{A_3} Q(w_3, A_1)Q'(w_3, W^N)P_*(dw_3) = \hat{P}(A_1 \times A_3),$$

which implies (i) and similarly, (ii).

(3) We claim that the map $w \to Q(w, A)$ is $\mathcal{B}_t(W^M)$-measurable for each $t > 0$ and $A \in \mathcal{B}_t(W^N)$. By the same arguments as (1) and (2), there is a regular conditional probability $Q_t(w, A)$ such that the map $w \to Q_t(w, A)$ is $\mathcal{B}_t(W^M)$-measurable and

$$\hat{P}(A \times A_3) = \int_{A_3} Q_t(w, A)P_*(dw), \quad A_3 \in \mathcal{B}_t(W^M).$$

Let $D \in \mathcal{B}(W^M)$ be arbitrary. Then D can be rewritten as

$$D = \{w \in W^M : \rho_t w \in C_1,\ \eta_t w \in C_2\}, \quad C_1, C_2 \in \mathcal{B}(W^M),$$

where $(\rho_t w)(s) = w(t \wedge s)$ and $(\eta_t w)(s) = w(t+s) - w(t)$. By the definition of P_*, we note that $\eta_t w$ and $\mathcal{B}_t(W^M)$ are independent with respect to P_*. Hence

$$\begin{aligned}\int_D Q_t(w, A)P_*(dw) &= \int_{\{\rho_t w \in C_1\}} Q_t(w, A)P_*(\eta_t w \in C_2)P_*(dw)\\ &= P_*(\eta_t w \in C_2)\int_{\{\rho_t w \in C_1\}} Q_t(w, A)P_*(dw)\\ &= P_*(\eta_t w \in C_2)\hat{P}(A \times \{\rho_t w \in C_1\}).\end{aligned}$$

On the other hand,

$$\begin{aligned}\hat{P}(A \times D) &= P(X \in A, B \in D)\\ &= P(X \in A, \rho_t(B) \in C_1, \eta_t(B) \in C_2)\\ &= P(X \in A, \rho_t(B) \in C_1)P(\eta_t(B) \in C_2)\\ &= \hat{P}(A \times \{\rho_t w \in C_1\})P_*(\eta_t w \in C_2).\end{aligned}$$

Hence

$$\hat{P}(A \times D) = \int_D Q_t(w, A)P_*(dw).$$

Thus, we deduce by (2) that $Q_t(w_3, A) = Q(w_3, A)$, P_*-a.s., which is $\mathcal{B}_t(W^M)$-measurable.

(4) We recall that $\{\beta_t\}$ is a Brownian motion under P_*. Let $s < t$ and $A_1, A_2 \in \mathcal{B}_s(W^N)$, $A_3 \in \mathcal{B}_s(W^M)$. Then, by (3),

$$\begin{aligned}&E^{P^*}[\exp\{(i(\xi, \beta(t) - \beta(s))\}1_{A_1 \times A_2 \times A_3}]\\ &= \int_{A_3} \exp\{(i(\xi, \beta(t) - \beta(s))\}Q(w_3, A_1)Q'(w_3, A_2)P_*(dw_3)\\ &= \int_{A_3} E^{P_*}[\exp\{i(\xi, \beta(t) - \beta(s))\}|\mathcal{B}_s(W^M)]Q(w_3, A_1)Q'(w_3, A_2)P_*(dw_3)\\ &= \exp(-\frac{|\xi|^2}{2}(t-s))\int_{A_3} Q(w_3, A_1)Q'(w_3, A_2)P_*(dw_3)\\ &= \exp(-\frac{|\xi|^2}{2}(t-s))P^*(A_1 \times A_2 \times A_3), \quad \xi \in \mathbf{R}^M,\end{aligned}$$

which implies (iii).

(5) Taking into account (i) and (2.10), we see by virtue of Proposition 2.1.6 (2) that (x, β) satisfies

$$x_t = x_0 + \int_0^t b(x_s)ds + \int_0^t \sigma(x_s)d\beta_s, \quad P^* \text{ a.s.}$$

Thus, (x, β) is a weak solution to (2.10), and so is (x', β). ∎

Theorem 2.5.3 (Yamada–Watanabe). *Let* $(\Omega, \mathcal{F}, P, \{\mathcal{F}_t\}; X, B)$ *and* $(\Omega', \mathcal{F}', P', \{\mathcal{F}'_t\}; X', B')$ *be two weak solutions of* (2.10). *Suppose the pathwise uniqueness holds. Then the uniqueness in law holds, and* (2.10) *has a unique strong solution.*

Proof.

(1) By Lemma 2.5.2, (x, β) and (x', β) are two weak solutions to (2.10) on the same stochastic basis $(\Omega^*, \mathcal{F}^*, P^*, \{\mathcal{F}^*_t\})$. The property of pathwise uniqueness yields that

$$P^*(x_t = x'_t \quad \text{for all } t \geq 0) = 1.$$

Thus we deduce $x \overset{\mathcal{L}}{\approx} x'$, which implies $X \overset{\mathcal{L}}{\approx} X'$.

(2) We adopt the notations such as Q, Q' in Lemma 2.5.2. By pathwise uniqueness, we have $P^*(\{(w_1, w_2, w_3) : w_1 = w_2\}) = 1$. Define the product measure $\tilde{Q}_{w_3}(dw_1, dw_2) := Q(w_3, dw_1)Q'(w_3, dw_2)$ on $W^N \times W^N$ for $w_3 \in W^M$. Then we have

$$\int_{W^M} \tilde{Q}_{w_3}(w_1 = w_2)P_*(dw_3) = 1,$$

so that

$$\tilde{Q}_{w_3}(w_1 = w_2) = 1, \quad \forall w_3 \in \tilde{W} \quad \text{with } P_*(\tilde{W}) = 1.$$

Hence

$$\begin{aligned}\tilde{Q}_{w_3}(\{(w_1, w_1)\}) &= Q(w_3, \{w_1\})Q'(w_3, \{w_1\}) \\ &= \tilde{Q}_{w_3}(\{w_1\} \times W^N)\tilde{Q}_{w_3}(W^N \times \{w_1\}) \\ &= \tilde{Q}_{w_3}(\{(w_1, w_1)\})\tilde{Q}_{w_3}(\{(w_1, w_1)\}).\end{aligned}$$

This yields that $\tilde{Q}_{w_3}(\{(w_1, w_1)\}) = 0$ or 1, for $w_3 \in \tilde{W}$. Therefore, we find a unique point $(w_0, w_0) \in W^N \times W^N$ satisfying $\tilde{Q}_{w_3}(\{(w_0, w_0)\}) = 1$ for $w_3 \in \tilde{W}$. Now we can define the map $F : W^M \to W^N$ by $w_0 = F(w_3)$. Recall that

$$\tilde{Q}_{w_3}(\{(w_0, w_0)\}) = Q(w_3, \{w_0\})Q'(w_3, \{w_0\}), \quad w_3 \in \tilde{W}.$$

Then it is obvious that $Q(w_3, \{F(w_3)\}) = Q'(w_3, \{F(w_3)\}) = 1$ for $w_3 \in \tilde{W}$. Hence, for $A \in \mathcal{B}(W^N)$,

$$Q(w_3, A) = Q'(w_3, A) = 1_A(F(w_3)), \quad w_3 \in \tilde{W}.$$

Since $\{F(w_3) \in A\} = \{Q(w_3, A) = 1\} \in \mathcal{B}_t(W^M)$ for $A \in \mathcal{B}_t(W^N)$, F is measurable. We note that

$$\begin{aligned}P^*(\{x = F(\beta)\}) &= \int_{W^M} \int_{\{w_1 = F(w_3)\}} Q(w_3, dw_1)P_*(dw_3) \\ &= \int_{W^M} Q(w_3, \{F(w_3)\})P_*(dw_3) \\ &= \int_{W^M} P_*(dw_3) = 1.\end{aligned}$$

Define the adapted process $\{\bar{X}_t\}$ by $\bar{X} = F(\bar{B})$ for given Brownian motion $\{\bar{B}_t\}$. Since $(x, \beta) \overset{\mathcal{L}}{\approx} (\bar{X}, \bar{B})$, we see that $\{\bar{X}_t\}$ is a strong solution to (2.10). The uniqueness is obvious. ■

2.6 Markov Properties

We consider the Markov property of a strong solution $(\Omega, \mathcal{F}, P, \{\mathcal{F}_t\}; X)$ of

$$dX_t = b(X_t)dt + \sigma(X_t)dB_t, \quad X_0 = x \in \mathbf{R}^N, \tag{2.11}$$

also denoted by $\{X_t^x\}$, starting at x.

Definition 2.6.1.

(a) *Let $P_x(\cdot)$ be the probability measure on $(\Omega, \mathcal{G}, \{\mathcal{G}_t\})$ defined by*

$$P_x(X_{t_1} \in A_1, \ldots, X_{t_n} \in A_n) = P(X_{t_1}^x \in A_1, \ldots, X_{t_n}^x \in A_n)$$

for any $t_0 = 0 < t_1 < \cdots < t_n,\ A_i \in \mathcal{B}(\mathbf{R}^N),\ i = 1, 2, \ldots, n,$ *where* $\mathcal{G} = \sigma(X_s, s \geq 0)$ *and* $\mathcal{G}_t = \sigma(X_s, s \leq t)$.

(b) *We define $\mathcal{F}^X = \cap_x(\mathcal{G})^{P_x}$ and $\mathcal{F}_t^X = \cap_x(\mathcal{G}_t)^{P_x}$, where $(\cdot)^{P_x}$ denotes the completion with respect to P_x.*

(c) *The system $(\Omega, \mathcal{F}^X, P, \{\mathcal{F}_t^X\}; X)$ is a strong Markov process if the map $x \to P_x(X_t \in A)$ is Borel measurable for $A \in \mathcal{B}(\mathbf{R}^N)$, and it has the strong Markov property:*

$$E_x\left[f(X_{t+\tau})|\mathcal{F}_\tau^X\right] = E_{X_\tau}[f(X_t)], \quad P_x\ a.s. \quad t \geq 0,$$

for any bounded Borel function f on $\mathbf{R}^N$ and $\mathcal{F}_t^X$-bounded stopping time τ, where E_x denotes the expectation with respect to P_x.

(d) *We say that the Markov property holds if the above equality holds whenever $\tau = T$ is a nonrandom time.*

Theorem 2.6.2. *We assume that*

$$b, \sigma : \text{Lipschitz}. \tag{2.12}$$

Then $(\Omega, \mathcal{F}^X, P, \{\mathcal{F}_t^X\}; X)$ is a strong Markov process such that $\mathcal{F}_{t+}^X = \mathcal{F}_t^X$, for all $t \geq 0$.

Proof.

(1) By monotonicity, it is easy to see that $h(r) := E[\sup_{0 \leq t \leq r} |X_t^x - X_t^y|^2]$ is continuous. Also, by Theorem 1.7.2 and Gronwall's inequality, we observe that

$$h(r) \leq C_r|x - y|^2, \quad x, y \in \mathbf{R}^N$$

for some constant $C_r > 0$, depending on $r > 0$. Hence, we have that if $x_n \to x$ in $\mathbf{R}^N$, then $X_t^{x_n} \to X_t^x$ in probability. Let g be a bounded continuous function on $\mathbf{R}^N$. Then

$$E[g(X_t^{x_n})] \to E[g(X_t^x)],$$

which implies that the map $x \to E[g(X_t^x)]$ is continuous. Let A be an open subset of $\mathbf{R}^N$. We set

$$g_k(x) = \frac{kd(x, A^c)}{1 + kd(x, A^c)}$$

for the metric d of $\mathbf{R}^N$. It is clear that $g_k(x) \nearrow 1_A(x)$ as $k \to \infty$, and hence, the map $x \to P(X_t^x \in A)$ is Borel measurable. Thus, by the monotone class theorem, we conclude that this map is Borel measurable for $A \in \mathcal{B}(\mathbf{R}^N)$.

(2) We claim that the solution $\{x_t\}$ of (2.11) for the canonical representation of a Brownian motion $\{\beta_t\}$ has the Markov property under the probability measure, also denoted by P. Fix $T > 0$ and let $Q(\omega, \cdot)$ the regular conditional probability for $E[\cdot|\mathcal{F}_T](\omega)$. We define the system $(\Omega, \mathcal{F}, \tilde{P}, \{\tilde{\mathcal{F}}_t\}; \tilde{x}, \tilde{\beta})$ by

$$\tilde{\beta}_t = \beta_{t+T} - \beta_T, \ \tilde{x}_t = x_{t+T}, \ \tilde{\mathcal{F}}_t = \mathcal{F}_{t+T}, \ \tilde{P}(\cdot) = Q(\omega_0, \cdot),$$

for fixed $\omega_0 \in \Omega_0$ with $P(\Omega_0) = 1$. Let $A \in \tilde{\mathcal{F}}_t$, $\xi \in \mathbf{R}^N$ and $u \geq t$. It is easy to see that $\{(\beta_{t+T} - \beta_T, \mathcal{F}_{t+T})\}_{t \geq 0}$ is a Brownian motion under P. By Theorem 2.3.2, we have

$$\begin{aligned} E^{\tilde{P}}\big[e^{i(\xi, \tilde{\beta}_u - \tilde{\beta}_t)} 1_A\big] &= E\big[e^{i(\xi, \tilde{\beta}_u - \tilde{\beta}_t)} 1_A | \mathcal{F}_T\big] \\ &= E\big[E\big[e^{i(\xi, \beta_{u+T} - \beta_{t+T})} | \mathcal{F}_{t+T}\big] 1_A | \mathcal{F}_T\big] \\ &= e^{-\frac{|\xi|^2}{2}(u-t)} \tilde{P}(A), \quad \xi \in \mathbf{R}^M. \end{aligned}$$

Hence, $(\Omega, \mathcal{F}, \tilde{P}, \{\tilde{\mathcal{F}}_t\}; \tilde{\beta})$ is a Brownian motion. Furthermore, using a change of variables, we have

$$\begin{aligned} \tilde{x}_t &= x + \int_0^{t+T} b(x_u) du + \int_0^{t+T} \sigma(x_u) d\beta_u \\ &= x_T + \int_T^{t+T} b(x_u) du + \int_T^{t+T} \sigma(x_u) d\beta_u \\ &= x_T + \int_0^t b(\tilde{x}_r) dr + \int_0^t \sigma(\tilde{x}_r) d\tilde{\beta}_r, \quad P\text{-a.s.} \end{aligned}$$

By definition,

$$\tilde{P}(x_T = x_T(\omega_0)) = 1_{\{x_T = x_T(\omega_0)\}}(\omega_0) = 1.$$

Hence

$$\tilde{x}_t = x_T(\omega_0) + \int_0^t b(\tilde{x}_r) dr + \int_0^t \sigma(\tilde{x}_r) d\tilde{\beta}_r, \quad \tilde{P}\text{-a.s.}$$

Also, by the definition of P_x

$$x_t = x_T(\omega_0) + \int_0^t b(x_r) dr + \int_0^t \sigma(x_r) d\beta_r, \quad P_{x_T(\omega_0)}\text{-a.s.}$$

By Proposition 2.1.6, under (2.12), the uniqueness in law holds, that is, $x \overset{\mathcal{L}}{\approx} \tilde{x}$. Therefore, for any bounded Borel function f,

$$E^{\tilde{P}}[f(\tilde{x}_t)] = E_{x_T(\omega_0)}[f(x_t)].$$

On the other hand,

$$E_x[f(x_{t+T})|\mathcal{F}_T](\omega_0) = E^{Q(\omega_0,\cdot)}[f(x_{t+T})] = E^{\tilde{P}}[f(\tilde{x}_t)],$$

which proves the Markov property of $\{x_t\}$.

(3) Let $0 \le t_1 < t_2 < \cdots < t_m \le T$ and $A_i \in \mathcal{B}(\mathbf{R}^N)$, $i = 1, 2, \ldots, m$. From (2) it follows that

$$E_x[f(x_{t+T})1_{\{x_{t_1}\in A_1,\ldots,x_{t_m}\in A_m\}}] = E_x[E_{x_T}[f(x_t)]1_{\{x_{t_1}\in A_1,\ldots,x_{t_m}\in A_m\}}].$$

Since $(X, B) \overset{\mathcal{L}}{\approx} (x, \beta)$, we have

$$E_x[f(X_{t+T})1_{\{X_{t_1}\in A_1,\ldots,X_{t_m}\in A_m\}}]] = E_x[E_{X_T}[f(X_t)]1_{\{X_{t_1}\in A_1,\ldots,X_{t_m}\in A_m\}}].$$

Thus we obtain the Markov property of $\{X_t\}$.

(4) We will show the strong Markov property with respect to $\{\mathcal{F}^X_{t+}\}$. Let τ be an $\{\mathcal{F}^X_{t+}\}$-stopping time such that $\tau < T$ and define for $k = 0, 1, \ldots, 2^n T - 1; n = 1, 2, \ldots,$

$$\tau^{(n)}(\omega) = (k+1)2^{-n} \quad \text{if} \quad k2^{-n} \le \tau(\omega) < (k+1)2^{-n}, \quad \omega \in \Omega.$$

It is obvious that $\tau^{(n)} \searrow \tau$ for all ω. Since

$$\{\tau < t\} = \bigcup_n \{\tau \le t - \frac{1}{n}\} \in \bigcup_n \bigcap_{\varepsilon>0} \mathcal{F}^X_{(t-1/n)+\varepsilon} \subset \mathcal{F}^X_t,$$

we have

$$\{\tau^{(n)} < t\} = \bigcup_{k=0}^{2^nT-1} (\{(k+1)2^{-n} < t\} \cap \{k2^{-n} \le \tau < (k+1)2^{-n}\}) \in \mathcal{F}^X_t, \quad t \ge 0.$$

Hence each $\tau^{(n)}$ is an $\{\mathcal{F}^X_{t+}\}$-stopping time. Furthermore, we note that

$$\{\tau^{(n)} = k2^{-n}\} = \{(k-1)2^{-n} \le \tau < k2^{-n}\} \in \mathcal{F}^X_{k2^{-n}}.$$

Then, by the Markov property,

$$\begin{aligned} E_x[g(X_{t+\tau^{(n)}})] &= \sum_k E_x[g(X_{t+k2^{-n}})1_{\{\tau^{(n)}=k2^{-n}\}}] \\ &= \sum_k E_x[E_{X_{k2^{-n}}}[g(X_t)]1_{\{\tau^{(n)}=k2^{-n}\}}] \\ &= E_x[E_{X_{\tau^{(n)}}}[g(X_t)]]. \end{aligned}$$

Letting $n \to \infty$, by continuity, we get

$$E_x[g(X_{t+\tau})] = E_x[E_{X_\tau}[g(X_t)]].$$

By the same argument as (1), we have

$$P_x(X_{t+\tau} \in A) = E_x[P_{X_\tau}(X_t \in A)], \qquad \text{for } A \in \mathcal{B}(\mathbf{R}^N).$$

For any $D \in \mathcal{F}^X_{\tau+}$, we define

$$\tau_D = \tau 1_D + T 1_{D^c}.$$

Clearly, τ_D is an $\{\mathcal{F}^X_{t+}\}$- stopping time. Therefore,

$$E_x[f(X_{t+\tau_D})] = E_x[f(X_{t+\tau})1_D + E_x[f(X_{t+T})|\mathcal{F}^X_T]1_{D^c}],$$
$$E_x[E_{X_{\tau_D}}[f(X_t)]] = E_x[E_{X_\tau}[f(X_t)]1_D + E_{X_T}[f(X_t)]1_{D^c}].$$

Consequently, we obtain

$$E_x[f(X_{t+\tau})|\mathcal{F}^X_{\tau+}] = E_{X_\tau}[f(X_t)], \quad P_x\text{-a.s.}$$

(5) Let $f_j, j = 1, 2, \ldots, n$, be bounded Borel functions and $0 \le t_1 < \cdots < t_i \le t < t_{i+1} < \cdots < t_n$. Choosing $t_{i+1} = t + s_1, t_{i+2} = t + s_1 + s_2, \ldots, t_n = t + s_1 + \cdots + s_k$, by (3) and (4), we have

$$\begin{aligned} E_x[\prod_{j=1}^n f_j(X_{t_j})|\mathcal{F}^X_{t+}] &= \prod_{j=1}^i f_j(X_{t_j})E_x[f_{i+1}(X_{s_1+t})\cdots f_n(X_{s_1+\cdots+s_k+t})|\mathcal{F}^X_{t+}] \\ &= \prod_{j=1}^i f_j(X_{t_j})E_{X_t}[f_{i+1}(X_{s_1})E_{X_{s_1}} \\ &\quad \times [\cdots f_{n-1}(X_{s_{k-1}})E_{X_{s_{k-1}}}[f_n(X_{s_k})]]] \\ &= E_x[\prod_{j=1}^n f_j(X_{t_j})|\mathcal{F}^X_t], \quad P_x\text{-a.s.} \end{aligned}$$

Hence

$$E_x\big[1_A|\mathcal{F}^X_{t+}\big] = E_x\big[1_A|\mathcal{F}^X_t\big], \quad P_x\text{-a.s.,}$$

for all $A \in \mathcal{G}$ and then $A \in \mathcal{F}^X$. If we take $A \in \mathcal{F}^X_{t+}$, then 1_A is $\mathcal{F}^X_t$-measurable P_x a.s., and so, $A \in \mathcal{F}^X_t$. Therefore, $\mathcal{F}^X_{t+} = \mathcal{F}^X_t$, for all $t \ge 0$. ■

3

Dynamic Programming

In many optimization problems, *dynamic programming* is known as a computational method to break up a complex problem into a sequence of easier subproblems by means of a recursive relation, which was found by R. Bellman [10].

In stochastic control problems, we usually minimize the cost function with the discount factor $\alpha > 0$:

$$J_x(p) = E[\int_0^\infty e^{-\alpha t} g(X(t), p(t))dt], \qquad \text{for } p = \{p(t)\}$$

subject to an N-dimensional stochastic differential system:

$$dX(t) = \{AX(t) + p(t)\}dt + \sigma dB(t), \quad X(0) = x, \quad p(t) \in \Gamma : \text{ compact, convex}$$

(cf. Section 2.5.4). The Hamilton–Jacobi–Bellman (HJB) equation associated with this problem is given by

$$\alpha u = \frac{1}{2} tr(\sigma\sigma^\top D^2 u) + (Ax, Dv) + \min_{q\in\Gamma}\{g(x, q) + (q, Dv)\}, \quad x \in \mathbf{R}^N.$$

Suppose we have a classical solution $u \in C^2$ of the HJB equation. Then Itô's formula gives

$$u(x) \le E[\int_0^\tau e^{-\alpha t} g(X(t), p(t))dt + e^{-\alpha\tau} u(X(\tau))], \quad \forall\tau \ge 0, \quad \forall p(t) : \text{ control}$$

$$u(x) = E[\int_0^\tau e^{-\alpha t} g(X^*(t), p^*(t))dt + e^{-\alpha\tau} u(X^*(\tau))], \quad \forall\tau \ge 0, \quad p^*(t) : \text{ optimal.}$$

This yields that the dynamic programming principle holds for u, that is,

$$u(x) = \inf_p E[\int_0^\tau e^{-\alpha t} g(X(t), p(t))dt + e^{-\alpha\tau} u(X(\tau))], \quad \forall\tau \ge 0.$$

However, in stochastic control theory, the difficulty stems from the smoothness of the solution of the HJB equation. Our main goal is to show that the dynamic programming principle holds for the value function $v(x) = \inf_p J_x(p)$ in the weak sense shown in Chapter 2, without its smoothness. In view of the Nisio nonlinear semigroup, this is valid if v is bounded and uniformly continuous on $\mathbf{R}^N$ [65, sections III.10, IV.7]. Using the proof of the Yamada–Watanabe Theorem, we here extend these conditions to the case when v has the polynomial growth with k-degree and the uniform continuity with k-weight, which will be needed for the viscosity properties of v in Part II.

W. H. Fleming and R. W. Rishel [63], W. H. Fleming and H. M. Soner [65], M. Nisio [129], N. V. Krylov [99], and I. I. Gihman and A. V. Skorokhod [73] are the basic references for this chapter.

3.1 Dynamic Programming Principle: Deterministic Case

We start with a simple example of dynamic programming principle (DPP) in deterministic control. We consider the linear differential equation,

$$\begin{cases} \dot{x}(t) = Ax(t) + Bp(t), \quad t > 0, \\ x(0) = x \in \mathbf{R}^N \end{cases} \tag{3.1}$$

for control functions $p = p(\cdot)$ in the class

$$\mathcal{A} := \{p(\cdot) : [0, \infty) \to \mathbf{R}^M, \text{ measurable, } p(t) \in \Gamma \text{ for all } t \geq 0\},$$

where A and B are, respectively, $N \times N$ and $N \times M$ matrices, and Γ is a compact convex subset of $\mathbf{R}^M$. The objective is to minimize the cost with discount rate $\alpha > 0$:

$$J_x(p) = \int_0^\infty e^{-\alpha t} g(x(t), p(t))dt$$

for the running cost $g(x, p)$ satisfying

$$\begin{cases} g : \mathbf{R}^N \times \mathbf{R}^M \to \mathbf{R}, \text{ continuous,} \\ |g(x, p) - g(y, p)| \leq C|x - y|, \\ 0 \leq g(x, p) \leq C, \end{cases} \tag{3.2}$$

where $C > 0$ is a constant independent of p. Here we notice that the discount factor $e^{-\alpha t}$ provides a dynamic force to derive the integrand down toward zero over time.

Definition 3.1.1. *Let* $v(x) = \inf_{p \in \mathcal{A}} J_x(p)$. *We say that the dynamic programming principle (DPP) holds for* $v(x)$ *if*

$$v(x) = \inf_{p \in \mathcal{A}} \{\int_0^T e^{-\alpha t} g(x(t), p(t))dt + e^{-\alpha T} v(x(T))\}, \quad \textit{for any } T > 0. \tag{3.3}$$

Proposition 3.1.2. *We assume* (3.2). *Then the DPP holds for* $v(x)$.

Proof. We denote by $\bar{v}(x)$ the right-hand side of (3.3).

(1) Let $x(t)$ be the solution to (3.1) for $p \in \mathcal{A}$. Let $\tilde{x}(t) := x(t+T)$ and $\tilde{p}(t) := p(t+T)$ for any fixed $T > 0$. Since

$$\begin{aligned}\tilde{x}(t) &= x + \int_0^{t+T} Ax(r)dr + \int_0^{t+T} Bp(r)dr \\ &= x(T) + \int_T^{t+T} Ax(r)dr + \int_T^{t+T} Bp(r)dr \\ &= x(T) + \int_0^t A\tilde{x}(r)dr + \int_0^t B\tilde{p}(r)dr,\end{aligned}$$

we see that $\tilde{x}(t)$ is a solution to (3.1) with $\tilde{x}(0) = x(T)$ for $\tilde{p}(t)$. By definition, we note that

$$\int_0^\infty e^{-\alpha t} g(\tilde{x}(t), \tilde{p}(t))dt \ge v(x(T)).$$

Then

$$\begin{aligned}J_x(p) &= \int_0^T e^{-\alpha t} g(x(t), p(t))dt + \int_T^\infty e^{-\alpha t} g(x(t), p(t))dt \\ &= \int_0^T e^{-\alpha t} g(x(t), p(t))dt + e^{-\alpha T} \int_0^\infty e^{-\alpha t} g(x(t+T), p(t+T))dt \\ &\ge \int_0^T e^{-\alpha t} g(x(t), p(t))dt + e^{-\alpha T} v(x(T)).\end{aligned}$$

Taking the infimum over $p \in \mathcal{A}$, we have $v(x) \ge \bar{v}(x)$.

(2) For any $\varepsilon > 0$, there exists $\bar{p} \in \mathcal{A}$ such that the corresponding solution $\bar{x}(t)$ of (3.1) with $\bar{x}(0) = x(T)$ satisfies

$$v(x(T)) + \varepsilon \ge \int_0^\infty e^{-\alpha t} g(\bar{x}(t), \bar{p}(t))dt.$$

Define

$$\hat{p}(t) = \begin{cases} p(t) & \text{if } t \le T, \\ \bar{p}(t-T) & \text{if } t > T, \end{cases}$$

and let $\hat{x}(t)$ be the solution of (3.1) corresponding to $\hat{p}$, which belongs to $\mathcal{A}$. Then it is easy to see that $(\hat{x}(t+T), \hat{p}(t+T)) = (\bar{x}(t), \bar{p}(t))$. Thus

$$\begin{aligned}
v(x) \le J_x(\hat{p}) &= \int_0^T e^{-\alpha t} g(\hat{x}(t), \hat{p}(t))dt + \int_T^\infty e^{-\alpha t} g(\hat{x}(t), \hat{p}(t))dt \\
&= \int_0^T e^{-\alpha t} g(x(t), p(t))dt + e^{-\alpha T} \int_0^\infty e^{-\alpha t} g(\hat{x}(t+T), \hat{p}(t+T))dt \\
&= \int_0^T e^{-\alpha t} g(x(t), p(t))dt + e^{-\alpha T} \int_0^\infty e^{-\alpha t} g(\bar{x}(t), \bar{p}(t))dt \\
&\le \int_0^T e^{-\alpha t} g(x(t), p(t))dt + e^{-\alpha T} v(x(T)) + \varepsilon.
\end{aligned}$$

Letting $\varepsilon \to 0$, we get $v(x) \le \bar{v}(x)$. ■

We examine the relation between the DPP and the HJB equation:

$$-\alpha v(x) + (Ax, Dv(x)) + \min_{p\in\Gamma}\{g(x,p) + (Bp, Dv(x))\} = 0, \quad x \in \mathbf{R}^N. \tag{3.4}$$

Proposition 3.1.3. *Let $v(x)$ satisfy the DPP. Suppose that $v \in C^1$. Then we have* (3.4).

Proof.

(1) Since

$$d(e^{-\alpha t} v(x(t))) = e^{-\alpha t}\{-\alpha v(x(t)) + (Dv(x(t)), \dot{x}(t))\}dt,$$

(3.3) can be rewritten as

$$\begin{aligned}
0 &= \inf_{p\in\mathcal{A}}\{\int_0^T e^{-\alpha t} g(x(t), p(t))dt + e^{-\alpha T} v(x(T)) - v(x)\} \\
&= \inf_{p\in\mathcal{A}} \int_0^T e^{-\alpha t}\{g(x(t), p(t)) - \alpha v(x(t)) + (Ax(t) + Bp(t), Dv(x(t)))\}dt.
\end{aligned}$$

(2) Let $x(t)$ be the solution to (3.1) for any constant control $p(t) = p \in \Gamma$, $t \ge 0$. By (1), we have

$$0 \le \int_0^T e^{-\alpha t}\{g(x(t), p) - \alpha v(x(t)) + (Ax(t) + Bp, Dv(x(t)))\}dt.$$

Dividing both sides by T and letting $T \to 0$, we have

$$0 \le g(x,p) - \alpha v(x) + (Ax + Bp, Dv(x)).$$

Thus

$$0 \le -\alpha v(x) + (Ax, Dv(x)) + \min_{p\in\Gamma}\{g(x,p) + (Bp, Dv(x))\}.$$

(3) Let $\varepsilon > 0$ and $T > 0$ be arbitrary. By (1), there exists $p = p(\cdot) \in \mathcal{A}$ such that

$$\varepsilon T \geq \int_0^T e^{-\alpha t}\{g(x(t), p(t)) - \alpha v(x(t)) + (Ax(t) + Bp(t), Dv(x(t)))\}dt.$$

We note by (3.2) that

$$\int_0^T e^{-\alpha t} g(x(t), p(t))dt = \int_0^T e^{-\alpha t} g(x, p(t))dt + o(T) \quad \text{as } T \to 0.$$

Then

$$\begin{aligned}\varepsilon T &\geq \int_0^T e^{-\alpha t}\{g(x, p(t)) - \alpha v(x(t)) + (Ax(t) + Bp(t), Dv(x(t)))\} \\ &\quad \times dt + o(T) \\ &\geq \int_0^T e^{-\alpha t} \min_p\{g(x, p) - \alpha v(x(t)) + (Ax(t) + Bp, Dv(x(t)))\} \\ &\quad \times dt + o(T).\end{aligned}$$

Dividing both sides by T and letting $T \to 0$, and then $\varepsilon \to 0$, we get

$$0 \geq -\alpha v(x) + (Ax, Dv(x)) + \min_{p\in\Gamma}\{g(x, p) + (Bp, Dv(x))\}. \qquad \blacksquare$$

3.2 Dynamic Programming Principle: Stochastic Case

Let us study the DPP for the stochastic case, based on the Nisio nonlinear semigroup [129]. We consider the stochastic differential equation,

$$\begin{cases} dX_t = b(X_t, p_t)dt + \sigma(X_t)dB_t, & t > 0, \\ X_0 = x \in \mathbf{R}^N \end{cases} \tag{3.5}$$

for control processes $p = \{p_t\}$ such that

$$p_t \in \Gamma \ a.s. \text{ for each } t \geq 0, \ \mathcal{F}_t^B\text{-progressively measurable}, \tag{3.6}$$

where Γ is a compact convex subset of $\mathbf{R}^N$, $b : \mathbf{R}^N \times \Gamma \to \mathbf{R}^N$ and $\sigma : \mathbf{R}^N \to \mathbf{R}^{N\times M}$ are continuous functions such that

$$|b(x, p) - b(y, p)| + \|\sigma(x) - \sigma(y)\| \leq L|x - y|, \tag{3.7}$$

for some constant $L > 0$ independent of $p \in \Gamma$, and $\{B_t\}$ is an M-dimensional Brownian motion on a stochastic basis $(\Omega, \mathcal{F}, P, \{\mathcal{F}_t\})$. We fix $T > 0$ and define $\{(S_p z_t, \mathcal{F}_t)\}$ by

$$S_p z_t = x + \int_0^t b(z_s, p_s)ds + \int_0^t \sigma(z_s)dB_s, \quad 0 \leq t \leq T,$$

for $\{z_t\} \in \mathcal{H}$, where $\mathcal{H}$ denotes the Banach space of continuous square integrable processes in Theorem 1.7.2 (1). Under (3.6) and (3.7), by the same line as

Theorem 1.7.2, we observe that S_p has a fixed point in $\mathcal{H}$ for each $p = \{p_t\}$, and equation (3.5) admits a unique solution $\{X_t\}$.

The objective is to minimize the cost with discount rate $\alpha > 0$:

$$J_x(p) = E[\int_0^\infty e^{-\alpha t} g(X_t, p_t)dt] \tag{3.8}$$

for the running cost $g(x, p)$ satisfying (3.2).

In order to obtain the dynamic programming principle, we consider the weak solution $\{X_t\}$ of (3.5) in the class $\mathcal{A}$ of all systems $\pi := (\Omega, \mathcal{F}, P, \{\mathcal{F}_t\}; \{B_t\}, \{p_t\})$ such that

(a) $(\Omega, \mathcal{F}, P, \{\mathcal{F}_t\})$: a stochastic basis,
(b) $\{B_t\}$: a Brownian motion on $(\Omega, \mathcal{F}, P, \{\mathcal{F}_t\})$,
(c) $\{p_t\}$ satisfies (3.6) for $\mathcal{F}_t^B$.

We set

$$J_x(\pi) = J_x(p), \quad \text{for } \pi \in \mathcal{A}.$$

Definition 3.2.1. *Let $v(x)$ be the value function defined by $v(x) = \inf_{\pi \in \mathcal{A}} J_x(\pi)$. We say that the dynamic programming principle (DPP) holds for $v(x)$ if*

$$v(x) = \inf_{\pi \in \mathcal{A}} E[\int_0^\tau e^{-\alpha t} g(X_t, p_t)dt + e^{-\alpha\tau} v(X_\tau)], \quad \textit{for any } \tau \in \mathcal{S}_b, \tag{3.9}$$

where $\mathcal{S}_b$ denotes the collections of all bounded $\{\mathcal{F}_t\}$-stopping times.

Proposition 3.2.2. *Let $\{X_t\}$ be a weak solution of (3.5) with $X_0 = x$ for $\pi = (\Omega, \mathcal{F}, P, \{\mathcal{F}_t\}; \{B_t\}, \{p_t\}) \in \mathcal{A}$. Then there exists a weak solution $\{(x_t, \beta_t)\}$ of (3.5):*

$$dx_t = b(x_t, q_t)dt + \sigma(x_t)d\beta_t, \quad t > 0, \quad x_0 = x$$

for some control $\{q_t\}$ satisfying (3.6) on the stochastic basis $(\Omega^*, \mathcal{F}^*, P^*, \{\mathcal{F}_t^*\})$ with the Polish space Ω^* such that

$$(X, \{\int_0^\cdot p_s ds\}, B) \overset{\mathcal{L}}{\approx} (x, \{\int_0^\cdot q_s ds\}, \beta).$$

Proof.

(1) Let $\hat{p} := \{\int_0^t p_s ds\}$, and let $(X, \hat{p}, B)$ be the random variable $(\Omega, \mathcal{F}, P) \to W^N \times W^N \times W^M$ defined by $\{(X_t, \hat{p}_t, B_t)\}$. Since $(X, \hat{p})$ is $\sigma(B)$-measurable, we see by Theorem 2.1.4 that $(X, \hat{p}) = \eta(B)$ for some measurable function $\eta : W^M \to W^N \times W^N$. Let $(\Omega^*, \mathcal{F}^*, P^*, \{\mathcal{F}_t^*\}; \{\beta_t\})$ be the canonical representation of $\{B_t\}$ and define

$$(x, \hat{q}) = \eta(\beta), \quad (x_t, \hat{q}_t) = \pi_t \circ (x, \hat{q}),$$

where π_t is as in Definition 2.1.1. Then

$$(X, \hat{p}, B) \overset{\mathcal{L}}{\approx} (x, \hat{q}, \beta).$$

By the same arguments as Proposition 2.1.6 (1), $\hat{q} = \{\hat{q}_t\}$ is $\mathcal{F}_t^\beta$-adapted and

$$\begin{aligned} &E^{P^*}[\exp\{i(\xi, x_t - G(x, q, \beta)(t))\}] \\ &= E[\exp\{i(\xi, X_t - G(X, p, B)(t))\}] = 1, \end{aligned}$$

where $i = \sqrt{-1}, \xi \in \mathbf{R}^N$ and $G(x, q, \beta)(t) = x + \int_0^t b(x_s, q_s)ds + \int_0^t \sigma(x_s)d\beta_s$ for $q = \{q_t\}$ given in (2) below. Thus (x, q, β) satisfies

$$dx_t = b(x_t, q_t)dt + \sigma(x_t)d\beta_t, \quad x_0 = x, \quad P^* \text{ a.s.}$$

(2) We claim that $\{\hat{q}_t\}$ can be written as $\hat{q}_t = \int_0^t q_s ds$ a.s. for an $\mathcal{F}_t^\beta$-progressively measurable process $\{q_t\}$ with $P^*(q_t \in \Gamma) = 1$ for all $t \geq 0$. Let

$$\begin{aligned} \Omega_{\hat{q}} = &\Big(\bigcap_{r,s \in \mathbf{Q}_+} \{\omega : |\hat{q}_r(\omega) - \hat{q}_s(\omega)| \leq \sup_{p \in \Gamma} |p||r - s|\}\Big) \\ &\cap \{\omega : \hat{q}_t(\omega) \text{ continuous}\}. \end{aligned}$$

By (1), $P^*(\Omega_{\hat{q}}) = 1$. For fixed $\omega \in \Omega_{\hat{q}}$, the map $t \in [0, \infty) \to \hat{q}_t(\omega)$ is continuous and Lipschitz. Since each component of $\hat{q}_t$ is an absolutely continuous function, it can be written as

$$\hat{q}_t = \hat{q}_t - \hat{q}_0 = \int_0^t \psi_s ds.$$

We set $\Delta_n \hat{q}_t = n(\hat{q}_t - \hat{q}_{(t-1/n)^+})$ for $n \in \mathbf{N}$. By Lebesgue's theorem

$$\lim_{n \to \infty} \Delta_n \hat{q}_t(\omega) = \psi_t(\omega), \quad a.e.\ t \in [0, T],\ \omega \in \Omega_{\hat{q}},$$

for every $T > 0$. By the dominated convergence theorem,

$$\begin{aligned} \int_0^T |\Delta_n \hat{q}_t(\omega) - \Delta_m \hat{q}_t(\omega)| dt &\leq \int_0^T |\Delta_n \hat{q}_t(\omega) - \psi_t(\omega)| dt \\ &\quad + \int_0^T |\Delta_m \hat{q}_t(\omega) - \psi_t(\omega)| dt \\ &\to 0 \quad \text{as } n, m \to \infty, \quad \omega \in \Omega_{\hat{q}}. \end{aligned}$$

Hence, by Fatou's lemma, we have

$$\begin{aligned} &\lim_{N \to \infty} \sup_{n,m \geq N} E^{P^*}\Big[\int_0^T |\Delta_n \hat{q}_t - \Delta_m \hat{q}_t| dt\Big] \\ &\leq E^{P^*}\Big[\lim_{N \to \infty} \sup_{n,m \geq N} \int_0^T |\Delta_n \hat{q}_t - \Delta_m \hat{q}_t| dt\Big] = 0. \end{aligned}$$

By completeness, there is a $\mathcal{B}([0,T]) \otimes \mathcal{F}_T^\beta$-measurable function $\varphi^T(t,\omega)$ such that

$$\lim_{n\to\infty} E^{P^*}[\int_0^T |\Delta_n \hat{q}_t - \varphi^T(t)|dt] = 0.$$

Since $\varphi^{\bar{T}}(t,\omega)1_{\{t\le T\}} = \varphi^T(t,\omega)$ a.e. for $T < \bar{T}$, we can define the $\{\mathcal{F}_t^\beta\}$-progressively measurable process $\{r_t\}$ by $r_t = \varphi^T(t,\cdot)$ for $t \le T$. For each $T > 0$, it is clear that

$$E^{P^*}[\int_0^T |d(\Delta_n \hat{q}_t, \Gamma) - d(r_t, \Gamma)|]dt \le E^{P^*}[\int_0^T |\Delta_n \hat{q}_t - r_t|dt] \to 0,$$

and similarly,

$$\begin{aligned} &E[\int_0^T |d(\Delta_n \hat{p}_t, \Gamma) - d(p_t, \Gamma)|]dt \\ &\le E[\int_0^T |\Delta_n \hat{p}_t - p_t|dt] \to 0 \quad \text{as } n \to \infty, \end{aligned}$$

where d denotes the distance. By (1), we have

$$E^{P^*}[d(\Delta_n \hat{q}_t, \Gamma)] = E[d(\Delta_n \hat{p}_t, \Gamma)].$$

Thus, by (3.6)

$$E^{P^*}[\int_0^T d(r_t, \Gamma)dt] = \int_0^T E[d(p_t, \Gamma)]dt = 0,$$

which implies that $E^{P^*}[\int_0^T 1_{\{r_t \notin \Gamma\}}dt] = 0$. Define $\{q_t\}$ by $q_t = r_t$ if $r_t \in \Gamma$, $q_t =$ an element in Γ otherwise. Then

$$E^{P^*}[\int_0^T |q_t - r_t|dt] = E^{P^*}[\int_0^T |q_t - r_t|1_{\{r_t \in \Gamma\}}dt] = 0.$$

Extracting a subsequence of $\{\Delta_n \hat{q}_t\}$, we get

$$\begin{aligned} &\int_0^T |\psi_t - q_t|dt \\ &\le \int_0^T \{|\psi_t - \Delta_n \hat{q}_t| + |\Delta_n \hat{q}_t - \varphi^T(t)| + |r_t - q_t|\}dt \to 0, \quad P^*\text{-a.s.} \end{aligned}$$

Therefore, we see that $\{q_t\}$ has the desired properties. ■

Proposition 3.2.3. *Let* $\{X_t^j\}$, $j = 1, 2, \ldots, n$, *be weak solutions of* (3.5) *with* $X_0^j = x^j$, *for* $\pi^j = (\Omega^j, \mathcal{F}^j, P^j, \{\mathcal{F}_t^j\}; \{B_t^j\}, \{p_t^j\}) \in \mathcal{A}$. *Then there exist weak solutions* $\{(x_t^j, \beta_t)\}$ *of* (3.5):

$$dx_t^j = b(x_t^j, q_t^j)dt + \sigma(x_t^j)d\beta_t, \quad t > 0, \quad x_0^j = x^j$$

for some controls $\{q_t^j\}$ *satisfying* (3.6) *on the same stochastic basis* $(\Omega^*, \mathcal{F}^*, P^*, \{\mathcal{F}_t^*\})$ *such that*

$$(X^j, \{\int_0^\cdot p_s^j ds\}, B^j) \overset{\mathcal{L}}{\approx} (x^j, \{\int_0^\cdot q_s^j ds\}, \beta), \quad j = 1, 2, \ldots, n.$$

Proof.

(1) Let $n = 2$. We set $\hat{p}^j = \{\int_0^t p_s^j ds\}$. By the proof of Lemma 2.5.2 with $(X^1, \hat{p}^1, B^1)$ and $(X^2, \hat{p}^2, B^2)$ replacing (X, B) and (X', B'), we get

$$(X^j, \hat{p}^j, B^j) \overset{\mathcal{L}}{\approx} (x^j, \hat{q}^j, \beta), \quad j = 1, 2,$$

where $(x^j, \hat{q}^j, \beta)$, $j = 1, 2$, are defined on the same stochastic basis. By virtue of Proposition 3.2.2 (2), we have

$$\hat{q}_t^j = \int_0^t q_s^j ds,$$

for some $\{q_t^j\}$. Thus the assertion holds for $n = 2$.

(2) For $n = 3$, let $(\bar{x}^j, \bar{q}^j, \bar{\beta})$, $j = 1, 2$, denote the weak solution $(x^j, \hat{q}^j, \beta)$ of (1). As in (1), we apply Lemma 2.5.2 to $(\bar{x}^1, \bar{q}^1, \bar{x}^2, \bar{q}^2, \bar{\beta})$ and $(X^3, \hat{p}^3, B^3)$. Then

$$(\bar{x}^1, \bar{q}^1, \bar{x}^2, \bar{q}^2, \bar{\beta}) \overset{\mathcal{L}}{\approx} (x^1, \hat{q}^1, x^2, \hat{q}^2, \beta), \quad (X^3, \hat{p}^3, B^3) \overset{\mathcal{L}}{\approx} (x^3, \hat{q}^3, \beta).$$

where $(x^j, \hat{q}^j, \beta)$, $j = 1, 2, 3$, are defined on the same stochastic basis. Therefore,

$$(X^j, \hat{p}^j, B^j) \overset{\mathcal{L}}{\approx} (x^j, \hat{q}^j, \beta), \quad j = 1, 2, 3.$$

Finally, by induction, we get the assertion. ∎

Proposition 3.2.4. *Suppose that*

$$\alpha > L + L^2/2. \tag{3.10}$$

Then, under (3.2) *and* (3.7), *we have:*

(i) $v(x)$ *is bounded.*
(ii) $v(x)$ *is Lipschitz.*

Proof.

(1) By (3.2), we have

$$0 \le J_x(\pi) = E[\int_0^\infty e^{-\alpha t} g(X_t, p_t) dt] \le C/\alpha,$$

which implies (i).

(2) Let $\{X_t\}$ and $\{X'_t\}$ be two weak solutions of (3.5) for $\pi = (\Omega, \mathcal{F}, P, \{\mathcal{F}_t\}; \{B_t\}, \{p_t\}) \in \mathcal{A}$ with $X_0 = x$ and $X'_0 = x'$, respectively. It is easy to see that

$$v(x) \le J_x(\pi) \le \sup_\pi |J_x(\pi) - J_{x'}(\pi)| + J_{x'}(\pi).$$

Taking the infimum of both sides, we have

$$v(x) - v(x') \le \sup_\pi |J_x(\pi) - J_{x'}(\pi)|.$$

Interchanging x and x', we obtain

$$|v(x) - v(x')| \le \sup_\pi |J_x(\pi) - J_{x'}(\pi)|.$$

(3) To prove (ii), we will estimate the right-hand side of this relation. By (2), we have

$$\begin{aligned} d(X_t - X'_t) = [b(X_t, p_t) - b(X'_t, p_t)]dt \\ + [\sigma(X_t) - \sigma(X'_t)]dB_t, \quad X_0 - X'_0 = x - x'. \end{aligned}$$

By Itô's formula and (3.7)

$$E[|X_t - X'_t|^2] \le |x - x'|^2 + (2L + L^2)E[\int_0^t |X_s - X'_s|^2 ds].$$

By Gronwall's inequality,

$$E[|X_t - X'_t|^2] \le |x - x'|^2 e^{(2L+L^2)t}.$$

Thus, by (3.2) and (3.10)

$$\begin{aligned} |J_x(\pi) - J_{x'}(\pi)| &= |E[\int_0^\infty e^{-\alpha t} g(X_t, p_t)dt] - E[\int_0^\infty e^{-\alpha t} g(X'_t, p_t)dt]| \\ &\le E[\int_0^\infty e^{-\alpha t} |g(X_t, p_t) - g(X'_t, p_t)|dt] \\ &\le CE[\int_0^\infty e^{-\alpha t} |X_t - X'_t|dt] \\ &\le C\int_0^\infty e^{-\alpha t} |x - x'| e^{(L+L^2/2)t} dt \\ &\le C|x - x'|/\{\alpha - (L + L^2/2)\}. \end{aligned}$$

■

Theorem 3.2.5. *We assume* (3.2), (3.7), *and* (3.10). *Then the DPP holds for* $v(x)$.

Proof. We denote by $\bar{v}(x)$ the right-hand side of (3.9).

(1) Let $\{X_t\}$ be the solution of (3.5) corresponding to any $\pi = (\Omega, \mathcal{F}, P, \{\mathcal{F}_t\}; \{B_t\}, \{p_t\}) \in \mathcal{A}$. By Proposition 3.2.2, we may consider that Ω is a Polish space and the regular conditional probability $Q(\omega, \cdot)$ for

$E[\cdot|\mathcal{F}_r](\omega)$ and any fixed $r \geq 0$ is defined on it. Let $\tilde{P}(\cdot) = Q(\omega_0, \cdot)$, for fixed $\omega_0 \in \Omega_0$ with $P(\Omega_0) = 1$, and

$$\tilde{X}_t = X_{t+r}, \ \tilde{\mathcal{F}}_t = \mathcal{F}_{t+r}, \ \tilde{p}_t = p_{t+r}, \ \tilde{B}_t = B_{t+r} - B_r.$$

Note that

$$\begin{aligned}\tilde{X}_t &= x + \int_0^{t+r} b(X_s, p_s)ds + \int_0^{t+r} \sigma(X_s)dB_s \\ &= X_r + \int_r^{t+r} b(X_s, p_s)ds + \int_r^{t+r} \sigma(X_s)dB_s \\ &= X_r + \int_0^t b(\tilde{X}_s, \tilde{p}_s)ds + \int_0^t \sigma(\tilde{X}_s)d\tilde{B}_s, \quad P\text{-a.s.}\end{aligned}$$

By the same arguments as Theorem 2.6.2 (2), $\{\tilde{X}_t\}$ is a weak solution of (3.5) with $\tilde{X}_0 = X_r(\omega_0)$ for $\tilde{\pi} := (\Omega, \mathcal{F}, \tilde{P}, \{\tilde{\mathcal{F}}_t\}; \{\tilde{B}_t\}, \{\tilde{p}_t\}) \in \mathcal{A}$. Then we have

$$J_{X_r(\omega_0)}(\tilde{\pi}) = E^{\tilde{P}}[\int_0^\infty e^{-\alpha s} g(\tilde{X}_s, \tilde{p}_s)dt].$$

Thus, by the definition of $\tilde{P}$

$$\begin{aligned}J_{X_r}(\tilde{\pi}) &= E[\int_0^\infty e^{-\alpha s} g(\tilde{X}_s, \tilde{p}_s)dt|\mathcal{F}_r] \\ &= e^{\alpha r} E[\int_r^\infty e^{-\alpha s} g(X_s, p_s)ds|\mathcal{F}_r], \quad P\text{-a.s.}\end{aligned}$$

(2) Let $\tau \in \mathcal{S}_b$ be random. Let $\{\tau_n\}$ be a sequence of countably valued stopping times approximating τ. Then, by (1), we have

$$E[e^{-\alpha\tau_n} J_{X_{\tau_n}}(\tilde{\pi})] = E[\int_{\tau_n}^\infty e^{-\alpha t} g(X_t, p_t)dt].$$

Letting $n \to \infty$, we get

$$E[e^{-\alpha\tau} J_{X_\tau}(\tilde{\pi})] = E[\int_\tau^\infty e^{-\alpha t} g(X_t, p_t)dt].$$

Thus

$$\begin{aligned}J_x(\pi) &= E[\int_0^\tau e^{-\alpha t} g(X_t, p_t)dt + \int_\tau^\infty e^{-\alpha t} g(X_t, p_t)dt] \\ &\geq E[\int_0^\tau e^{-\alpha t} g(X_t, p_t)dt + e^{-\alpha\tau} v(X_\tau)].\end{aligned}$$

Taking the infimum over $\pi \in \mathcal{A}$, we have $v(x) \geq \bar{v}(x)$.

(3) To prove the reverse inequality, we assume that τ is nonrandom, $\tau = r$. Let $\delta > 0$ and $\{O_i : i = 1, 2, \dots\}$ be a sequence of open subsets of $\mathbf{R}^N$ such that

$$\bigcup_{i=1}^\infty O_i = \mathbf{R}^N, \quad \text{diameter}(O_i) < \delta.$$

We take a sequence of pairwise disjoint sets $S_i \in \mathcal{B}(\mathbf{R}^N)$, $i = 1, 2, \ldots, n$, such that

$$\{x : |x| \leq R\} \subset \overset{n}{\underset{j=1}{\cup}} O_j = \overset{n}{\underset{i=1}{\cup}} S_i, \quad S_i \subset O_i,$$

and set $S_{n+1} = (\cup_{i=1}^{n} S_i)^c$ for $R > 0$ chosen later. For any i, we take $x_i \in S_i$. Then there is $\pi^{(i)} = (\Omega^{(i)}, \mathcal{F}^{(i)}, P^{(i)}, \{\mathcal{F}_t^{(i)}\}; \{B_t^{(i)}\}, \{p_t^{(i)}\}) \in \mathcal{A}$, for $\varepsilon > 0$, such that

$$J_{x_i}(\pi^{(i)}) \leq v(x_i) + \varepsilon.$$

Also, we can find $\pi = (\Omega, \mathcal{F}, P, \{\mathcal{F}_t\}; \{B_t\}, \{p_t\}) \in \mathcal{A}$ such that

$$E[\int_0^\tau e^{-\alpha t} g(X_t, p_t)dt + e^{-\alpha\tau} v(X_\tau)] \leq \bar{v}(x) + \varepsilon.$$

By Proposition 3.2.3, we may consider that $\{p_t^{(i)}\}$ and $\{p_t\}$ are defined on the same stochastic basis $(\Omega^*, \mathcal{F}^*, P^*, \{\mathcal{F}_t^*\})$, which depends on a choice of $\{p_t^{(i)}\}$ and $\{p_t\}$, carrying the Brownian motion $\{\beta_t\}$. Define

$$p_t^\tau = p_t 1_{\{t<\tau\}} + \left(\sum_{i=1}^{n+1} p_{t-\tau}^{(i)} 1_{\{X_\tau \in S_i\}}\right) 1_{\{t \geq \tau\}}.$$

Clearly, $\pi^\tau = (\Omega^*, \mathcal{F}^*, P^*, \{\mathcal{F}_t^*\}; \{\beta_t\}, \{p_t^\tau\}) \in \mathcal{A}$. Let $\{X_t^\tau\}$ be the solution of (3.5) for π^τ:

$$dX_t^\tau = b(X_t^\tau, p_t^\tau)dt + \sigma(X_t^\tau)d\beta_t, \quad X_0^\tau = x.$$

Then, by (1)

$$d\tilde{X}_t^\tau = b(\tilde{X}_t^\tau, \tilde{p}_t^\tau)dt + \sigma(\tilde{X}_t^\tau)d\tilde{\beta}_t, \quad \tilde{X}_0^\tau = X_\tau.$$

It is easy to see that $X_t^\tau = X_t$ if $t < \tau$, and $\tilde{X}_t^\tau = X_t^{(i)}$, $\tilde{p}_t^\tau = p_t^{(i)}$ if $X_\tau \in S_i$. Thus, by (1), we deduce that

$$\begin{aligned} J_{X_\tau}(\tilde{\pi^\tau}) &= E^{P^*}[\int_0^\infty e^{-\alpha t} g(\tilde{X}_t^\tau, \tilde{p}_t^\tau)dt|\mathcal{F}_\tau^*] \\ &= E^{P^*}[\int_0^\infty e^{-\alpha t} g(X_t^{(i)}, p_t^{(i)})dt|\mathcal{F}_\tau^*] \\ &= J_{X_\tau}(\pi^{(i)}), \quad \text{a.s.} \quad \text{if } X_\tau \in S_i, \quad i = 1, 2, \ldots, n+1. \end{aligned}$$

(4) By Proposition 3.2.4 (2), we can take $\delta > 0$ of (3), for any $\varepsilon > 0$, such that whenever $|x - x'| < \delta$,

$$|J_x(\pi) - J_{x'}(\pi)| + |v(x) - v(x')| < \varepsilon, \quad \text{for all } \pi \in \mathcal{A}.$$

For any $x \in S_i$, $i = 1, 2, \ldots, n$, we have

$$\begin{aligned} J_x(\pi^{(i)}) &\leq J_{x_i}(\pi^{(i)}) + |J_x(\pi^{(i)}) - J_{x_i}(\pi^{(i)})| \leq v(x_i) + 2\varepsilon \\ &\leq v(x) + 3\varepsilon. \end{aligned}$$

Also, by the same line as Proposition 1.8.3, we have

$$E[\sup_{0\le t\le \tau} |X_t - x|^2] \le C_0(1+|x|^2),$$

for some constant C_0, independent of $\{p_t\}$. By Chebyshev's inequality,

$$\begin{aligned} P(|X_\tau| > R) &\le \frac{1}{R^2} E[|X_\tau|^2] \le \frac{1}{R^2} E[\sup_{0\le t\le\tau} |X_t|^2] \\ &\le \frac{2}{R^2}\{C_0(1+|x|^2) + |x|^2\}, \quad \forall \pi \in \mathcal{A}. \end{aligned}$$

Thus, by Proposition 3.2.4, we fix x and choose $R > 0$ such that

$$\sup_{\pi\in\mathcal{A}} E[|J_{X_\tau}(\pi)| 1_{\{|X_\tau|>R\}}] < \varepsilon.$$

Now, applying (1) and (3), we have

$$\begin{aligned} E[\int_\tau^\infty e^{-\alpha t} g(X_t^\tau, p_t^\tau)dt] &= E[E[\int_\tau^\infty e^{-\alpha t} g(X_t^\tau, p_t^\tau)dt | \mathcal{F}_\tau^*]] \\ &= E[e^{-\alpha\tau} J_{X_\tau}(\tilde{\pi^\tau})] \\ &= E[e^{-\alpha\tau} \sum_{i=1}^{n+1} J_{X_\tau}(\pi^{(i)}) 1_{\{X_\tau\in S_i\}}] \\ &\le E[e^{-\alpha\tau} \sum_{i=1}^{n} \{v(X_\tau) + 3\varepsilon\} 1_{\{X_\tau\in S_i\}}] \\ &\quad + E[e^{-\alpha\tau} J_{X_\tau}(\pi^{(n+1)}) 1_{\{|X_\tau|>R\}}] \\ &\le E[e^{-\alpha\tau} v(X_\tau)] + 3\varepsilon + \varepsilon. \end{aligned}$$

Hence, by (3) again,

$$\begin{aligned} v(x) &\le E[\int_0^\tau e^{-\alpha t} g(X_t^\tau, p_t^\tau)dt + \int_\tau^\infty e^{-\alpha t} g(X_t^\tau, p_t^\tau)dt] \\ &\le E[\int_0^\tau e^{-\alpha t} g(X_t, p_t)dt + e^{-\alpha\tau} v(X_\tau)] + 4\varepsilon \\ &\le \bar{v}(x) + 5\varepsilon. \end{aligned}$$

Therefore, letting $\varepsilon \to 0$, we get $v(x) \le \bar{v}(x)$.

(5) Let $\tau \in \mathcal{S}_b$ be random. We know by (4) that

$$v(x) \le \inf_{\pi\in\mathcal{A}} E[\int_0^t e^{-\alpha s} g(X_s, p_s)ds + e^{-\alpha t} v(X_t)], \quad t \ge 0.$$

Applying (1), we have

$$\begin{aligned} v(X_r) = v(\tilde{X}_0) &\le E^{\tilde{P}}[e^{-\alpha t} v(\tilde{X}_t) + \int_0^t e^{-\alpha s} g(\tilde{X}_s, \tilde{p}_s)ds] \\ &= E[e^{-\alpha t} v(X_{t+r}) + \int_0^t e^{-\alpha s} g(X_{r+s}, p_{r+s})ds | \mathcal{F}_r], \quad P\text{-a.s.} \end{aligned}$$

Hence $\{e^{-\alpha t}v(X_t) + \int_0^t e^{-\alpha s}g(X_s, p_s)ds\}$ is a submartingale. By the optional sampling theorem,

$$v(X_0) \leq E[e^{-\alpha\tau}v(X_\tau) + \int_0^\tau e^{-\alpha s}g(X_s, p_s)ds|\mathcal{F}_0], \quad P\text{-a.s.}$$

Taking the expectation and then the infimum over $\mathcal{A}$, we conclude that $v(x) \leq \bar{v}(x)$, which implies $v(x) = \bar{v}(x)$. ■

3.3 Dynamic Programming Principle: Polynomial Growth

We consider the DPP for the stochastic control problem (3.5)–(3.8), where the running cost $g(x, p)$ is assumed to satisfy the polynomial growth conditions with k-degree and the uniform continuity with k-weight, that is,

There exist $C > 0$ and $C_\rho > 0$ (independent of $p \in \Gamma$), for any $\rho > 0$, such that

$$0 \leq g(x, p) \leq C(1 + |x|^k), \quad \text{for some } k \in \mathbf{N}, \tag{3.11}$$

$$|g(x, p) - g(y, p)| \leq C_\rho|x - y|^k + \rho(1 + |x|^k + |y|^k), \quad x, y \in \mathbf{R}^N. \tag{3.12}$$

We notice that Propositions 3.2.2 and 3.2.3 are still valid, and we improve the sufficient conditions of Theorem 3.2.5.

We assume that

$$\alpha > kL(1 + (k - 1)L/2), \quad k \geq 2 : \text{even}, \tag{3.13}$$

and $\xi(x) := C(1 + |x|^k)$ satisfies

$$-\alpha\xi(x) + \mathcal{L}\xi(x) + \max_{p\in\Gamma}(b(x, p), D\xi(x)) \leq -\xi(x), \tag{3.14}$$

where $\mathcal{L}\xi(x) = \frac{1}{2}tr(\sigma(x)\sigma^\top(x)D^2\xi(x))$. Define the value function

$$v(x) = \inf_{\pi\in\mathcal{A}} J_x(\pi).$$

Proposition 3.3.1. *Suppose* (3.7), (3.11), (3.12), (3.13), *and* (3.14) *hold. Then we have*

(i) $0 \leq v(x) \leq \xi(x)$,
(ii) $\forall\rho > 0, \exists C_\rho > 0 : |v(x) - v(y)| \leq C_\rho|x - y|^k + \rho(1 + |x|^k + |y|^k), x,$ $y \in \mathbf{R}^N$.

Proof.

(1) By Itô's formula

$$\begin{aligned} e^{-\alpha T}\xi(X_T) = \xi(x) + \int_0^T e^{-\alpha t}\{-\alpha\xi(X_t) + \mathcal{L}\xi(X_t) \\ + (b(X_t, p_t), D\xi(X_t))\}dt \\ + \int_0^T e^{-\alpha t} D\xi(X_t)\sigma(X_t)dB_t, \quad \text{a.s.} \end{aligned}$$

By virtue of Proposition 1.8.3, we see that

$$E[\sup_{0\le t\le T} |X_t - x|^{2k}] \le CT^k, \tag{3.15}$$

and then

$$E[\int_0^T |e^{-\alpha t} D\xi(X_t)\sigma(X_t)|^2 dt] \le CE[\int_0^T (1 + |X_t|^{2k})dt] < \infty.$$

This yields that $\{\int_0^t e^{-\alpha s} D\xi(X_s)\sigma(X_s)dB_s\}$ is a martingale. Taking the expectation, we deduce by (3.14) that

$$E[e^{-\alpha T}\xi(X_T) + \int_0^T e^{-\alpha t}\xi(X_t)dt] \le \xi(x). \tag{3.16}$$

Letting $T \to \infty$, by (3.11), we have

$$J_x(\pi) \le E[\int_0^\infty e^{-\alpha t}\xi(X_t)dt] \le \xi(x), \quad \pi \in \mathcal{A}, \tag{3.17}$$

which implies (i). By (3.17), we remark that $\liminf_{t\to\infty} E[e^{-\alpha t}\xi(X_t)] = 0$.

(2) Let $\{Y_t\}$ be the solution to (3.5) with $Y_0 = y$ for given Brownian motion $\{B_t\}$. Since

$$\begin{aligned} d(X_t - Y_t) = [b(X_t, p_t) - b(Y_t, p_t)]dt + [\sigma(X_t) - \sigma(Y_t)]dB_t, \\ X_0 - Y_0 = x - y, \end{aligned}$$

we have by Itô's formula and (3.7)

$$E[|X_t - Y_t|^k] \le |x - y|^k + k(L + (k-1)L^2/2)E[\int_0^t |X_s - Y_s|^k ds].$$

By Gronwall's inequality,

$$E[|X_t - Y_t|^k] \le |x - y|^k \exp\{k(L + (k-1)L^2/2)t\}.$$

By (3.11), (3.12), (3.13) and (3.16), we get

$$\begin{aligned}
|J_x(\pi) - J_y(\pi)| &= |E[\int_0^\infty e^{-\alpha t}\{g(X_t, p_t) - g(Y_t, p_t)\}dt] \\
&\le E[\int_0^\infty e^{-\alpha t}\{C_\rho |X_t - Y_t|^k + \rho(1 + |X_t|^k + |Y_t|^k)\}dt] \\
&\le C_\rho |x - y|^k/(\alpha - kL(1 + (k-1)L/2)) \\
&\quad + \rho C_0(1 + \xi(x) + \xi(y)),
\end{aligned}$$

for some $C_0 > 0$. By the same argument as Proposition 3.2.4 (2), we obtain (ii). ■

Theorem 3.3.2. *We assume* (3.7), (3.11), (3.12), (3.13), *and* (3.14). *Then the DPP holds for* $v(x)$, *that is,*

$$v(x) = \inf_{\pi \in \mathcal{A}} E[\int_0^\tau e^{-\alpha t} g(X_t, p_t)dt + e^{-\alpha\tau} v(X_\tau)], \quad \tau \in \mathcal{S}_b.$$

Proof. We note that $E[\sup_{0 \le t \le T} e^{-\alpha t}\xi(X_t)] < \infty$ and the proof of Theorem 3.2.5 is valid except for (4). So it is enough to verify (4) with nonrandom τ, by using the same notations as Theorem 3.2.5.

By Proposition 3.3.1, we can take $0 < \delta \le 1$, for any $\varepsilon > 0$, such that whenever $|x - y| < \delta$,

$$\begin{aligned}
|J_x(\pi) - J_y(\pi)| + |v(x) - v(y)| &\le C_\rho \delta^k + \rho\{1 + |x|^k + 2^k(|x|^k + \delta^k)\} \\
&\le \{C_\rho \delta^k + \rho(1 + 2^k)\}\xi(x)/C \\
&< \left(\frac{\varepsilon}{2} + \frac{\varepsilon}{2}\right)\xi(x) = \varepsilon\xi(x), \qquad \text{for all } \pi \in \mathcal{A}.
\end{aligned}$$

For any $x \in S_i$, $i = 1, 2, \ldots, n$, we have

$$\begin{aligned}
J_x(\pi^{(i)}) &\le J_{x_i}(\pi^{(i)}) + |J_x(\pi^{(i)}) - J_{x_i}(\pi^{(i)})| \\
&\le v(x_i) + \varepsilon(1 + \xi(x)) \\
&\le v(x) + \varepsilon(1 + 2\xi(x)).
\end{aligned}$$

Also,

$$\begin{aligned}
E[|X_\tau|^k 1_{\{|X_\tau| > R\}}]^2 &\le E[|X_\tau|^{2k}]P(|X_\tau| > R) \\
&\le E[\sup_{0 \le t \le T} |X_t|^{2k}] \cdot \frac{C}{R^2}(1 + |x|^2).
\end{aligned}$$

Thus, by (3.15) and (3.17), we fix x and choose $R > 0$ such that

$$\sup_{\pi \in \mathcal{A}} E[|J_{X_\tau}(\pi)| 1_{\{|X_\tau| > R\}}] < \varepsilon.$$

Now, taking into account (3.16), we have

$$\begin{aligned}
E[\int_{\tau}^{\infty} e^{-\alpha t} g(X_t^{\tau}, p_t^{\tau}) dt] &= E[E[\int_{\tau}^{\infty} e^{-\alpha t} g(X_t^{\tau}, p_t^{\tau}) dt | \mathcal{F}_{\tau}]] \\
&= E[e^{-\alpha \tau} J_{X_\tau}(\tilde{\pi^{\tau}})] \\
&= E[e^{-\alpha \tau} \sum_{i=1}^{n+1} J_{X_\tau}(\pi^{(i)}) 1_{\{X_\tau \in S_i\}}] \\
&\le E[e^{-\alpha \tau} \sum_{i=1}^{n} \{v(X_\tau) + \varepsilon(1 + 2\xi(X_\tau))\} 1_{\{X_\tau \in S_i\}}] \\
&\quad + E[e^{-\alpha \tau} J_{X_\tau}(\pi^{(n+1)}) 1_{\{|X_\tau| > R\}}] \\
&\le E[e^{-\alpha \tau} v(X_\tau)] + \varepsilon(1 + 2\xi(x)) + \varepsilon.
\end{aligned}$$

Hence

$$\begin{aligned}
v(x) &\le E[\int_0^{\tau} e^{-\alpha t} g(X_t^{\tau}, p_t^{\tau}) dt + \int_{\tau}^{\infty} e^{-\alpha t} g(X_t^{\tau}, p_t^{\tau}) dt] \\
&\le E[\int_0^{\tau} e^{-\alpha t} g(X_t, p_t) dt + e^{-\alpha \tau} v(X_\tau)] + 2\varepsilon(1 + \xi(x)) \\
&\le \bar{v}(x) + \varepsilon + 2\varepsilon(1 + \xi(x)).
\end{aligned}$$

Therefore, letting $\varepsilon \to 0$, we get $v(x) \le \bar{v}(x)$. ∎

Example 3.3.3. *We can show that $f(x) := |x|^k$ fulfills (3.12). Let $1/k + 1/k' = 1$. By Young's inequality, we have*

$$\begin{aligned}
|f(x) - f(y)| &= |(|x| - |y|)(|x|^{k-1} + |x|^{k-2}|y| + \cdots + |y|^{k-1})| \\
&\le \frac{1}{a}|x - y| \cdot a(|x| + |y|)^{k-1} \\
&\le \frac{1}{k}(\frac{1}{a}|x - y|)^k + \frac{1}{k'}\{a(|x| + |y|)^{k-1}\}^{k'} \\
&\le C_\rho |x - y|^k + \rho(1 + |x|^k + |y|^k),
\end{aligned}$$

for a suitable choice of $a > 0$.

3.4 The HJB Equations: Stochastic Case

We examine the relation between the DPP in Section 3.3 and the HJB equation:

$$-\alpha v(x) + \mathcal{L} v(x) + \min_{p \in \Gamma} h(x, p, Dv) = 0, \qquad x \in \mathbf{R}^N, \tag{3.18}$$

where

$$h(x, p, Dv) = g(x, p) + (b(x, p), Dv(x)).$$

Proposition 3.4.1. *Let $v(x)$ satisfy the DPP. Suppose that $v \in C^2$ and*

$$|v(x)| + |Dv(x)| + |D^2 v(x)| + |\mathcal{L}v(x)| \le C(1 + |x|^k), \tag{3.19}$$

for some constant $C > 0$, independent of $p \in \Gamma$. Then we have (3.18).

Proof.

(1) By Itô's formula, we have

$$\begin{aligned} d(e^{-\alpha t} v(X_t)) = e^{-\alpha t}\{&-\alpha v(X_t) + \mathcal{L}v(X_t) + (b(X_t, p_t), Dv(X_t))\}dt \\ &+ Dv(X_t)\sigma(X_t)dB_t. \end{aligned}$$

By (3.15), we see that

$$E[\sup_{0\le t\le T} |X_t - x|^{2m}] < \infty, \quad E[\int_0^T |X_t|^{2m} dt] < \infty, \quad \forall m \ge 0.$$

Also, by (3.19)

$$E[\int_0^t e^{-\alpha s} |Dv(X_s)\sigma(X_s)|^2 ds] < \infty.$$

This yields that $\{\int_0^t e^{-\alpha s} Dv(X_s)\sigma(X_s)dB_s\}$ is a martingale. Thus the DPP can be rewritten as

$$\begin{aligned} 0 &= \inf_{\pi\in\mathcal{A}} E[\int_0^T e^{-\alpha t} g(X_t, p_t)dt + e^{-\alpha T} v(X_T) - v(x)] \\ &= \inf_{\pi\in\mathcal{A}} E[\int_0^T e^{-\alpha t}\{-\alpha v(X_t) + \mathcal{L}v(X_t) + h(X_t, p_t, Dv)\}dt]. \end{aligned}$$

(2) Let $\{X_t\}$ be the solution to (3.5) for any constant control $p(t) = p$, for all $t \ge 0$. By (1), we have

$$0 \le E[\int_0^T e^{-\alpha t}\{-\alpha v(X_t) + \mathcal{L}v(X_t) + h(X_t, p, Dv)\}dt].$$

Divide both sides by T and let $T \to 0$. Then, taking into account (3.19), we have

$$0 \le -\alpha v(x) + \mathcal{L}v(x) + \min_{p\in\Gamma} h(x, p, Dv).$$

(3) Let $\varepsilon > 0$ and $T > 0$ be arbitrary. By (1), there exists $\pi = (\Omega, \mathcal{F}, P, \{\mathcal{F}_t\}; \{B_t\}, \{p_t\}) \in \mathcal{A}$ such that

$$\varepsilon T \ge E[\int_0^T e^{-\alpha t}\{-\alpha v(X_t) + \mathcal{L}v(X_t) + h(X_t, p_t, Dv\}dt].$$

Under (3.19), we note by (3.11), (3.12) that

$$\begin{aligned}|h(x,p,Dv)-h(y,p,Dv)| &\le |g(x,p)-g(y,p)| \\ &\quad + |b(x,p)\|Dv(x)-Dv(y)| \\ &\quad + |b(x,p)-b(y,p)||Dv(y)| \\ &\le C_\rho|x-y|^k+\rho(1+|x|^k \\ &\quad +|y|^k)+2L|x-y|C(1+|x|^k).\end{aligned}$$

Hence

$$E[\int_0^T e^{-\alpha t}h(X_t,p_t,Dv)dt]=E[\int_0^T e^{-\alpha t}h(x,p_t,Dv)dt]+o(T) \quad \text{as } T\to 0.$$

Thus

$$\begin{aligned}\varepsilon T &\ge E[\int_0^T e^{-\alpha t}\{-\alpha v(X_t)+\mathcal{L}v(X_t)+h(x,p_t,Dv\}dt]+o(T) \\ &\ge E[\int_0^T e^{-\alpha t}\min_{p\in\Gamma}\{-\alpha v(X_t)+\mathcal{L}v(X_t)+h(x,p,Dv)\}dt]+o(T).\end{aligned}$$

Dividing both sides by T and letting $T\to 0$ and then $\varepsilon\to 0$, we get

$$0\ge -\alpha v(x)+\mathcal{L}v(x)+\min_{p\in\Gamma}h(x,p,Dv).$$

■

We present a synthesis of optimal control in terms of a classical solution u of the HJB equation (3.18). The following result is called a *verification theorem.*

Theorem 3.4.2. *Let $u\in C^2$ be a solution of* (3.18) *satisfying* (3.19). *We make the assumptions of Theorem 3.3.2 and assume that there exists a Γ-valued continuous function $\hat{p}(x)$ on $\mathbf{R}^N$ such that*

(i) $h(x,\hat{p}(x),Du)=\min_{p\in\Gamma}h(x,p,Du)$,

(ii) $\begin{cases} dz_t=b(z_t,\hat{p}(z_t))dt+\sigma(z_t)dB_t, \\ z_0=x\end{cases}$

admits a unique solution $\{z_t\}$.

Then $u(x)\le v(x)$ and the feedback control $\{\hat{p}(z_t)\}$ is optimal, that is, $u(x)=J_x(\hat{p}(z_\cdot))$.

Proof. By (3.18), it is clear that

$$\begin{aligned}0 &= -\alpha u(x)+\mathcal{L}u(x)+h(x,\hat{p}(x),Du) \\ &\le -\alpha u(x)+\mathcal{L}u(x)+h(x,p,Du), \quad p\in\Gamma.\end{aligned}$$

Itô's formula gives

$$e^{-\alpha T}u(X_T) = u(x) + \int_0^T e^{-\alpha t}\{-\alpha u(X_t) + \mathcal{L}u(X_t) + (b(X_t, p_t), Du(X_t))\}dt$$
$$+ \int_0^T e^{-\alpha t} Du(X_t)\sigma(X_t)dB_t, \quad \text{a.s.},$$

and hence

$$E[e^{-\alpha T}u(X_T)] \geq u(x) - E[\int_0^T e^{-\alpha t} g(X_t, p_t)dt].$$

By (3.19) and Proposition 3.3.1 (1), we get

$$\liminf_{T\to\infty} E[e^{-\alpha T}|u(X_T)|] \leq \liminf_{T\to\infty} E[e^{-\alpha T}\xi(X_T)]/C_0 = 0$$

for some constant $C_0 > 0$. Hence

$$E[\int_0^\infty e^{-\alpha t} g(X_t, p_t)dt] \geq u(x),$$

so that $v(x) \geq u(x)$. By the same calculation as above, we deduce

$$J(\hat{p}(z.)) = E[\int_0^\infty e^{-\alpha t} g(z_t, \hat{p}(z_t))dt] = u(x). \qquad \blacksquare$$

Example 3.4.3. *Let $N = 1$, $\Gamma = [-1, 1]$, $b(x, p) = xp$, $\sigma(x) = x$, $g(x, p) = x^2$, $L = 1$, $k = 2$ in Section 3.3. Assume $\alpha \geq 4$. Then the sufficient conditions of Theorem 3.3.2 are fulfilled. The associated HJB equation is*

$$-\alpha u + \frac{1}{2}x^2 u'' + \min_{p\in\Gamma} xu'p + x^2 = 0, \quad x \in \mathbf{R}.$$

Substituting $u(x) = Ax^2$ into this equation, we can show that $A = 1/(\alpha + 1) > 0$ and

$$xu'\hat{p} = -|xu'| = \min_{p\in\Gamma} xu'p \quad \text{with} \quad \hat{p}(x) = -\text{sgn}(xu'(x)) = -1,$$

and there exists a unique solution $\{z_t\}$ of

$$\begin{cases} dz_t = -z_t dt + z_t dB_t, \\ z_0 = x. \end{cases}$$

4

Viscosity Solutions of Hamilton–Jacobi–Bellman Equations

The dynamic programming principle,

$$v(x) = \inf_{p} E[\int_0^\tau e^{-\alpha t} g(X(t), p(t))dt + e^{-\alpha\tau} v(X(\tau))], \quad \tau \geq 0,$$

is closely related to the HJB equation,

$$\alpha u = \frac{1}{2} tr(\sigma\sigma^\top D^2 u) + (Ax, Du) + \min_{q \in \Gamma}\{g(x, q) + (q, Du)\}, \quad x \in \mathbf{R}^N,$$

associated with the stochastic control problem to minimize the cost function,

$$J_x(p) = E[\int_0^\infty e^{-\alpha t} g(X(t), p(t))dt], \qquad \text{for } p = \{p(t)\},$$

subject to

$$dX(t) = \{AX(t) + p(t)\}dt + \sigma dB(t), \quad X(0) = x, \quad p(t) \in \Gamma.$$

Suppose $v \in C^2$. Then, by Itô's formula, the dynamic programming principle implies that v is a classical solution of the HJB equation. What is the meaning of its solution if the smoothness of v is dropped ? The answer is that v is a viscosity solution of the HJB equation. In this chapter, we present the notion of viscosity solutions, that was introduced by M. G. Crandall and P. L. Lions in 1981. Our main goal is to show that this answer is true and also the uniqueness holds. The relationship between viscosity solutions and Markov processes is also given.

M. Bardi and I. Cappuzzo Dolcetta [8], W. H. Fleming and H. M. Soner [65], S. Koike [97], M. G. Crandall, H. Ishii and P. L. Lions [36], H. Ishii [78], L. C. Evans [59], and P. L. Lions [102] are basic references for this chapter.

4.1 Definition of Viscosity Solutions

We consider the second-order partial differential equations

$$F(x, u, Du, D^2u) = 0 \quad \text{in } Q, \tag{4.1}$$

where Q is an open domain of $\mathbf{R}^N$, $F : Q \times \mathbf{R} \times \mathbf{R}^N \times \mathbf{S}^N \to \mathbf{R}$ is continuous, $\mathbf{S}^N$ is the collection of $N \times N$ symmetric matrices, and F is degenerate elliptic, that is,

$$F(x, r, \lambda, X) \leq F(x, r, \lambda, Y),$$

for all $(x, r, \lambda) \in Q \times \mathbf{R} \times \mathbf{R}^N$, $X, Y \in \mathbf{S}^N$ provided $X \geq Y$, that is, $X - Y$ is nonnegative definite.

Definition 4.1.1. *$u \in C(Q)$ is a viscosity solution of* (4.1) *if the following assertions are satisfied:*

(a) *u is a viscosity subsolution of* (4.1)*, that is, for any $\varphi \in C^2(Q)$ and any local maximum point $z \in Q$ of $u - \varphi$,*

$$F(z, u(z), D\varphi(z), D^2\varphi(z)) \leq 0,$$

and

(b) *u is a viscosity supersolution of* (4.1)*, that is, for any $\varphi \in C^2(Q)$ and any local minimum point $\bar{z} \in Q$ of $u - \varphi$,*

$$F(\bar{z}, u(\bar{z}), D\varphi(\bar{z}), D^2\varphi(\bar{z})) \geq 0.$$

Theorem 4.1.2. *Let u be a solution of* (4.1) *in C^2. Then u is a viscosity solution of* (4.1).

Proof. Let $\varphi \in C^2(Q)$ and let $z \in Q$ be a local maximum point of $u - \varphi$. It is well known that

$$D(u - \varphi)(z) = 0, \quad D^2(u - \varphi)(z) \leq 0.$$

By degenerate ellipticity, we have

$$0 = F(z, u(z), Du(z), D^2u(z)) \geq F(z, u(z), D\varphi(z), D^2\varphi(z)).$$

This yields that u is a subsolution of (4.1). Replacing z by the local minimum point $\bar{z}$, we see that u is a supersolution of (4.1). ■

Theorem 4.1.3. *Let $u \in C(Q)$ be a viscosity solution of* (4.1)*. If $u \in C^2$, then it is a classical solution of* (4.1).

Proof. Let $z \in Q$ be arbitrary. Taking $\varphi = u$, we see that $u(x) - \varphi(x) = u(z) - \varphi(z) = 0$, for all $x \in Q$. Thus, by definition,

$$F(z, u(z), Du(z), D^2u(z)) = F(z, u(z), D\varphi(z), D^2\varphi(z)) = 0. \quad ■$$

4.2 The HJB Equations: First-Order Case

Consider the first-order partial differential equation:

$$F(x, u, Du, D^2u) := \alpha u(x) - (Du(x), Ax) \\ - \min_{p \in \Gamma}\{(Du(x), Bp) + g(x, p)\} = 0 \text{ in } \mathbf{R}^N. \quad (4.2)$$

This is the HJB equation (3.4) of Proposition 3.1.3 to minimize the cost

$$J_x(p) = \int_0^\infty e^{-\alpha t} g(x(t), p(t))dt,$$

over $p \in \mathcal{A}$, subject to

$$\dot{x}(t) = Ax(t) + Bp(t), \quad x(0) = x \in \mathbf{R}^N, \quad (4.3)$$

where $\mathcal{A}$ is as in Section 3.3.1 and $g : \mathbf{R}^N \times \mathbf{R}^M \to \mathbf{R}$ is a continuous function satisfying

$$|g(x, p) - g(y, p)| \le C|x - y|, \quad 0 \le g(x, p) \le C, \quad (4.4)$$

for some constant $C > 0$.

Theorem 4.2.1. *Let $v(x)$ be the value function, that is $v(x) = \inf_{p \in \mathcal{A}} J_x(p)$. We assume* (4.4). *Then v is a viscosity solution of* (4.2).

Proof.

(1) We claim that v is uniformly continuous on $\mathbf{R}^N$. We choose $T > 0$, for any $\varepsilon > 0$, such that

$$2C \int_T^\infty e^{-\alpha t} dt < \varepsilon.$$

Let $y(t)$ be the solution of (4.3) for $p \in \mathcal{A}$ with $y(0) = y$. Then

$$|x(t) - y(t)| = |e^{At}(x - y)| \le e^{|A|t}|x - y|.$$

Taking $C_\varepsilon = C \displaystyle\int_0^T e^{(-\alpha + |A|)t} dt$, we have

$$\begin{aligned} |v(x) - v(y)| &\le \sup_{p \in \mathcal{A}} \int_0^\infty e^{-\alpha t}|g(x(t), p(t)) - g(y(t), p(t))|dt \\ &\le C \sup_{p \in \mathcal{A}} \int_0^T e^{-\alpha t}|x(t) - y(t)|dt + 2C \int_T^\infty e^{-\alpha t} dt \\ &< C_\varepsilon|x - y| + \varepsilon. \end{aligned}$$

This yields the uniform continuity of v on $\mathbf{R}^N$.

(2) Subsolution property. Let $\varphi \in C^1(\mathbf{R}^N)$ and z be a local maximum point of $v - \varphi$. Then, there is a ball $B_r(z)$ such that $v(x) - v(z) \le \varphi(x) - \varphi(z)$, for all $x \in B_r(z)$. Let $x(t)$ be the solution of (4.3) for any constant control $p(t) = p$ with $x(0) = z$. By the DPP in Proposition 3.1.2

$$v(z) \le \int_0^T e^{-\alpha t} g(x(t), p)dt + e^{-\alpha T} v(x(T)).$$

Hence, for sufficiently small $T > 0$,

$$\begin{aligned}
0 &\ge -\int_0^T e^{-\alpha t} g(x(t), p)dt - e^{-\alpha T} v(x(T)) + v(z) \\
&\ge -\int_0^T e^{-\alpha t} g(x(t), p)dt - e^{-\alpha T}\{\varphi(x(T)) - \varphi(z)\} - (e^{-\alpha T} - 1)v(z) \\
&= -\int_0^T e^{-\alpha t} g(x(t), p)dt - e^{-\alpha T} \\
&\quad \times \int_0^T (D\varphi(x(t)), Ax(t) + Bp)dt - (e^{-\alpha T} - 1)v(z).
\end{aligned}$$

Dividing by T and letting $T \to 0$, we get

$$-g(z, p) - (D\varphi(z), Az + Bp) + \alpha v(z) \le 0.$$

Since

$$\sup_{p \in \Gamma}\{-(D\varphi(z), Bp) - g(x, p)\} = -\min_{p \in \Gamma}\{(D\varphi(z), Bp) + g(x, p)\},$$

we deduce

$$\alpha v(z) - (D\varphi(z), Az) - \min_{p \in \Gamma}\{(D\varphi(z), Bp) + g(x, p)\} \le 0.$$

This implies that v is a viscosity subsolution of (4.2).

(3) Supersolution property. Let $\varphi \in C^1(\mathbf{R}^N)$ and $\bar{z}$ be a local mimimum point of $v - \varphi$. Then, there is a ball $B_r(\bar{z})$ such that $v(x) - v(\bar{z}) \ge \varphi(x) - \varphi(\bar{z})$ for all $x \in B_r(\bar{z})$. For any $\varepsilon > 0$ and $T > 0$, by the DPP of Proposition 3.1.2, there exists $p \in \mathcal{A}$ such that

$$v(\bar{z}) + \varepsilon T \ge \int_0^T e^{-\alpha t} g(x(t), p(t))dt + e^{-\alpha T} v(x(T)),$$

where $x(t)$ is the solution of (4.3) for $p(t)$ with $x(0) = \bar{z}$. Hence, for sufficiently small $T > 0$

$$\begin{aligned}
-\varepsilon T &\leq -\int_0^T e^{-\alpha t} g(x(t), p(t))dt - e^{-\alpha T} v(x(T)) + v(\bar{z}) \\
&\leq -\int_0^T e^{-\alpha t} g(x(t), p(t))dt - e^{-\alpha T}\{\varphi(x(T)) - \varphi(\bar{z})\} \\
&\quad - (e^{-\alpha T} - 1)v(\bar{z}) \\
&= -\int_0^T e^{-\alpha t} g(x(t), p(t))dt - e^{-\alpha T} \\
&\quad \times \int_0^T (D\varphi(x(t)), Ax(t) + Bp(t))dt - (e^{-\alpha T} - 1)v(\bar{z}).
\end{aligned}$$

By (4.4), we note that

$$\int_0^T e^{-\alpha t} g(x(t), p(t))dt = \int_0^T g(\bar{z}, p(t))dt + o(T).$$

Moreover,

$$\int_0^T (D\varphi(x(t)), Ax(t) + Bp(t))dt = \int_0^T (D\varphi(\bar{z}), A\bar{z} + Bp(t))dt + o(T).$$

Hence

$$\begin{aligned}
-\varepsilon T &\leq -\int_0^T g(\bar{z}, p(t))dt \\
&\quad - \int_0^T (D\varphi(\bar{z}), A\bar{z} + Bp(t))dt - (e^{-\alpha T} - 1)v(\bar{z}) + o(T) \\
&\leq \int_0^T -(D\varphi(\bar{z}), A\bar{z}) + \max_{p\in\Gamma}\{-g(\bar{z}, p) - (D\varphi(\bar{z}), Bp)\}dt \\
&\quad - (e^{-\alpha T} - 1)v(\bar{z}) + o(T).
\end{aligned}$$

Dividing by T, letting $T \to 0$ and then $\varepsilon \to 0$, we get

$$0 \leq -(D\varphi(\bar{z}), A\bar{z}) - \min_{p\in\Gamma}\{(D\varphi(\bar{z}), Bp) + g(\bar{z}, p)\} + \alpha v(\bar{z}),$$

that is, v is a viscosity supersolution of (4.2). ■

Theorem 4.2.2. *Let $v_i(x) \in C(\mathbf{R}^N)$, $i = 1, 2$, be two bounded and uniformly continuous viscosity solutions of* (4.2). *We assume* (4.4). *Then we have*

$$v_1(x) = v_2(x), \quad \textit{for all } x \in \mathbf{R}^N.$$

Proof. We will show that

$$v_1(x) \leq v_2(x), \quad \text{for } x \in \mathbf{R}^N.$$

Then, interchanging v_1 and v_2, we obtain the assertion.

(1) Suppose it were not true. Then there is $\bar{x} \in \mathbf{R}^N$ such that $v_1(\bar{x}) > v_2(\bar{x})$. Hence we find $\zeta_0 > 0$ such that

$$\delta := \sup_{x \in \mathbf{R}^N} v_1(x) - v_2(x) - 2\zeta_0 \psi(x) > 0,$$

where $\psi(x) := \dfrac{1}{2} \log(1 + |x|^2)$. Define

$$\Psi(x, y) = v_1(x) - v_2(y) - \frac{n}{2}|x - y|^2 - \zeta\{\psi(x) + \psi(y)\},$$

for each $n \in \mathbf{N}$ and $0 < \zeta \leq \zeta_0$ chosen later. Since Ψ is continuous and $\lim_{|x|+|y| \to \infty} \Psi(x, y) = -\infty$, there exists $(x_n, y_n) \in \mathbf{R}^N \times \mathbf{R}^N$ such that

$$\Psi(x_n, y_n) = \sup_{(x,y) \in \mathbf{R}^N \times \mathbf{R}^N} \Psi(x, y),$$

and then

$$\Psi(x_n, y_n) \geq \sup_{x \in \mathbf{R}^N} v_1(x) - v_2(x) - 2\zeta\psi(x) \geq \delta.$$

Since $\Psi(x_n, x_n) + \Psi(y_n, y_n) \leq 2\Psi(x_n, y_n)$, we have, by a simple calculation,

$$n|x_n - y_n|^2 \leq v_1(x_n) - v_1(y_n) + v_2(x_n) - v_2(y_n).$$

By the boundedness of v_1 and v_2,

$$|x_n - y_n| \leq \frac{C}{\sqrt{n}}.$$

Thus, by the uniform continuity of v_1 and v_2

$$\lim_{n \to \infty} n|x_n - y_n|^2 \leq \lim_{n \to \infty} \sup_{|x-y| \leq \frac{C}{\sqrt{n}}} \{|v_1(x) - v_1(y)| + |v_2(x) - v_2(y)|\} = 0.$$

(2) We note that

$$\Psi(x, y_n) = v_1(x) - \varphi_1(x),$$
$$\Psi(x_n, y) = -v_2(y) + \varphi_2(y),$$

where

$$\varphi_1(x) := v_2(y_n) + \frac{n}{2}|x - y_n|^2 + \zeta\{\psi(x) + \psi(y_n)\} \in C^1(\mathbf{R}^N),$$
$$\varphi_2(y) := v_1(x_n) - \frac{n}{2}|x_n - y|^2 - \zeta\{\psi(x_n) + \psi(y)\} \in C^1(\mathbf{R}^N).$$

It is easily seen that

$$D\varphi_1(x) = n(x - y_n) + \zeta\frac{x}{1 + |x|^2}, \quad D\varphi_2(y) = n(x_n - y) - \zeta\frac{y}{1 + |y|^2}.$$

Since $v_1 - \varphi_1$ attains its maximum at x_n and $v_2 - \varphi_2$ attains its mimimum at y_n, the viscosity property gives

$$\alpha v_1(x_n) - (D\varphi_1(x_n), Ax_n) - \min_{p\in\Gamma}\{(D\varphi_1(x_n), Bp) + g(x_n, p)\} \leq 0,$$
$$\alpha v_2(y_n) - (D\varphi_2(y_n), Ay_n) - \min_{p\in\Gamma}\{(D\varphi_2(y_n), Bp) + g(y_n, p)\} \geq 0.$$

Putting together these inequalities, we get

$$\begin{aligned}
\alpha(v_1(x_n) - v_2(y_n)) &\leq (D\varphi_1(x_n), Ax_n) - (D\varphi_2(y_n), Ay_n) \\
&\quad + \max_{p\in\Gamma}\{(D\varphi_1(x_n) - D\varphi_2(y_n), Bp) \\
&\quad + g(x_n, p) - g(y_n, p)\} \\
&\leq n|x_n - y_n|^2\|A\| + 2\zeta\|A\| + 2\zeta\|B\| \cdot \max_{p\in\Gamma}|p| \\
&\quad + C|x_n - y_n| \\
&\leq n|x_n - y_n|^2\|A\| + \alpha\delta/2 + C|x_n - y_n|,
\end{aligned}$$

where we choose $0 < \zeta \leq \zeta_0$ such that $2\zeta(\|A\| + \|B\| \cdot \max_{p\in\Gamma}|p|) < \alpha\delta/2$. Also, by virtue of (1),

$$\alpha\delta \leq \alpha\Psi(x_n, y_n) \leq \alpha(v_1(x_n) - v_2(y_n)).$$

Letting $n \to \infty$, we have $\delta \leq 0$, which is a contradiction. ■

4.3 The HJB Equations: Second-Order Case

Consider the second-order partial differential equation:

$$\begin{aligned}
F(x, u, Du, D^2u) :&= \alpha u - \frac{1}{2}tr(\sigma(x)\sigma^\top(x)D^2u) \\
&\quad - \min_{p\in\Gamma}\{(b(x, p), Du) + g(x, p)\} = 0 \text{ in } \mathbf{R}^N. \quad (4.5)
\end{aligned}$$

This is the HJB equation (3.18) of Proposition 3.4.1 associated with the optimization problem of Section 3.3.2 to minimize the cost

$$J_x(\pi) = E[\int_0^\infty e^{-\alpha t} g(X_t, p_t)dt],$$

over $\pi = (\Omega, \mathcal{F}, P, \{\mathcal{F}_t\}; \{B_t\}, \{p_t\}) \in \mathcal{A}$, subject to

$$dX_t = b(X_t, p_t)dt + \sigma(X_t)dB_t, \quad X_0 = x \in \mathbf{R}^N. \quad (4.6)$$

Here $\mathcal{A}$ is a class of all control processes $p = \{p_t\}$ given in Section 3.3.2, $g : \mathbf{R}^N \times \mathbf{R}^M \to \mathbf{R}$, $b : \mathbf{R}^N \times \Gamma \to \mathbf{R}^N$, $\sigma : \mathbf{R}^N \to \mathbf{R}^{N\times M}$ are continuous functions such that

$$|g(x, p) - g(y, p)| \leq C|x - y|, \quad 0 \leq g(x, p) \leq C, \quad (4.7)$$

and

$$|b(x, p) - b(y, p)| + \|\sigma(x) - \sigma(y)\| \leq L|x - y|, \quad (4.8)$$

for the compact convex subset Γ of $\mathbf{R}^N$ and the constant $L > 0$ satisfying

$$\alpha > L + L^2/2. \tag{4.9}$$

Theorem 4.3.1. *Let $v(x)$ be the value function, that is, $v(x) = \inf_{\pi \in \mathcal{A}} J_x(\pi)$. We assume* (4.7), (4.8), *and* (4.9). *Then v is a viscosity solution of* (4.5).

Proof.

(1) We claim that for $(x, r, \lambda, X) \in \mathbf{R}^N \times \mathbf{R} \times \mathbf{R}^N \times \mathbf{S}^N$,

$$F(x, r, \lambda, X) := \alpha r - \frac{1}{2} tr(A \cdot X) - \min_{p \in \Gamma}\{(b(x, p), \lambda) + g(x, p)\} \text{ is degenerate elliptic,}$$

where $A = \sigma(x)\sigma^\top(x)$. Since A is symmetric and nonnegative definite, there exists an orthogonal matrix M such that

$$M^\top AM = \text{diag}(\mu_1, \dots, \mu_N), \quad M^\top M = I, \quad \mu_i \geq 0, \quad i = 1, \dots, N.$$

Let $Y \in \mathbf{S}^N$ and $X \geq Y$. Then $(\eta_{ij}) := M^\top(Y - X)M \leq 0$. Therefore,

$$\begin{aligned} 2\{F(x, r, \lambda, X) - F(x, r, \lambda, Y)\} &= tr(A \cdot (Y - X)) \\ &= tr(\text{diag}(\mu_1, \dots, \mu_N)(\eta_{ij})) \\ &= \mu_1\eta_{11} + \dots + \mu_N\eta_{NN} \leq 0. \end{aligned}$$

(2) By (4.6) and Gronwall's inequality, we have

$$\kappa := \sup_{p \in \mathcal{A}} \sup_{0 \leq s \leq 1} E[|X_s|^2] < \infty.$$

Since

$$\sup_{0 \leq s \leq h} E[|X_s|^2] \leq \kappa, \quad 0 < h < 1,$$

we see, by (4.8), that

$$\begin{aligned} E[\sup_{0 \leq s \leq h} |X_s - X_0|^2] &= E[\sup_{0 \leq s \leq h} |\int_0^s b(X_t, p_t)dt + \int_0^s \sigma(X_t)dB_t|^2] \\ &\leq CE[\int_0^h (1 + |X_t|)^2 dt] \\ &\leq Ch(1 + \kappa) \to 0 \quad \text{as } h \to 0, \quad \forall p \in \mathcal{A}. \end{aligned}$$

Let $\mathcal{O}$ be a bounded neighborhood of x and $\rho = \rho_x$ be the first exit time from $\mathcal{O}$. Then

$$\begin{aligned} P(\rho \leq h) &\leq P(\sup_{0 \leq s \leq h} |X_s - x| \geq d(x, \partial\mathcal{O})) \\ &\leq (E[\sup_{0 \leq s \leq h} |X_s - x|^2])(d(x, \partial\mathcal{O}))^{-2}, \end{aligned}$$

from which

$$\frac{1}{h}E[h - h \wedge \rho] \le \frac{1}{h}E[(h-\rho)1_{\{\rho \le h\}}] \le P(\rho \le h) \to 0 \quad \text{as} \quad h \to 0.$$

(3) Subsolution property. Let $\varphi \in C^2$ and z be the local maximum point of $v - \varphi$. Then, there is a bounded neighborhood $\mathcal{O}$ of z such that $v(x) - \varphi(x) \le v(z) - \varphi(z)$, for all $x \in \mathcal{O}$. We may consider $v(z) = \varphi(z)$, adding a constant if necessary.

Let $\{X_t\}$ be the solution to (4.6) with $X_0 = z$ for any constant control $p_t = p$. By the DPP in Theorem 3.2.5,

$$v(z) \le E[\int_0^\tau e^{-\alpha t} g(X_t, p)dt + e^{-\alpha\tau} v(X_\tau)], \quad \tau \in \mathcal{S}_b.$$

Then, for $\rho = \rho_z$,

$$\begin{aligned}\varphi(z) &\le E[\int_0^{h\wedge\rho} e^{-\alpha t} g(X_t, p)dt + e^{-\alpha(h\wedge\rho)} v(X_{h\wedge\rho})] \\ &\le E[\int_0^{h\wedge\rho} e^{-\alpha t} g(X_t, p)dt + e^{-\alpha(h\wedge\rho)} \varphi(X_{h\wedge\rho})].\end{aligned}$$

We subtract $\varphi(z)$ from both sides and apply Itô's formula to obtain

$$\begin{aligned}0 &\le E[\int_0^{h\wedge\rho} e^{-\alpha t} g(X_t, p)dt + \{e^{-\alpha(h\wedge\rho)}\varphi(X_{h\wedge\rho}) - \varphi(z)\}] \\ &= E[\int_0^{h\wedge\rho} e^{-\alpha t} g(X_t, p)dt \\ &\quad + \int_0^{h\wedge\rho} e^{-\alpha t}\{-\alpha\varphi + \frac{1}{2}tr(\sigma(x)\sigma^\top(x)D^2\varphi) + (b(x,p), D\varphi)|_{x=X_t}\}dt].\end{aligned}$$

Dividing by h and letting $h \to 0$, by (2) and the uniform continuity of $\varphi, D\varphi, D^2\varphi$ on $\bar{\mathcal{O}}$, we deduce

$$-\alpha\varphi(z) + \frac{1}{2}tr(\sigma(z)\sigma^\top(z)D^2\varphi) + (b(z,p), D\varphi) + g(z,p) \ge 0.$$

Therefore,

$$F(z, v(z), D\varphi(z), D^2\varphi(z)) \le 0.$$

(4) By the DPP in Theorem 3.2.5, we have

$$v(x) = \inf_{\pi\in\mathcal{A}} E[\int_0^\tau e^{-\alpha t} g(X_t, p_t)dt + e^{-\alpha\tau} v(X_\tau)], \quad \tau \in \mathcal{S}_b.$$

For any $0 < h < 1$, there exists $\pi^h = (\Omega^h, \mathcal{F}^h, P^h, \{\mathcal{F}_t^h\}; \{B_t^h\}, \{p_t^h\}) \in \mathcal{A}$ such that

$$v(x) + h^2 \ge E^h[\int_0^\tau e^{-\alpha t} g(X_t^h, p_t^h)dt + e^{-\alpha\tau} v(X_\tau^h)], \quad \forall \tau \in \mathcal{S}_b,$$

where E^h denotes the expectation with respect to P^h, and

$$dX_t^h = b(X_t^h, p_t^h)dt + \sigma(X_t^h)dB_t^h, \quad X_0^h = x.$$

By the same line as (2), we have

$$E^h[\sup_{0\le s\le h} |X_s^h - X_0^h|^2] \to 0 \text{ as } h \to 0.$$

Taking $\tau = h \wedge \rho_x^{(h)}$ for the first exit time $\rho_x^{(h)}$ of $\{X_t^h\}$ from $\mathcal{O}$, we obtain

$$\frac{1}{h}E^h[h - h \wedge \rho_x^{(h)}] \le P^h(\rho_x^{(h)}) \le h) \to 0.$$

$$\begin{aligned}
&\sup_{0\le t\le h} E^h[|(b(X_t^h, p_t^h) - b(x, p_t^h), D\varphi(X_t^h))|1_{\{t\le\rho_x^{(h)}\}}] \\
&\quad \le \sup_{0\le t\le h} E^h[|b(X_t^h, p_t^h) - b(x, p_t^h)|] \sup_{x\in\bar{\mathcal{O}}} |D\varphi(x)| \to 0,
\end{aligned}$$

and

$$\sup_{0\le t\le h} E^h[|g(X_t^h, p_t^h) - g(x, p_t^h)|] \to 0 \quad \text{as} \quad h \to 0.$$

(5) Supersolution property. Let $\varphi \in C^2$ and z be the minimizer of $v - \varphi$. Then, by (4)

$$v(X_\tau^h) - \varphi(X_\tau^h) \ge v(z) - \varphi(z) = 0, \quad \text{a.s.},$$

$$\varphi(z) \ge E^h\Big[\int_0^\tau e^{-\alpha t} g(X_t^h, p_t^h)dt + e^{-\alpha\tau}\varphi(X_\tau^h)\Big] - h^2,$$

where $\tau = h \wedge \rho_z^{(h)}$ and $X_0^h = z$. We subtract $\varphi(z)$ from both sides and apply Itô's formula to obtain

$$\begin{aligned}
0 \ge \frac{1}{h}E^h\Big[&\int_0^\tau e^{-\alpha t} g(X_t^h, p_t^h)dt + \int_0^\tau e^{-\alpha t}\{-\alpha\varphi + \frac{1}{2}tr(\sigma(x)\sigma^\top(x)D^2\varphi) \\
&+ (b(x, p_t^h), D\varphi)\Big|_{x=X_t^h}\}dt\Big] - h.
\end{aligned}$$

By (4) and the uniform continuity of φ, $D\varphi$, $D^2\varphi$ on $\bar{\mathcal{O}}$, we see that

$$\begin{aligned}
0 \ge \liminf_{h\to 0} \frac{1}{h}E^h\Big[&\int_0^\tau e^{-\alpha t} g(z, p_t^h)dt + \int_0^\tau e^{-\alpha t}\{-\alpha\varphi(z) \\
&+ \frac{1}{2}tr(\sigma(z)\sigma^\top(z)D^2\varphi(z)) + (b(z, p_t^h), D\varphi(z))\}dt\Big]
\end{aligned}$$

and thus,

$$
\begin{aligned}
0 &\geq \liminf_{h\to 0} \frac{1}{h} E^h\Big[\int_0^h e^{-\alpha t} g(z, p_t^h)dt + \int_0^h e^{-\alpha t}\{-\alpha\varphi(z) \\
&\quad + \frac{1}{2} tr(\sigma(z)\sigma^\top(z)D^2\varphi(z)) + (b(z, p_t^h), D\varphi(z))dt\Big] \\
&\geq \lim_{h\to 0} \frac{1}{h} \int_0^h e^{-\alpha t}\{-\alpha\varphi(z) + \frac{1}{2} tr(\sigma(z)\sigma^\top(z)D^2\varphi(z)) \\
&\quad + \min_{p\in\Gamma}(g(z, p) + (b(z, p), D\varphi(z))\}dt \\
&= -F(z, v(z), D\varphi(z), D^2\varphi(z)).
\end{aligned}
$$

■

4.4 Uniqueness of Viscosity Solutions

4.4.1 An Equivalent Definition

Definition 4.4.1. *$u \in C(Q)$ is a viscosity solution of* (4.1) *in Section 4.1 if the following assertions are satisfied:*

(a) $F(x, u(x), \lambda, X) \leq 0, \quad \forall(\lambda, X) \in J^{2,+}u(x), \quad \forall x \in Q,$

and

(b) $F(x, u(x), \lambda, X) \geq 0, \quad \forall(\lambda, X) \in J^{2,-}u(x), \quad \forall x \in Q,$

where $J^{2,+}$ and $J^{2,-}$ are the second-order superjet and subjet defined by

$$
\begin{aligned}
&J^{2,+}u(x) = \{(\lambda, X) \in \mathbf{R}^N \times \mathbf{S}^N : \\
&\quad \limsup_{y\to x} \frac{u(y) - u(x) - \langle\lambda, y - x\rangle - \frac{1}{2}\langle X(y - x), y - x\rangle}{|y - x|^2} \leq 0\},
\end{aligned}
$$

$$
\begin{aligned}
&J^{2,-}u(x) = \{(\lambda, X) \in \mathbf{R}^N \times \mathbf{S}^N : \\
&\quad \liminf_{y\to x} \frac{u(y) - u(x) - \langle\lambda, y - x\rangle - \frac{1}{2}\langle X(y - x), y - x\rangle}{|y - x|^2} \geq 0\},
\end{aligned}
$$

with the inner product $\langle\ ,\ \rangle$.

Proposition 4.4.2. *We have*

$$
\begin{aligned}
&J^{2,+}u(z) = \{(D\varphi(z), D^2\varphi(z)) : \\
&\quad \varphi \in C^2(Q), z \in Q \text{ is a local maximum point of } u - \varphi\}, \\
&J^{2,-}u(z) = \{(D\varphi(z), D^2\varphi(z)) : \\
&\quad \varphi \in C^2(Q), z \in Q \text{ is a local minimum point of } u - \varphi\},
\end{aligned}
$$

and Definitions 4.1.1 and 4.4.1 are equivalent.

Proof. It suffices to show that $J^{2,+}u$ coincides with the right-hand side, denoted by J.

(1) Let $(D\varphi(z), D^2\varphi(z)) \in J$. Then

$$u(x) - \varphi(x) \le u(z) - \varphi(z) \quad \text{as } x \to z.$$

Hence, by Taylor's formula,

$$\begin{aligned} u(x) &\le u(z) + \varphi(x) - \varphi(z) \\ &\le u(z) + \langle D\varphi(z), x - z\rangle \\ &\quad + \frac{1}{2}\langle D^2\varphi(z)(x-z), x - z\rangle + o(|x-z|^2) \quad \text{as } x \to z. \end{aligned}$$

This implies $(D\varphi(z), D^2\varphi(z)) \in J^{2,+}u(z)$.

(2) Let $(\lambda, X) \in J^{2,+}u(z)$. We set

$$h(y) = \begin{cases} \dfrac{(u(z+y) - u(z) - \langle\lambda, y\rangle - \frac{1}{2}\langle Xy, y\rangle)^+}{|y|^2} & \text{if } y \neq 0, \\ 0 & \text{if } y = 0. \end{cases}$$

By extension, we may consider $u \in C(\mathbf{R}^N)$. By the definition of $J^{2,+}u(z)$, h is continuous on $\mathbf{R}^N$. For $r \ge 0$, we define

$$j(r) = \sup\{h(y) : |y| \le r\}.$$

Then j is a nondecreasing continuous function on $[0, \infty)$ with $j(0) = 0$, and

$$\begin{aligned} u(x) &\le u(z) + \langle\lambda, x - z\rangle + \frac{1}{2}\langle X(x-z), x - z\rangle \\ &\quad + j(|x-z|)|x-z|^2 \quad \text{as } x \to z. \end{aligned}$$

Furthermore, for $r > 0$, we set $\eta(r) = \int_r^{\sqrt{3}r} dt \int_t^{2t} j(s)ds$. It is clear that

$$\begin{aligned} &\eta \in C^2[0,\infty), \quad \eta(0) = \eta'(0) = \eta''(0) = 0, \\ &\eta(r) \ge \int_r^{\sqrt{3}r} t\, j(t)dt \ge r^2 j(r). \end{aligned}$$

Define

$$\varphi(x) = \langle\lambda, x - z\rangle + \frac{1}{2}\langle X(x-z), x - z\rangle + \eta(|x-z|).$$

We observe that

$$\begin{aligned} &\varphi \in C^2(Q), \\ &(D\varphi(z), D^2\varphi(z)) = (\lambda, X), \\ &(u - \varphi)(x) \le u(z) + j(|x-z|)|x-z|^2 - \eta(|x-z|) \le u(z) \\ &\quad = (u-\varphi)(z) \ \text{ as } x \to z, \end{aligned}$$

which implies $(\lambda, X) \in J$. ∎

4.4.2 Alexandrov's Theorem and Ishii's Lemma

Definition 4.4.3.

(a) *A map $f: \mathbf{R}^N \to \mathbf{R}^M$ is differentiable on $\mathbf{R}^N$ if, for any $x \in \mathbf{R}^N$, there exists an $M \times N$ matrix L, depending on x, such that*

$$\lim_{y \to x} \frac{|f(y) - f(x) - L(y-x)|}{|y-x|} = 0,$$

or, equivalently,

$$f(y) = f(x) + L(y-x) + o(|y-x|) \quad as \ y \to x.$$

We write $Df(x)$ for L if it exists, and call $Df(x)$ the derivative of f at x.

(b) *$f : \mathbf{R}^N \to \mathbf{R}$ is said to be twice differentiable on $\mathbf{R}^N$ if, for any $x \in \mathbf{R}^N$, there exists $(\lambda, X) \in \mathbf{R}^N \times \mathbf{S}^N$, depending on x, such that*

$$f(y) = f(x) + \langle \lambda, y-x \rangle + \frac{1}{2}\langle X(y-x), y-x \rangle$$
$$+o(|y-x|^2) \quad as \ y \to x.$$

Theorem 4.4.4 (Alexandrov). *Let $f : \mathbf{R}^N \to \mathbf{R}$ be convex. Then f is twice differentiable almost everywhere on $\mathbf{R}^N$.*

Proof. We show that for almost all $x \in \mathbf{R}^N$, there exists $(\lambda, X) \in \mathbf{R}^N \times \mathbf{S}^N$ satisfying Definition 4.4.3 (b). We notice that such (λ, X) belongs to $J^{2,+}f(x) \cap J^{2,-}f(x)$.

(1) Let $x, y \in \bar{B}_\delta(0), x \neq y$ for any $\delta > 0$. Take $z = y + \delta \frac{y-x}{|y-x|} \in B_{2\delta}(0)$. Clearly,

$$y = \xi z + (1-\xi)x, \quad \xi := \frac{|y-x|}{\delta + |y-x|}.$$

Since f is convex, we have

$$f(y) = f(\xi z + (1-\xi)x) \leq \xi f(z) + (1-\xi) f(x).$$

Hence

$$f(y) - f(x) \leq \xi\{f(z) - f(x)\} \leq \frac{1}{\delta}|y-x|\{\max_{u \in \bar{B}_{2\delta}(0)} f(u) - \min_{u \in \bar{B}_{2\delta}(0)} f(u)\}.$$

Exchanging x and y, we have a constant $C > 0$, depending on δ, such that

$$|f(y) - f(x)| \leq C|x-y|, \quad x, y \in \bar{B}_\delta(0).$$

This implies that f is locally Lipschitz on $\mathbf{R}^N$. By Rademacher's theorem (Appendix D), such f is differentiable almost everywhere. This means that the set $F_1 := \{x \in \mathbf{R}^N : Df(x) \text{ exists}\}$ has full measure, that is, $m^*(F_1^c) = 0$, where m^* denotes the Lebesgue measure.

(2) For each $x \in \mathbf{R}^N$, the subdifferential ∂f of f is defined by

$$\partial f(x) = \{q \in \mathbf{R}^N : f(y) \geq f(x) + \langle q, y-x \rangle, \quad \text{for all } y \in \mathbf{R}^N\}.$$

Let $\mathcal{E}(f) = \{(y, r) \in \mathbf{R}^{N+1} : r \geq f(y)\}$. Clearly, $\mathcal{E}(f)$ is convex and $(x, f(x))$ belongs to the boundary of $\mathcal{E}(f)$. By the classical separation theorem, there is $(q^*, r^*) \neq 0$ in $\mathbf{R}^{N+1}$ such that the hyperplane $\Pi = \{(y, r) \in \mathbf{R}^{N+1} : \langle (q^*, r^*), (y, r) \rangle = 0\}$ separates $\mathcal{E}(f)$ and $(x, f(x))$. This means that

$$\langle (q^*, r^*), (y, r) \rangle \geq \langle (q^*, r^*), (x, f(x)) \rangle = 0, \qquad \text{for all } (y, r) \in \mathcal{E}(f),$$

from which

$$r^* r \geq \langle -q^*, y - x \rangle + r^* f(x).$$

Since $|q^*|^2 + (r^*)^2 \neq 0$ and $(x, f(x) + 1) \in \mathcal{E}(f)$, we have $r^* > 0$. Therefore, taking $r = f(y)$, we deduce that $\partial f(x)$ is nonempty. It is not difficult to see that if f is differentiable at x, then

$$\langle Df(x), tw \rangle = \langle q, tw \rangle + o(t) \quad \text{as } t \to 0, \quad \text{for } q \in \partial f(x), \ |w| = 1,$$

and thus $\partial f(x) = \{Df(x)\}$. Conversely, it is known in convex analysis (see F. H. Clark [35, Theorem 2.5.1]; M. Bardi and I. Capuzzo-Dolcetta [8, chapter 2, Proposition 4.7]) that

① $\partial f(x) =$ the convex hull of $D^* f(x)$
$$:= \{p \in \mathbf{R}^N : p = \lim_{n \to \infty} Df(x_n), \ x_n \to x\}.$$
② If $\partial f(x)$ is a singleton, then f is differentiable at x.

In particular, $\partial f(x) = \{Df(x)\}$ for all $x \in F_1$ and, by definition, the map $x \to Df(x)$ is Borel measurable.

(3) Let H be the set-valued map on $\mathbf{R}^N$ defined by $H(x) = x + \partial f(x)$. Then H satisfies the following:

① $|y_1 - y_2| \geq |x_1 - x_2| \quad$ for $y_1 \in H(x_1), \ y_2 \in H(x_2)$,
② For any $y \in \mathbf{R}^N$, there exists $\hat{x} \in \mathbf{R}^N$ uniquely such that $y \in H(\hat{x})$.

Indeed, let $y_1 \in H(x_1)$ and $y_2 \in H(x_2)$. Then there are $q_1 \in \partial f(x_1)$ and $q_2 \in \partial f(x_2)$ such that $y_1 = x_1 + q_1$, $y_2 = x_2 + q_2$. Hence

$$\langle q_1 - q_2, x_2 - x_1 \rangle \leq \{f(x_2) - f(x_1)\} + \{f(x_1) - f(x_2)\} = 0.$$

Thus

$$\begin{aligned} |x_2 - x_1|^2 &\leq \langle x_2 - x_1 + q_2 - q_1, x_2 - x_1 \rangle \\ &= \langle y_2 - y_1, x_2 - x_1 \rangle \\ &\leq |y_2 - y_1||x_2 - x_1|, \end{aligned}$$

which implies ①. For any $y \in \mathbf{R}^N$, we set $\eta(x) = \frac{1}{2}|x|^2 + f(x) - \langle x, y \rangle$. Then,

$$\eta(x) \geq \frac{1}{2}|x|^2 + f(x_0) + \langle q, x - x_0 \rangle - |x||y| \quad \to \infty \quad \text{as } |x| \to \infty,$$

for fixed $x_0 \in \mathbf{R}^N$ and $q \in \partial f(x_0)$. Since η is continuous, η has a global minimum at some $\hat{x} \in \mathbf{R}^N$ so that $\eta(x) \geq \eta(\hat{x})$, for all $x \in \mathbf{R}^N$. By (2) ①, we have

$$\partial\eta(x) = x + \partial f(x) - y, \quad \forall x \in \mathbf{R}^N.$$

Thus

$$0 \in \partial\eta(\hat{x}) = \hat{x} + \partial f(\hat{x}) - y = H(\hat{x}) - y,$$

which implies ②. Uniqueness is immediate from ①. By ① and ②, we observe that the inverse $J := H^{-1} = (I + \partial f)^{-1} : \mathbf{R}^N \to \mathbf{R}^N$ is a single-valued and Lipschitz function with Lipschitz constant 1.

(4) Let M be a null set. By the definition of the Lebesgue measure, we can select a sequence $\{Q_j\}$ of open cells, for any $\varepsilon > 0$, such that $M \subset Q := \cup_j Q_j$ and $m^*(Q) \leq \sum_j m^*(Q_j) < \varepsilon$. Let $\mathcal{F} = \{B : B = \bar{B}_r(x) \subset Q, r < \varepsilon\}$. By Vitali's covering theorem (see Appendix E), there exists a countable disjoint family $\mathcal{G} := \{\bar{B}_{r_i}(a_i)\} \subset \mathcal{F}$ such that

$$Q \subset \bigcup_{B\in\mathcal{F}} B \subset \bigcup_{B\in\mathcal{G}} \hat{B} \equiv \bigcup_i \bar{B}_{s_i}(a_i), \quad \hat{B} := \bar{B}_{5r}(x), \quad s_i := 5r_i.$$

If $x \in \bar{B}_{s_i}(a_i)$, for some i, then we have by (3),

$$|J(x) - J(a_i)| \leq |x - a_i| \leq s_i,$$

so that $J(\bar{B}_{s_i}(a_i)) \subset \bar{B}_{s_i}(J(a_i))$. Thus

$$\begin{aligned} m^*(J(M)) \leq m^*(J(Q)) &\leq \sum_i m^*(\bar{B}_{s_i}(J(a_i))) = \sum_i m^*(\bar{B}_{s_i}(a_i)) \\ &= 5^N \sum_i m^*(\bar{B}_{r_i}(a_i)) \leq 5^N m^*(Q) < 5^N \varepsilon, \end{aligned}$$

which implies $m^*(J(M)) = 0$. Therefore, J maps null sets to null sets.

(5) By the area formula (see Appendix F), we have

$$\int_{J(A)} dy \leq \int_A |\det DJ(x)| dx,$$

for $A \subset F_1$. Define $F_2 = \{x \in \mathbf{R}^N : DJ(x) \text{ exists and } \det DJ(x) \neq 0\}$. Then

$$\int_{J(F_1 \cap F_2^c)} dy \leq \int_{F_1 \cap F_2^c} |\det DJ(x)|\, dx = \int_{F_1 \cap \{\det DJ = 0\}} |\det DJ(x)|\, dx = 0.$$

It is easy to see that

$$J(F_2)^c \subset J(F_1^c \cup F_2^c) = J(F_1^c \cup \{F_1 \cap F_2^c\}) \subset J(F_1^c) \cup J(F_1 \cap F_2^c).$$

Therefore, taking into account (4), we deduce that $m^*(J(F_2)^c) = 0$.

(6) For fixed $x \in F_2$, we claim that there exist $r, R > 0$ such that for each $|k| \leq r$ there is $h \in \mathbf{R}^N$ satisfying $J(x + h) = J(x) + k$ and $|h| \leq R|k|$.

As in the inverse function theorem, we associate a function ξ_k defined by

$$\xi_k(z) = z - K^{-1}\{J(z) - (J(x) + k)\}, \qquad \text{for } z \in \mathbf{R}^N,$$

where $K := DJ(x)$ and $k \in \mathbf{R}^N$. Choose $0 < \rho \leq 1$ as $2\rho\|K^{-1}\| \leq 1$. By definition, there is $s > 0$ such that if $|z - x| \leq s$, then

$$|J(z) - J(x) - K(z - x)| \leq \rho|z - x|.$$

Taking $r = \rho s$, we have

$$\begin{aligned}
|\xi_k(z) - x| &= |z - x - K^{-1}\{J(z) - J(x)\} + K^{-1}k| \\
&\leq \|K^{-1}\||K(z - x) - \{J(z) - J(x)\}| + \|K^{-1}\||k| \\
&\leq \|K^{-1}\|(\rho|z - x| + |k|) \\
&\leq 2\|K^{-1}\|\rho s \leq s \qquad \text{for } |k| \leq r.
\end{aligned}$$

This yields that the continuous function ξ_k maps $\bar{B}_s(x)$ into itself for $|k| \leq r$. By the Brouwer fixed-point theorem (see Appendix G), there exists $z_0 \in \bar{B}_s(x)$ such that $\xi_k(z_0) = z_0$. Therefore, taking $h = z_0 - x$ and $R = \|K^{-1}\|/(1 - \rho\|K^{-1}\|)$, we obtain this assertion.

(7) Let $y = J(x) \in F_3 := J(F_2) \cap F_1$. By (5), F_3 has full measure. Since $x \in F_2$, it is obvious that

$$J(x + h) - J(x) = DJ(x)h + o(|h|),$$

or, equivalently,

$$DJ(x)^{-1}(J(x + h) - J(x)) = h + o(|h|) \qquad \text{as } h \to 0.$$

By (3), we have $|k| = |J(x + h) - J(x)| \leq |h|$ for all h and by (6), $|k| \leq r$, $|h| \leq R|k|$ for some h. Hence

$$h = DJ(x)^{-1}k + o(|k|) \qquad \text{as } k \to 0.$$

Since $m^*(F_1^c) = 0$, we can take $y + k \in F_1$ as $k \to 0$, extracting a subsequence. By the definition of J in (3),

$$Df(J(x)) = H(y) - y = x - J(x).$$

Then

$$\begin{aligned}
Df(y + k) &= Df(J(x + h)) = x + h - J(x + h) \\
&= Df(y) + J(x) + h - J(x + h) \\
&= Df(y) + (I - DJ(x))h + o(|h|) \\
&= Df(y) + Yk + o(|k|),
\end{aligned}$$

where $Y := DJ(x)^{-1} - I$. Let $u_k \in \partial f(y + k)$, for any $|k| \leq r, k \to 0$. By (2), we have $\partial f(y + k) = Df(y + k)$ if $y + k \in F_1$. Thus we conclude that

$$|u_k - Df(y) - Yk|1_{F_1}(y + k) = o(|k|) \qquad \text{as } k \to 0.$$

(8) Let $y = J(x) \in F_3$ and $u(t) \in \partial f(y + tw)$, $t \geq 0$, for fixed $w \in \partial B_1(0) := \{x : |x| = 1\}$. By (7), there is $\delta > 0$ (independent of w), for any $\varepsilon > 0$, such that

$$|u(t) - Df(y) - tYw|1_{F_1}(y + tw) \leq \varepsilon t,$$

whenever $t \in [0, \delta]$. By (1), we see that

$$\int_0^\delta 1_{F_1^c}(y + tw)dt = 0, \quad w \notin N_0,$$

where $N_0 \subset \partial B_1(0)$ is null. Also, by definition,

$$f(y + (h + s)w) - f(y + sw) \geq \langle u(s), hw \rangle, \quad \forall h \in \mathbf{R}, \quad s \geq 0.$$

By convexity, the map $s \to f(y + sw)$ is locally Lipschitz, and absolutely continuous on each compact interval. We apply Lebesgue's theorem to obtain

$$f(y + (h + s)w) - f(y + sw) = \int_s^{h+s} \langle u(t), w \rangle dt.$$

Hence

$$\begin{aligned}
&|f(y + sw) - f(y) - s\langle Df(y), w\rangle - \frac{s^2}{2}\langle Yw, w\rangle| \\
&= |\int_0^s \{\langle u(t), w\rangle - \langle Df(y), w\rangle - t\langle Yw, w\rangle\}dt| \\
&\leq \int_0^s |\langle u(t), w\rangle - \langle Df(y), w\rangle - t\langle Yw, w\rangle|dt \\
&\leq \int_0^s |u(t) - Df(y) - tYw||w|dt \\
&= \int_0^s |u(t) - Df(y) - tYw|1_{F_1}(y + tw)dt \leq \varepsilon s^2/2, \\
&\qquad\qquad \text{for } s \in [0, \delta], \quad w \notin N_0.
\end{aligned}$$

Since the map $w \to f(y + sw)$ is continuous, this is valid for all $w \in \partial B_1(0)$. Thus

$$f(y + k) = f(y) + \langle Df(y), k\rangle + \frac{1}{2}\langle Yk, k\rangle + o(|k|^2) \quad \text{as } k \to 0.$$

Setting $X = (Y + Y^\top)/2 \in \mathbf{S}^N$, we get

$$f(y + k) = f(y) + \langle Df(y), k\rangle + \frac{1}{2}\langle Xk, k\rangle + o(|k|^2) \quad \text{as } k \to 0.$$

The uniqueness of X is immediate. ∎

Remark 4.4.5. *When $N = 1$, we can prove Theorem 4.4.4 by the following facts (see H. L. Royden [139]):*

① *For all $x \in (a, b)$, there exist the left and right derivatives, denoted by $d^- f(x)$, $d^+ f(x)$, and $d^- f(x) = d^+ f(x)$ except on a countable set.*

② *$d^- f(x)$ is nondecreasing, and then differentiable a.e. on each $[a, b]$.*

Theorem 4.4.6 (Ishii's lemma). *Let G be an open and bounded subset of $\mathbf{R}^N$. Given $u, v \in C(\bar{G})$ and $k > 0$, we set*

$$\psi(x, y) = u(x) - v(y) - \frac{k}{2}|x - y|^2.$$

Suppose that ψ has a local maximum at $(\hat{x}, \hat{y}) \in G \times G$. Then there exist $X, Y \in \mathbf{S}^N$ such that

(i) $(k(\hat{x} - \hat{y}), X) \in \bar{J}^{2,+}u(\hat{x})$,
(ii) $(k(\hat{x} - \hat{y}), Y) \in \bar{J}^{2,-}v(\hat{y})$,
(iii) $\begin{pmatrix} X & 0 \\ 0 & -Y \end{pmatrix} \le 3k \begin{pmatrix} I & -I \\ -I & I \end{pmatrix}, \qquad I = \textit{identity},$

where

$$\bar{J}^{2,\pm}u(x) = \{(\lambda, X) : \exists x_n \to x, \exists(\lambda_n, X_n) \in J^{2,\pm}u(x_n),$$
$$(u(x_n), \lambda_n, X_n) \to (u(x), \lambda, X)\}.$$

For the proof, we need the following lemma.

Lemma 4.4.7 (Jensen). *Let $f : \mathbf{R}^N \to \mathbf{R}$ be semiconvex, that is, the map $x \to f(x) + C|x|^2$ is convex for some constant $C \in \mathbf{R}$, and let $f(x) < f(0)$ hold for all $x \in \mathbf{R}^N$, $x \neq 0$. Then, for any $r > 0$, there exists $0 < \delta < r$ such that*

$$K_{r,\delta} = \{x \in B_r(0) : \text{there exists } p \in \bar{B}_\delta(0) \text{ for which } \sup_{y \in B_r(0)} f_p(y) \le f_p(x)\}$$

has positive measure, where $f_p(x) = f(x) - \langle p, x \rangle$.

Proof.

(1) Let us consider the case when $f \in C^2(\mathbf{R}^N)$. Fix $r > 0$ and we take sufficiently small $0 < \delta < r$ such that $\max_{x \in \partial B_r(0)} f(x) + \delta r < f(0)$. For $|p| \le \delta$, we have

$$\max_{x \in B_r(0)} \{f(x) - \langle p, x \rangle\} \ge f(0).$$

This shows that $f_p(x)$ attains the maximum at some $\tilde{x} \in B_r(0)$. Hence $Df(\tilde{x}) = p$, so that $B_\delta(0) \subset Df(K_{r,\delta})$. Also, by semiconvexity, we get $D^2 f + 2CI \ge 0$. Moreover $D^2 f(x) = D^2 f_p(x) \le 0$, for $x \in K_{r,\delta}$, and then

$$-2CI \le D^2 f(x) \le 0, \quad \text{for } x \in K_{r,\delta}.$$

In particular, by the symmetry of $D^2 f(x)$, we see that $|\det D^2 f(x)| \leq (2C)^N$ for $x \in K_{r,\delta}$. Thus, by the area formula,

$$m^*(B_\delta(0)) \leq \int_{Df(K_{r,\delta})} dy \leq \int_{K_{r,\delta}} |\det D^2 f(x)| dx \leq (2C)^N m^*(K_{r,\delta}),$$

which implies

$$m^*(K_{r,\delta}) \geq \omega_N(\delta/2C)^N > 0,$$

where ω_N is the volume of the unit ball in $\mathbf{R}^N$.

(2) In the general case, in which f is not smooth, we proceed by approximation. Let $\rho \in C^\infty(\mathbf{R}^N)$ be the mollifier (see Definition 5.1.3 of Chapter 5). For any $n \in \mathbf{N}$, we denote by f^n for $n \in \mathbf{N}$ the convolution:

$$f^n(x) = (1/n)^{-N} \int \rho(\frac{x-y}{1/n}) f(y) dy.$$

We see that $f^n \in C^\infty(\mathbf{R}^N)$, $f^n(x) + C|x|^2$ is convex for the same constant C of f, and f^n converges to f uniformly on $\bar{B}_r(0)$. Then, for sufficiently large n, there is $0 < \delta < r$ such that $\max_{x \in \partial B_r(0)} f^n(x) + \delta r < f(0)$. Applying (1) to f^n, we have

$$m^*(K^n_{r,\delta}) \geq \omega_N(\delta/2C)^N,$$

where $K^n_{r,\delta}$ denotes $K_{r,\delta}$ for f^n replacing f. Setting $A_k := \cup_{n=k}^\infty K^n_{r,\delta}$, we have $\cap_{k=1}^\infty A_k \subset K_{r,\delta}$. Indeed, let $x \in \cap_{k=1}^\infty A_k$. We choose $p_k \in \bar{B}_\delta(0)$ and $n_k \to \infty$ such that $\sup_{y \in \bar{B}_r(0)} f^{n_k}_{p_k}(y) \leq f^{n_k}_{p_k}(x)$. By extracting a subsequence, we see that $p_k \to p \in \bar{B}_\delta(0)$ and $\sup_{y \in \bar{B}_r(0)} f_p(y) \leq f_p(x)$. Thus $x \in K_{r,\delta}$. Consequently,

$$\begin{aligned} m^*(K_{r,\delta}) &\geq m^*(\cap_{k=1}^\infty A_k) = \lim_{k\to\infty} m^*(A_k) \geq \limsup_{k\to\infty} m^*(K^k_{r,\delta}) \\ &\geq \omega_N(\delta/2C)^N. \end{aligned}$$

■

Proof of Theorem 4.4.6. By translation and extension, it suffices to prove the theorem in case that $G = \mathbf{R}^N$, $\hat{x} = \hat{y} = 0$, $u(0) = v(0) = 0$, $\max_{(x,y)\in\mathbf{R}^{2N}} \psi(x, y) = \psi(0, 0)$, ψ is bounded in $\mathbf{R}^N$.

(1) We may suppose that for all $x, y \in \mathbf{R}^N$

$$u(x) + w(y) \leq \frac{1}{2} < A \begin{pmatrix} x \\ y \end{pmatrix}, \begin{pmatrix} x \\ y \end{pmatrix} >, \tag{4.10}$$

where $w := -v$ and

$$A = k \begin{pmatrix} I & -I \\ -I & I \end{pmatrix}.$$

Therefore, (i), (ii), and (iii) can be rewritten as

(i′) $(0, X) \in \bar{J}^{2,+}u(0)$,

(ii′) $(0, Y) \in \bar{J}^{2,+}w(0)$,

(iii′) $\begin{pmatrix} X & 0 \\ 0 & Y \end{pmatrix} \le 3A.$

By Schwarz's inequality, we have

$$\begin{aligned}\langle Aa, a\rangle &= \langle Ab, b\rangle + \langle A(a-b), a-b\rangle + 2\langle a-b, Ab\rangle \\ &\le \langle Ab, b\rangle + \|A\||a-b|^2 + k|a-b|^2 + \frac{1}{k}|Ab|^2 \\ &= \langle (A + \frac{1}{k}A^2)b, b\rangle + (k + \|A\|)|a-b|^2, \quad \forall\, a, b \in \mathbf{R}^{2N}.\end{aligned}$$

Setting $B = 3A = A + \frac{1}{k}A^2$ and $m = k + \|A\|$, by (4.10), we get

$$u(x) + w(y) - \frac{m}{2}\{|x-\xi|^2 + |y-\eta|^2\} \le \frac{1}{2}\langle B\begin{pmatrix}\xi\\ \eta\end{pmatrix}, \begin{pmatrix}\xi\\ \eta\end{pmatrix}\rangle,$$

for $\xi, \eta \in \mathbf{R}^N$.

(2) Define the sup-convolution of u by

$$\hat{u}(\xi) = \sup_x (u(x) - \frac{m}{2}|x-\xi|^2).$$

By (1), it is clear that

$$\hat{u}(\xi) + \hat{w}(\eta) \le \frac{1}{2}\langle B\theta, \theta\rangle, \quad \theta := \begin{pmatrix}\xi\\ \eta\end{pmatrix}.$$

Since

$$0 = u(0) + w(0) \le \hat{u}(0) + \hat{w}(0) \le \frac{1}{2}\langle B\tilde{0}, \tilde{0}\rangle = 0, \quad \tilde{0} := \begin{pmatrix}0\\ 0\end{pmatrix},$$

we have

$$\hat{u}(0) + \hat{w}(0) = \max_{\xi,\eta}\left(\hat{u}(\xi) + \hat{w}(\eta) - \frac{1}{2}\langle B\theta, \theta\rangle\right).$$

(3) We observe that

$$\hat{u}(\xi) + \frac{m}{2}|\xi|^2 = \sup_x (u(x) - \frac{m}{2}|x|^2 + m\langle x, \xi\rangle).$$

Since the map $\xi \to u(x) - \frac{m}{2}|x|^2 + m\langle x, \xi\rangle$ is affine, we have that

$$\hat{u}(\xi) + \frac{m}{2}|\xi|^2 \quad \text{is convex,}$$

so that $\hat{u}$ is semiconvex.

(4) Let $f(\theta) = \hat{u}(\xi) + \hat{w}(\eta)$ and $f_r(\theta) = f(\theta) - \frac{1}{2}\langle B\theta, \theta\rangle - r|\theta|^2$, for any $r > 0$. Clearly, by (2), $f_r(\theta)$ attains the strict maximum at $\theta = \tilde{0}$. By (3) and the definition of B, $f_r(\theta) + \frac{1}{2}\langle B\theta, \theta\rangle$ is semiconvex.

By Theorem 4.4.4 and Lemma 4.4.7, there exist $\theta_r, p_r \in B_r(\tilde{0})$ such that the map $\theta \to f_r(\theta) + \langle p_r, \theta\rangle$ is twice differentiable and has the local maximum at θ_r. Therefore, we find an element in $J^{2,+}f_r(\theta_r) \cap J^{2,-}f_r(\theta_r)$, denoted by $(Df_r(\theta_r), D^2 f_r(\theta_r))$. It is clear that $(Df(\theta_r), D^2 f(\theta_r)) \in J^{2,+} f(\theta_r) \cap J^{2,-} f(\theta_r)$. By maximality and the defintion of $J^{2,\pm}$,

$$\begin{aligned} Df_r(\theta_r) + p_r &= \tilde{0}, \\ D^2 f_r(\theta_r) &= D^2 f(\theta_r) - (B + 2rI) \le 0. \end{aligned}$$

Also, since $g(\theta) := f(\theta) + \frac{m}{2}|\theta|^2$ is convex, there is $\nu_r \in \mathbf{R}^{2N}$ such that $g(\theta) - g(\theta_r) \ge \langle \nu_r, \theta - \theta_r \rangle$ for all $\theta \in \mathbf{R}^{2N}$. Hence, applying Theorem 4.4.4, we get $-mI \le D^2 f(\theta_r)$ and thus

$$\|D^2 f(\theta_r)\| = \sup_{|z|=1} |\langle D^2 f(\theta_r) z, z\rangle| \le \|B\| + 2r + m.$$

Note that $\theta_r, p_r \to 0$ and then $Df(\theta_r) \to \tilde{0}$ as $r \to 0$. Extracting a subsequence, we obtain $D^2 f(\theta_r) \to Z \in \mathbf{S}^{2N}$. Passing to the limit, we have

$$(\tilde{0}, Z) \in \bar{J}^{2,+} f(\tilde{0}) \cap \bar{J}^{2,-} f(\tilde{0}), \quad Z \le B.$$

Thus, we deduce that there exist $X, Y \in \mathbf{S}^N$ such that

$$\begin{aligned} &(0, X) \in \bar{J}^{2,+}\hat{u}(0) \cap \bar{J}^{2,-}\hat{u}(0), \\ &(0, Y) \in \bar{J}^{2,+}\hat{w}(0) \cap \bar{J}^{2,-}\hat{w}(0), \\ &Z = \begin{pmatrix} X & 0 \\ 0 & Y \end{pmatrix} \le B. \end{aligned}$$

(5) To prove (i′), (ii′), and (iii′), it suffices to show that if $(0, X) \in \bar{J}^{2,+}\hat{u}(0)$, then $(0, X) \in \bar{J}^{2,+}u(0)$. Let $(\mu, X) \in J^{2,+}\hat{u}(\eta)$, for $\eta \in \mathbf{R}^N$. Since $\psi(x, \eta) + v(\eta) - \frac{\|A\|}{2}|x - \eta|^2 \to -\infty$ as $|x| \to \infty$, we find $z \in \mathbf{R}^N$ such that

$$\hat{u}(\eta) = u(z) - \frac{m}{2}|z - \eta|^2.$$

By the definition of $J^{2,+}$, we see that, for any $x \in \mathbf{R}^N$,

$$\begin{aligned} u(x) - \frac{m}{2}|x - \zeta|^2 \le \hat{u}(\zeta) & \\ &\le \hat{u}(\eta) + \langle \mu, \zeta - \eta\rangle + \frac{1}{2}\langle X(\zeta - \eta), \zeta - \eta\rangle + o(|\zeta - \eta|^2) \\ &= u(z) - \frac{m}{2}|z - \eta|^2 + \langle \mu, \zeta - \eta\rangle \\ &\quad + \frac{1}{2}\langle X(\zeta - \eta), \zeta - \eta\rangle + o(|\zeta - \eta|^2) \quad \text{as } \zeta \to \eta. \end{aligned}$$

Taking $\zeta = x - z + \eta$, we have

$$u(x) \le u(z) + \langle \mu, x - z\rangle + \frac{1}{2}\langle X(x - z), x - z\rangle + o(|x - z|^2) \text{ as } x \to z,$$

which implies $(\mu, X) \in J^{2,+}u(z)$.

Now, choosing $x = z$ and $\zeta = \eta - \varepsilon\{m(\eta - z) + \mu\}$, for $\varepsilon > 0$ in the above, we have

$$\langle m(\eta - z) + \mu, \eta - \zeta\rangle - \frac{m}{2}|\eta - \zeta|^2 \leq O(|\zeta - \eta|^2),$$

and then

$$\varepsilon|m(\eta - z) + \mu|^2 = \frac{1}{\varepsilon}|\eta - \zeta|^2 \leq O(\varepsilon^2) \quad \text{as } \varepsilon \to 0,$$

which implies $m(\eta - z) + \mu = 0$ and thus $(\mu, X) \in J^{2,+}u(\eta + \mu/m)$. In particular, putting $\eta = \mu = 0$, we obtain the desired result. ■

Remark 4.4.8. *If $u, v \in C^2$, then the maximum principle in G gives*

$$D\psi(\hat{x}, \hat{y}) = 0, \quad D^2\psi(\hat{x}, \hat{y}) \leq 0.$$

Note that

$$Du(\hat{x}) = k(\hat{x} - \hat{y}), \qquad Dv(\hat{y}) = -k(\hat{x} - \hat{y}),$$

and

$$D^2\psi(\hat{x}, \hat{y}) = \begin{pmatrix} D^2u(\hat{x}) & 0 \\ 0 & -D^2v(\hat{y}) \end{pmatrix} - k\begin{pmatrix} I & -I \\ -I & I \end{pmatrix}.$$

Taking $X = D^2u(\hat{x})$, $Y = D^2v(\hat{y})$, we obtain the desired properties, which are extended in Ishii's lemma.

4.4.3 Uniqueness for the HJB Equations

We consider the uniqueness for the HJB equation:

$$\begin{aligned} F(x, u, Du, D^2u) := \;& \alpha u - \frac{1}{2}tr(\sigma(x)\sigma^\top(x)D^2u) \\ & - \min_{p\in\Gamma}\{(b(x, p), Du) + g(x, p)\} = 0 \text{ in } \mathbf{R}^N, \end{aligned} \tag{4.11}$$

discussed in Section 4.3.

Theorem 4.4.9. *We make the assumptions of Theorem 4.3.1. Let $u_i \in C(\mathbf{R}^N)$, $i = 1, 2$, be two bounded, uniformly continuous viscosity solutions of* (4.11), *that is, $u_i \in U_b(\mathbf{R}^N)$. Then we have*

$$u_1 = u_2.$$

Proof. It is sufficient to show that $u_1 \leq u_2$.

(1) Suppose there exists x_0 such that $u_1(x_0) - u_2(x_0) > 0$. Then we find $\eta > 0$ such that

$$\varrho := \sup_{x\in\mathbf{R}^N} [u_1(x) - u_2(x) - 2\eta\psi(x)] > 0,$$

where $\psi(x) := (1 + |x|^2)^{1/2} + \psi_0$ for some constant $\psi_0 > 0$ chosen later. By boundedness, we note that

$$u_1(x) - u_2(x) - 2\eta\psi(x) \quad \to \quad -\infty \quad \text{as } |x| \to \infty.$$

Thus there exists $\bar{x} \in \mathbf{R}^N$ such that $\varrho = u_1(\bar{x}) - u_2(\bar{x}) - 2\eta\psi(\bar{x})$.

(2) Define

$$\Psi_k(x, y) = u_1(x) - u_2(y) - \frac{k}{2}|x - y|^2 - \eta\{\psi(x) + \psi(y)\},$$

for $k > 0$. Since $\Psi_k(x, y) \to -\infty$ as $|x| + |y| \to \infty$, there exists $(x_k, y_k) \in \mathbf{R}^{2N}$ such that

$$\Psi_k(x_k, y_k) = \sup \Psi_k(x, y) \geq \Psi_k(\bar{x}, \bar{x}) = \varrho,$$

from which

$$\frac{k}{2}|x_k - y_k|^2 < u_1(x_k) - u_2(y_k) - \eta\{\psi(x_k) + \psi(y_k)\}.$$

Thus we deduce that the sequences $\{x_k\}$, $\{y_k\}$ and $\{k|x_k - y_k|^2\}$ are bounded by some constant $C > 0$, and

$$|x_k - y_k| \leq C/\sqrt{k} \quad \to \quad 0 \quad \text{as } k \to \infty.$$

(3) By the definition of (x_k, y_k), we have

$$\Psi_k(x_k, y_k) \geq \Psi_k(x_k, x_k).$$

Hence, by (2),

$$\frac{k}{2}|x_k - y_k|^2 \leq u_2(x_k) - u_2(y_k) + \eta\{\psi(x_k) - \psi(y_k)\} \quad \to \quad 0 \text{ as } k \to \infty.$$

By extracting a subsequence, we have

$$x_k \to \tilde{x}, \quad y_k \to \tilde{y} \quad \text{as } k \to \infty.$$

Thus, by (2), we get $\tilde{x} = \tilde{y}$. Furthermore, passing to the limit, we deduce

$$u_1(\tilde{x}) - u_2(\tilde{x}) - 2\eta\psi(\tilde{x}) > 0.$$

(4) Let $u(x) = u_1(x) - \eta\psi(x)$ and $v(y) = u_2(y) + \eta\psi(y)$. Applying Ishii's lemma to

$$\Psi_k(x, y) = u(x) - v(y) - \frac{k}{2}|x - y|^2,$$

we obtain $X, Y \in \mathbf{S}^N$ such that

$$(k(x_k - y_k), X) \in \bar{J}^{2,+}u(x_k),$$
$$(k(x_k - y_k), Y) \in \bar{J}^{2,-}v(y_k),$$
$$\begin{pmatrix} X & 0 \\ 0 & -Y \end{pmatrix} \leq 3k \begin{pmatrix} I & -I \\ -I & I \end{pmatrix}.$$

From the definition of $J^{2,+}u(x_k)$, $J^{2,-}v(y_k)$, it follows that

$$J^{2,+}u_1(x) = \{(\lambda, \hat{X}) + \eta(D\psi(x), D^2\psi(x)) : (\lambda, \hat{X}) \in J^{2,+}u(x)\},$$
$$J^{2,-}u_2(x) = \{(\lambda, \hat{X}) - \eta(D\psi(x), D^2\psi(x)) : (\lambda, \hat{X}) \in J^{2,-}v(x)\}.$$

Hence

$$(\lambda_1, \bar{X}) := (k(x_k - y_k), X) + \eta(D\psi(x_k), D^2\psi(x_k)) \in \bar{J}^{2,+}u_1(x_k),$$
$$(\lambda_2, \bar{Y}) := (k(x_k - y_k), Y) - \eta(D\psi(y_k), D^2\psi(y_k)) \in \bar{J}^{2,-}u_2(y_k).$$

By Definition 4.4.1, (4.11) gives

$$F(x_k, u_1(x_k), \lambda_1, \bar{X}) \le 0,$$
$$F(y_k, u_2(y_k), \lambda_2, \bar{Y}) \ge 0.$$

Putting these inequalities together, we get

$$\begin{aligned}\alpha(u_1(x_k) - u_2(y_k)) &\le \{tr(\sigma(x_k)\sigma^\top(x_k)\bar{X}) - tr(\sigma(y_k)\sigma^\top(y_k)\bar{Y})\}/2 \\ &\quad + \max_{p\in\Gamma} |(b(x_k, p), \lambda_1) - (b(y_k, p), \lambda_2)| \\ &\quad + \max_{p\in\Gamma} |g(x_k, p) - g(y_k, p)| \\ &\equiv I_1 + I_2 + I_3.\end{aligned}$$

(5) Since the matrix

$$\begin{pmatrix} AA^\top & AB^\top \\ BA^\top & BB^\top \end{pmatrix} := \begin{pmatrix} \sigma(x_k)\sigma^\top(x_k) & \sigma(x_k)\sigma^\top(y_k) \\ \sigma(y_k)\sigma^\top(x_k) & \sigma(y_k)\sigma^\top(y_k) \end{pmatrix}$$

is symmetric and nonnegative definite, we have

$$\begin{aligned} tr\begin{pmatrix} AA^\top & AB^\top \\ BA^\top & BB^\top \end{pmatrix}\begin{pmatrix} X & 0 \\ 0 & -Y \end{pmatrix} &\le 3k\, tr\begin{pmatrix} AA^\top & AB^\top \\ BA^\top & BB^\top \end{pmatrix}\begin{pmatrix} I & -I \\ -I & I \end{pmatrix} \\ &= 3k\, tr(AA^\top - AB^\top - BA^\top + BB^\top) \\ &= 3k\, tr(A - B)(A^\top - B^\top) \\ &= 3k\|A - B\|^2. \end{aligned}$$

Hence, by (4.8) and (3),

$$\begin{aligned} I_1 &\le \frac{1}{2}[3k\|\sigma(x_k) - \sigma(y_k)\|^2 + \eta\{tr(\sigma(x_k)\sigma^\top(x_k)D^2\psi(x_k)) \\ &\quad + tr(\sigma(y_k)\sigma^\top(y_k)D^2\psi(y_k))\}] \\ &\to \eta\, tr(\sigma(\tilde{x})\sigma^\top(\tilde{x})D^2\psi(\tilde{x})) \quad \text{as } k \to \infty. \end{aligned}$$

By (4.7), it is obvious that

$$I_3 \to 0 \quad \text{as } k \to \infty.$$

By (4.8) and (2), we see that

$$I_2 \le kL|x_k - y_k|^2 + \eta \max_{p\in\Gamma}\{|(b(x_k, p), D\psi(x_k))| + |(b(y_k, p), D\psi(y_k))|\}$$
$$\to 2\eta \max_{p\in\Gamma} |(b(\tilde{x}, p), D\psi(\tilde{x}))| \qquad \text{as } k \to \infty.$$

Taking into account (4.9) and Lipschitz continuity, by a direct calculation, we have

$$\begin{aligned}
\mathcal{J} &:= \frac{1}{2} tr(\sigma(x)\sigma^\top(x)D^2\psi(x)) + \max_{p\in\Gamma}|(b(x, p), D\psi(x))| \\
&\le \frac{1}{2}\|\sigma(x)\|^2/(1 + |x|^2)^{1/2} + \max_{p\in\Gamma}|b(x, p)||x|/(1 + |x|^2)^{1/2} \\
&\le \frac{1}{2}(L|x| + C)^2/(1 + |x|^2)^{1/2} + (L|x| + C)|x|/(1 + |x|^2)^{1/2} \\
&= \{(\frac{L^2}{2} + L)|x|^2 + (L + 1)C|x| + \frac{C^2}{2}\}/(1 + |x|^2)^{1/2} \\
&\le \alpha(1 + |x|^2)^{1/2} + (L + 1)C + C^2/2,
\end{aligned}$$

for some constant $C > 0$. Choose $\psi_0 > 0$ such that $(L + 1)C + C^2/2 \le \alpha\psi_0$. Then we get $\mathcal{J} \le \alpha\psi(x)$. Therefore, we deduce $u_1(\tilde{x}) - u_2(\tilde{x}) \le 2\eta\psi(\tilde{x})$. This is contrary to (3). ■

4.5 Stability

In this section, we consider the stability property of the limit of viscosity solutions u_n as $n \to \infty$.

Theorem 4.5.1. *Let F_n, $n \in \mathbf{N}$, be continuous degenerate elliptic in $Q \times \mathbf{R} \times \mathbf{R}^N \times \mathbf{S}^N$ and let $u_n \in C(Q)$ be a viscosity solution of*

$$F_n(x, u_n, Du_n, D^2u_n) = 0 \quad \textit{in } Q. \tag{4.12}$$

Assume that

$$u_n \to u \quad \textit{locally uniformly in } Q, \tag{4.13}$$
$$F_n \to F \quad \textit{locally uniformly in } Q \times \mathbf{R} \times \mathbf{R}^N \times \mathbf{S}^N. \tag{4.14}$$

Then u is a viscosity solution of

$$F(x, u, Du, D^2u) = 0 \quad \textit{in } Q. \tag{4.15}$$

Proof.

(1) Let $\varphi \in C^2(Q)$ and $z \in Q$ be the local maximum point of $u - \varphi$. Then

$$u(x) - \varphi(x) \le u(z) - \varphi(z), \quad \forall x \in \bar{B}_\delta(z) \subset Q,$$

for some $\delta > 0$. Define $\tilde{\varphi} \in C^2(Q)$ by

$$\tilde{\varphi}(x) = \varphi(x) + |x - z|^4.$$

It is easy to check that

$$u(x) - \tilde{\varphi}(x) < u(z) - \tilde{\varphi}(z), \quad \forall x \in \bar{B}_\delta(z),\ x \neq z.$$

Let z_n be the maximizer of $u_n - \tilde{\varphi}$ in $\bar{B}_\delta(z)$. By extracting a subsequence $\{z_{n_k}\}$ of $\{z_n\}$, we have

$$z_{n_k} \to z' \in \bar{B}_\delta(z).$$

By (4.13)

$$(u_{n_k} - \tilde{\varphi})(z_{n_k}) \to (u - \tilde{\varphi})(z') \quad \text{as } k \to \infty.$$

Furthermore,

$$\max_{x \in \bar{B}_\delta(z)} (u_{n_k}(x) - \tilde{\varphi}(x)) \to \max_{x \in \bar{B}_\delta(z)} (u(x) - \tilde{\varphi}(x)).$$

Hence

$$(u - \tilde{\varphi})(z') \geq (u - \tilde{\varphi})(x), \quad x \in \bar{B}_\delta(z).$$

In particular,

$$(u - \tilde{\varphi})(z') \geq (u - \tilde{\varphi})(z).$$

Thus we deduce that $z = z'$, and hence $z_n \to z$ as $n \to \infty$.

(2) Since $z_n \in B_\delta(z)$ for sufficiently large n, the viscosity property of (4.12) gives that

$$F_n(z_n, u_n(z_n), D\tilde{\varphi}(z_n), D^2\tilde{\varphi}(z_n)) \leq 0.$$

Note that $D\varphi(z) = D\tilde{\varphi}(z)$ and $D^2\varphi(z) = D^2\tilde{\varphi}(z)$. Letting $n \to \infty$, by (4.13) and (4.14), we get

$$F(z, u(z), D\varphi(z), D^2\varphi(z)) \leq 0.$$

This yields that u is a subsolution of (4.15). By a similar argument, we obtain the supersolution property. ∎

As an immediate consequence of the stability result, we obtain the following, called the vanishing viscosity method.

Proposition 4.5.2. *Let $\{G_\varepsilon,\ \varepsilon > 0\}$ be a family of continuous functions on $Q \times \mathbf{R} \times \mathbf{R}^N$ such that G_ε converges to some function G locally uniformly as $\varepsilon \to 0$. Suppose that $u_\varepsilon \in C^2(Q)$ is a solution of the equation*

$$-\varepsilon\, tr(D^2 u_\varepsilon) + G_\varepsilon(x, u_\varepsilon, Du_\varepsilon) = 0 \quad \text{in } Q,$$

and that u_ε converges to some $u \in C(Q)$ locally uniformly as $\varepsilon \to 0$. Then u is a viscosity solution of

$$G(x, u, Du) = 0 \quad \text{in } Q.$$

Proof. Put $F_\varepsilon(x, r, \lambda, X) = -\varepsilon tr(X) + G_\varepsilon(x, r, \lambda)$ and $F(x, r, \lambda, X) = G(x, r, \lambda)$. Let K be a compact subset of $Q \times \mathbf{R} \times \mathbf{R}^N \times \mathbf{S}^N$. Then we have

$$\begin{aligned} |F_\varepsilon(x, r, \lambda, X) - F(x, r, \lambda, X)| &\le \varepsilon \|X\| + |G_\varepsilon(x, r, \lambda) - G(x, r, \lambda)| \\ &\to 0 \quad \text{as } \varepsilon \to 0, \quad \forall (x, r, \lambda, X) \in K. \end{aligned}$$

By Theorem 4.5.1, we get the assertion. ■

Now, we present an application of the stability result to the HJB equation,

$$\begin{aligned} F(x, u, Du, D^2u) &:= \alpha u - \frac{1}{2} tr(\sigma(x)\sigma^\top(x) D^2 u) \\ &- \inf_{p \in \mathbf{R}^N} \{(b(x, p), Du) + g(x, p)\} = 0 \text{ in } \mathbf{R}^N. \end{aligned} \tag{4.16}$$

We make the following assumption:

$$\begin{cases} b : \mathbf{R}^N \times \mathbf{R}^N \to \mathbf{R}^N \text{ continuous}, \\ |b(x, p) - b(y, p)| + \|\sigma(x) - \sigma(y)\| \le L|x - y|, \quad x, y \in \mathbf{R}^N, \\ \sup_{p \in \mathbf{R}^N} |b(0, p)| < \infty. \end{cases} \tag{4.17}$$

Define

$$F_n(x, r, \lambda, X) = \alpha r - \frac{1}{2} tr(\sigma(x)\sigma^\top(x) X) - \min_{|p| \le n} \{(b(x, p), \lambda) + g(x, p)\}$$

for $(x, r, \lambda, X) \in \mathbf{R}^N \times \mathbf{R} \times \mathbf{R}^N \times \mathbf{S}^N$ and $n \in \mathbf{N}$. By Theorem 4.3.1, we observe that there exists a viscosity solution $u_n \in U_b(\mathbf{R}^N)$ of

$$F_n(x, u_n, Du_n, D^2u_n) = 0 \quad \text{in } \mathbf{R}^N.$$

Proposition 4.5.3. *We assume* (4.7), (4.9), *and* (4.17). *Then we have*

$$u_n \to u \in U_b(\mathbf{R}^N) \quad \text{locally uniformly as } n \to \infty,$$

and the limit u is a viscosity solution of (4.16).

Proof. It suffices to check (4.13) and (4.14) in Theorem 4.5.1.

(1) Recall that u_n is the value function for the optimization problem with $\Gamma = \bar{B}_n(0)$ in Section 4.3. Furthermore, by Proposition 3.2.4 (3),

$$|u_n(x) - u_n(x')| \le C|x - x'|/\{\alpha - (L + L^2/2)\},$$

for some $C > 0$, independent of n. It is clear that u_n is nonnegative and nonincreasing. Thus $u_n(x)$ converges to $u(x)$ for every $x \in \mathbf{R}^N$. Letting $n \to \infty$, we see that u is also Lipschitz. By Dini's theorem, we obtain (4.13).

(2) By (4.9) and (4.17), we have

$$\begin{aligned}
|F_n(x, r, \lambda, X) - F_n(x', r', \lambda', X')| &\leq \alpha |r - r'| \\
&+ \frac{1}{2}|tr(\sigma(x)\sigma^\top(x)X - \sigma(x')\sigma^\top(x')X')| \\
&+ \max_{|p| \leq n}\{|(b(x, p), \lambda) - (b(x', p), \lambda')| + |g(x, p) - g(x', p)|\} \\
&\leq C\{|r - r'| + |x - x'|(|x| + |x'| + 1)\|X\| + (1 + |x'|^2)\|X - X'\| \\
&+ (\sup_{p \in \mathbf{R}^N} |b(0, p)| + |x|)|\lambda - \lambda'| + |x - x'|(|\lambda'| + 1)\}.
\end{aligned}$$

Note that F_n is nondecreasing and F_n converges to F. Letting $n \to \infty$, we see that F is continuous. By Dini's theorem, (4.14) is fulfilled. ■

4.6 Viscosity Solutions and Markov Processes

We consider the linear elliptic equation of the form:

$$H_\alpha(x, u, Du, D^2u) := \alpha u - \mathcal{L}u - g = 0 \quad \text{in } \mathbf{R}^N, \tag{4.18}$$

for given $\alpha > 0$ and

$$g \in U_b(\mathbf{R}^N), \tag{4.19}$$

where

$$\mathcal{L}u = \frac{1}{2}tr(\sigma(x)\sigma^\top(x)D^2u) + (b(x), Du).$$

We assume that

$$b, \sigma : \text{ Lipschitz}, \tag{4.20}$$

and let $\{X_t\}$ denote the unique solution of

$$dX_t = b(X_t)dt + \sigma(X_t)dB_t, \quad X_0 = x. \tag{4.21}$$

Proposition 4.6.1. *Let $u \in C^2(\mathbf{R}^N)$ be a bounded solution of* (4.18). *Then, under* (4.19) *and* (4.20),

$$u(x) = E[\int_0^\infty e^{-\alpha t} g(X_t)dt].$$

Proof. Applying Itô's formula, we have

$$e^{-\alpha T}u(X_T) = u(x) + \int_0^T e^{-\alpha t}\{-\alpha u(X_t) + \mathcal{L}u(X_t)\}dt$$
$$+ \int_0^T e^{-\alpha t} Du(X_t)\,\sigma(X_t)dB_t, \quad \text{a.s.}$$

By a localizing argument, the expectation of the local martingale vanishes. Hence

$$E[e^{-\alpha T}u(X_T)] = u(x) + E[\int_0^T e^{-\alpha t}\{-\alpha u(X_t) + \mathcal{L}u(X_t)\}dt]$$
$$= u(x) - E[\int_0^T e^{-\alpha t} g(X_t)dt].$$

Letting $T \to \infty$, we get a representation of u, as claimed. ■

Conversely, we can state the following:

Theorem 4.6.2. *We assume* (4.19) *and* (4.20). *Then*

$$v(x) = E[\int_0^\infty e^{-\alpha t} g(X_t)dt] \tag{4.22}$$

is a unique viscosity solution of (4.18) *in* $U_b(\mathbf{R}^N)$.

Proof.

(1) For any $f \in U_b(\mathbf{R}^N)$ and $\varepsilon > 0$, we define the operator $\mathcal{T} = \mathcal{T}_\varepsilon$ by

$$\mathcal{T}f(x) = E[\int_0^\infty e^{-(\alpha+\frac{1}{\varepsilon})t}(\frac{1}{\varepsilon}f + g)(X_t)dt],$$

and we claim that $\mathcal{T}$ maps $U_b(\mathbf{R}^N)$ into itself. Clearly, $\|\mathcal{T}f\| < \infty$. By the latter part of Theorem 2.3.2 (2), we note that there exists $C_\rho > 0$, for any $\rho > 0$, such that

$$|f(x) - f(y)| \le C_\rho|x - y| + \rho, \quad x, y \in \mathbf{R}^N.$$

Let $\{Y_t\}$ be the solution to (4.21) with $Y_0 = y$. Since

$$|b(x) - b(y)| + \|\sigma(x) - \sigma(y)\| \le L|x - y|, \quad \text{for some} \quad L > 0,$$

By Itô's formula, we get

$$E[|X_t - Y_t|^2] \le |x - y|^2 e^{(2L+L^2)t}.$$

Taking $\varepsilon < 1/(L + L^2/2)$, we have

$$\begin{aligned}|\mathcal{T} f(x) - \mathcal{T} f(y)| &\le \frac{1}{\varepsilon} E[\int_0^\infty e^{-(\alpha+\frac{1}{\varepsilon})t} |f(X_t) - f(Y_t)| dt] \\ &\le \frac{1}{\varepsilon} E[\int_0^\infty e^{-(\alpha+\frac{1}{\varepsilon})t} \{C_\rho |X_t - Y_t| + \rho\} dt] \\ &\le \frac{C_\rho |x-y|}{(\alpha\varepsilon + 1) - \varepsilon(L + L^2/2)} + \frac{\rho}{\alpha\varepsilon + 1}.\end{aligned}$$

Thus $\mathcal{T} f$ is uniformly continuous and $\mathcal{T} f \in U_b(\mathbf{R}^N)$.

(2) For any $f_1, f_2 \in U_b(\mathbf{R}^N)$,

$$\begin{aligned}|\mathcal{T} f_1(x) - \mathcal{T} f_2(x)| &\le \frac{1}{\varepsilon} E[\int_0^\infty e^{-(\alpha+\frac{1}{\varepsilon})t} |f_1(X_t) - f_2(X_t)| dt] \\ &\le \frac{1}{\alpha\varepsilon + 1} \|f_1 - f_2\|.\end{aligned}$$

Therefore, by the contraction mapping theorem, $\mathcal{T}$ has a fixed point $v \in U_b(\mathbf{R}^N)$, that is, $\mathcal{T} v = v$. Therefore,

$$v(x) = E[\int_0^\infty e^{-(\alpha+\frac{1}{\varepsilon})t} (\frac{1}{\varepsilon} v + g)(X_t) dt] \quad \text{for sufficiently small } \varepsilon > 0.$$

(3) Let $G_\gamma f(x) = E[\int_0^\infty e^{-\gamma t} f(X_t) dt]$ for $f \in U_b(\mathbf{R}^N)$, $\gamma > 0$ and let $\tau \in \mathcal{S}_b$. By the strong Markov property of $\{X_t\}$,

$$\begin{aligned}E[\int_\tau^\infty e^{-\gamma t} f(X_t) dt | \mathcal{F}_\tau^X] &= e^{-\gamma\tau} E[\int_0^\infty e^{-\gamma t} f(X_{t+\tau}) dt | \mathcal{F}_\tau^X] \\ &= e^{-\gamma\tau} G_\gamma f(X_\tau), \quad \text{a.s.}\end{aligned}$$

Hence

$$G_\gamma f(x) = E[\int_0^\tau e^{-\gamma t} f(X_t) dt + e^{-\gamma\tau} G_\gamma f(X_\tau)].$$

Thus, by (2),

$$v(x) = E[\int_0^\tau e^{-(\alpha+\frac{1}{\varepsilon})t} (\frac{1}{\varepsilon} v + g)(X_t) dt + e^{-(\alpha+\frac{1}{\varepsilon})\tau} v(X_\tau)].$$

We notice that this relation corresponds to the DPP in Theorem 3.2.5. By the same line as the proof of Theorems 4.3.1 and 4.4.9, we observe that v is a unique viscosity solution of

$$(\alpha + \frac{1}{\varepsilon}) v - \mathcal{L} v - (\frac{1}{\varepsilon} v + g) = 0 \quad \text{in } \mathbf{R}^N,$$

or, equivalently,

$$\alpha v - \mathcal{L} v - g = 0 \quad \text{in } \mathbf{R}^N.$$

(4) Put $\bar{v}(x) = \mathcal{T}_\delta v$ for any $\delta > 0$. By the same argument as (3), $\bar{v}$ is a unique viscosity solution of

$$(\alpha + \frac{1}{\delta})\bar{v} - \mathcal{L}\bar{v} - (\frac{1}{\delta} v + g) = 0 \quad \text{in } \mathbf{R}^N.$$

By (3), we have

$$(\alpha + \frac{1}{\delta})v - \mathcal{L}v - (\frac{1}{\delta} v + g) = 0 \quad \text{in } \mathbf{R}^N.$$

Hence, by uniqueness,

$$v(x) = \bar{v}(x) = E[\int_0^\infty e^{-(\alpha+\frac{1}{\delta})t}(\frac{1}{\delta} v + g)(X_t)dt].$$

Letting $\delta \to \infty$, by the dominated convergence theorem, we obtain (4.22).

■

5

Classical Solutions of Hamilton–Jacobi–Bellman Equations

In this chapter, we study how the viscosity solutions of HJB equations in Chapter 4 turn smooth. We first observe that the DPP holds for the value function $v(x)$ with initial state x and v is a viscosity solution of the HJB equation associated with the problem. Next, the C^2-regularity of v is shown by the following procedure:

(a) Consider the boundary value problem on any fixed ball, also called the Dirichlet problem, with the boundary condition given by v.
(b) In the Dirichlet problem, we have a classical solution w, which is also a viscosity solution.
(c) By the uniqueness of viscosity solutions, we obtain $v = w$ and thus v is smooth.

This viscosity solutions technique is developed by D. Duffie et al. [51] in an optimal consumption/investment problem.

The plan of this chapter is as follows. We present a brief review of the Dirichlet problem for linear elliptic equations, and show the existence of classical solutions of the Dirichlet problem for HJB equations. Finally, we examine the important role of (a)–(c) in the stochastic linear-quadratic (LQ) control problems. As is well known in [40, 65] the control region for the stochastic linear regulator problem is unbounded and the associated HJB equation admits a classical solution. Based on the viscosity solutions technique, the optimal control policies of the LQ problems with bounded control regions are given. The technique (a)–(c) can be applied to economic problems in Part II.

A. Bensoussan [16], L. C. Evans [59], H. Brezis [30], D. Gilbarg and N. S. Trudinger [74], S. Koike [97], and Y. S. Chen and L. C. Wu [32] are basic references for this chapter.

5.1 Linear Elliptic Equations: Weak Solutions

5.1.1 Hölder Spaces

Definition 5.1.1. *Let U be an open subset of $\mathbf{R}^N$ and $0 < \gamma \le 1$.*

(a) *The mapping $u : \bar{U} \to \mathbf{R}$ is Hölder continuous with exponent γ if*

$$|u(x) - u(y)| \le C|x - y|^\gamma, \quad x, y \in \bar{U},$$

for some constant $C > 0$.

(b) *If u is bounded and continuous, the norm is $\|u\|_{C(\bar{U})} := \sup_{x \in \bar{U}} |u(x)|$.*

(c) *The Hölder seminorm is $[u]_{C^{0,\gamma}(\bar{U})} := \sup\{\frac{|u(x)-u(y)|}{|x-y|^\gamma} : x, y \in \bar{U}, x \neq y\}$.*

(d) *The Hölder norm is $\|u\|_{C^{0,\gamma}(\bar{U})} := \|u\|_{C(\bar{U})} + [u]_{C^{0,\gamma}(\bar{U})}$.*

(e) *We write*

$$D_i u = \frac{\partial u}{\partial x_i}, \quad D_{ij} u = \frac{\partial^2 u}{\partial x_i \partial x_j}, \quad Du = (D_1 u, \ldots, D_N u) = \textit{gradient of } u,$$

$$D^2 u = (D_{ij} u) = \textit{matrix of second derivatives } D_{ij} u, i, j = 1, 2, \ldots, N,$$

$$D^\beta u = \frac{\partial^{|\beta|} u}{\partial x_1^{\beta_1} \cdots \partial x_N^{\beta_N}} \quad \textit{for } \beta = (\beta_1, \ldots, \beta_N),\ \beta_i \in \mathbf{Z}_+,\ |\beta| = \sum_{i=1}^N \beta_i,$$

$$D^0 u = u.$$

(f) *The Hölder space $C^{k,\gamma}(\bar{U})$ is the collection of $u \in C^k(\bar{U})$ such that the kth-partial derivatives of u are Hölder continuous with exponent γ and $\|u\|_{C^{k,\gamma}(\bar{U})} < \infty$.*

(g) *We define the norms on the spaces on $C^k(\bar{U})$, $C^{k,\gamma}(\bar{U})$ by*

$$\|u\|_{C^k(\bar{U})} := \sum_{m=0}^k \|D^m u\|_{C(\bar{U})}, \quad \|D^m u\|_{C(\bar{U})} := \sum_{|\beta|=m} \|D^\beta u\|_{C(\bar{U})},$$

$$\|u\|_{C^{k,\gamma}(\bar{U})} := \|u\|_{C^k(\bar{U})} + [D^k u]_{C^{0,\gamma}(\bar{U})}, \quad [D^k u]_{C^{0,\gamma}(\bar{U})} := \sum_{|\beta|=k} [D^\beta u]_{C^{0,\gamma}(\bar{U})}.$$

It can be shown that $C^{k,\gamma}(\bar{U})$ is a Banach space.

Theorem 5.1.2 (Interpolation inequality). *Let $\mathcal{O}$ be a ball with radius R and $u \in C^{2,\gamma}(\bar{\mathcal{O}})$, for $0 < \gamma \le 1$. Then there exists a constant $C > 0$, for any $0 < \varepsilon \le R$, such that*

(i) $\|D^2 u\|_{C(\bar{\mathcal{O}})} \le \varepsilon^\gamma [D^2 u]_{C^{0,\gamma}(\bar{\mathcal{O}})} + \frac{C}{\varepsilon^2} \|u\|_{C(\bar{\mathcal{O}})}$,

(ii) $\|Du\|_{C(\bar{\mathcal{O}})} \le \varepsilon \|D^2 u\|_{C(\bar{\mathcal{O}})} + \frac{C}{\varepsilon} \|u\|_{C(\bar{\mathcal{O}})}$.

Proof.

(1) We claim that there exists a constant $C_\varepsilon > 0$, depending on $\varepsilon > 0$, such that

① $\|D^2u\|_{C(\bar{\mathcal{O}})} \le \varepsilon[D^2u]_{C^{0,\gamma}(\bar{\mathcal{O}})} + C_\varepsilon\|u\|_{C(\bar{\mathcal{O}})}$,

② $\|Du\|_{C(\bar{\mathcal{O}})} \le \varepsilon\|D^2u\|_{C(\bar{\mathcal{O}})} + C_\varepsilon\|u\|_{C(\bar{\mathcal{O}})}$.

To prove ①, dividing both sides by $\|u\|_{C^2(\bar{\mathcal{O}})}$, we may consider $\|u\|_{C^2(\bar{\mathcal{O}})} = 1$. If it were not true, then there exists $u_n \in C^{2,\gamma}(\bar{\mathcal{O}})$, for any $n \in \mathbf{N}$, such that

$$\|D^2u_n\|_{C(\bar{\mathcal{O}})} > \varepsilon[D^2u_n]_{C^{0,\gamma}(\bar{\mathcal{O}})} + n\|u_n\|_{C(\bar{\mathcal{O}})}.$$

Hence

$$\|u_n\|_{C^{2,\gamma}(\bar{\mathcal{O}})} = \|u_n\|_{C^2(\bar{\mathcal{O}})} + [D^2u_n]_{C^{0,\gamma}(\bar{\mathcal{O}})} < 1 + \frac{1}{\varepsilon}.$$

By the Ascoli–Arzelà theorem, extracting a subsequence, we have

$$u_n \to u \quad \text{in } C^2(\bar{\mathcal{O}}) \quad \text{as } n \to \infty.$$

Also,

$$\|u_n\|_{C(\bar{\mathcal{O}})} < \frac{1}{n}\|D^2u_n\|_{C(\bar{\mathcal{O}})} \le \frac{1}{n}\|u_n\|_{C^2(\bar{\mathcal{O}})} = \frac{1}{n} \to 0.$$

Thus we deduce that $u = 0$ and hence $\|u_n\|_{C^2(\bar{\mathcal{O}})} \to 0$, which is a contradiction.

By the same line as above, we get

$$\|Du\|_{C(\bar{\mathcal{O}})} \le \varepsilon[Du]_{C^{0,\gamma}(\bar{\mathcal{O}})} + C_\varepsilon\|u\|_{C(\bar{\mathcal{O}})}.$$

We note that

$$\begin{aligned}|Du(x) - Du(y)| &\le \|\int_0^1 D^2u(x + t(y-x))dt\||x-y| \\ &\le \|D^2u\|_{C(\bar{\mathcal{O}})}|x-y|,\end{aligned}$$

and hence

$$[Du]_{C^{0,\gamma}(\bar{\mathcal{O}})} \le \sup_{x,y\in\bar{\mathcal{O}}}(|x|+|y|)^{1-\gamma}\|D^2u\|_{C(\bar{\mathcal{O}})}.$$

Therefore, we obtain ②.

(2) For any $z \in \mathcal{O}$ and $0 < \varepsilon \le R$, there is a ball $B_\varepsilon := B_\varepsilon(x_0) \subset \mathcal{O}$ for some $x_0 \in \mathcal{O}$ such that $z \in \bar{B}_\varepsilon$. We set $y = x/\varepsilon + (1 - 1/\varepsilon)x_0$, $\tilde{u}(y) = u(\varepsilon y + (1-\varepsilon)x_0) = u(x)$ for $x \in \bar{B}_\varepsilon$. Then $\tilde{u} \in C^{2,\gamma}(\bar{B}_1)$. By ①, we find $C > 0$ such that

$$\|D_y^2\tilde{u}\|_{C(\bar{B}_1)} \le [D_y^2\tilde{u}]_{C^{0,\gamma}(\bar{B}_1)} + C\|\tilde{u}\|_{C(\bar{B}_1)}.$$

Hence

$$\varepsilon^2\|D_x^2 u\|_{C(\bar{B}_\varepsilon)} \le \varepsilon^{\gamma+2}[D_x^2 u]_{C^{0,\gamma}(\bar{B}_\varepsilon)} + C\|u\|_{C(\bar{B}_\varepsilon)}.$$

Thus

$$\begin{aligned}|D^2u(z)| \le \|D^2u\|_{C(\bar{B}_\varepsilon)} &\le \varepsilon^\gamma [D^2u]_{C^{0,\gamma}(\bar{B}_\varepsilon)} + \frac{C}{\varepsilon^2}\|u\|_{C(\bar{B}_\varepsilon)}\\ &\le \varepsilon^\gamma [D^2u]_{C^{0,\gamma}(\bar{O})} + \frac{C}{\varepsilon^2}\|u\|_{C(\bar{O})},\end{aligned}$$

which implies (i).

(3) Similarly, by ②, we find $C > 0$ such that

$$\|D_y\tilde{u}\|_{C(\bar{B}_1)} \le \|D_y^2\tilde{u}\|_{C(\bar{B}_1)} + C\|\tilde{u}\|_{C(\bar{B}_1)}.$$

Hence

$$\varepsilon\|D_x u\|_{C(\bar{B}_\varepsilon)} \le \varepsilon^2\|D_x^2 u\|_{C(\bar{B}_\varepsilon)} + C\|u\|_{C(\bar{B}_\varepsilon)}.$$

Thus

$$\begin{aligned}|Du(z)| \le \|Du\|_{C(\bar{B}_\varepsilon)} &\le \varepsilon\|D^2u\|_{C(\bar{B}_\varepsilon)} + \frac{C}{\varepsilon}\|u\|_{C(\bar{B}_\varepsilon)}\\ &\le \varepsilon\|D^2u\|_{C(\bar{O})} + \frac{C}{\varepsilon}\|u\|_{C(\bar{O})},\end{aligned}$$

which implies (ii). ■

Definition 5.1.3.

(a) *We define $\rho \in C^\infty(\mathbf{R}^N)$ by*

$$\rho(x) = \begin{cases} c\exp(\frac{1}{|x|^2-1}) & \text{for } |x| < 1,\\ 0 & \text{for } |x| \ge 1,\end{cases}$$

where $c > 0$ is chosen so that $\int \rho(x)dx = 1$. The function ρ is the mollifier.

(b) *For any locally integrable function u on $\mathbf{R}^N$ and $h > 0$, the function*

$$u_h(x) = h^{-N}\int_{\mathbf{R}^N} \rho(\frac{x-y}{h})u(y)dy = \int_{B_1(0)} \rho(z)u(x-hz)dz$$

is the mollification of u.

Lemma 5.1.4. *Let $u \in C(\mathbf{R}^N)$. Then*

(i) $u_h \to u$ *locally uniformly* as $h \to 0$,
(ii) $\sup_x |u_h(x)| \le \sup_x |u(x)|$,
(iii) $|D^k u_h(x)| \le Ch^{-k}\sup_{y\in B_h(x)}|u(y)|$ $(k = 1, 2, \ldots)$,
(iv) $|\frac{\partial}{\partial h}Du_h(x)| \le Ch^{-2}\sup_{y\in B_h(x)}|u(y)|$,
where the constant $C > 0$ depends only on N, k.

Lemma 5.1.5. *Let $u \in C^{0,\gamma}(\mathbf{R}^N)$ and $0 < \gamma \le 1$. Then*

(i) $|u_h(x) - u(x)| \le h^\gamma [u]_{C^{0,\gamma}(B_h(x))}$,
(ii) $[u_h]_{C^{0,\gamma}(\mathbf{R}^N)} \le [u]_{C^{0,\gamma}(\mathbf{R}^N)}$,
(iii) $|D^k u_h(x)| \le C h^{\gamma-k}[u]_{C^{0,\gamma}(B_h(x))}, \quad (k = 1, 2, \dots)$,
(iv) $|\frac{\partial u_h}{\partial h}(x)| \le C h^{\gamma-1}[u]_{C^{0,\gamma}(B_h(x))}$,
where the constant $C > 0$ depends only on N, k.

See Y. Z. Chen and L. C. Wu [32, chapter 2, pp. 20–23] for the proofs of Lemmas 5.1.4 and 5.1.5.

5.1.2 Sobolev Spaces and the Imbedding Theorem

Definition 5.1.6. *Let u be locally integrable in U, and $\beta = (\beta_1, \dots, \beta_N)$, $\beta_i \in \mathbf{Z}_+$. We say that a locally integrable function v on U is the βth-weak partial derivative of u if*

$$\int_U v(x)\phi(x)dx = (-1)^{|\beta|} \int_U u(x) D^\beta \phi(x) dx,$$

for all test functions $\phi \in C_c^\infty(U)$. Here $C_c^\infty(U)$ denotes the set of functions f in $C^\infty(U)$ whose support $\mathrm{supp}(f) := \overline{\{x : f(x) \neq 0\}}$ is compact in U. We write $v = D^\beta u$, which is uniquely defined up to a set of measure zero if it exists.

Definition 5.1.7. *Let U be an open subset of $\mathbf{R}^N$, $p \ge 1$ and $k \in \mathbf{Z}_+$.*

(a) *The Sobolev space is a Banach space defined by*

$$W^{k,p}(U) = \{u \in L^p(U) : D^\beta u \in L^p(U) \text{ for } |\beta| \le k\}.$$

(b) *The norm is* $\|u\|_{W^{k,p}(U)} = \left(\sum_{|\beta| \le k} \int_U |D^\beta u(x)|^p dx \right)^{1/p}$.
(c) *We denote by $W_0^{k,p}(U)$ the closure of $C_c^\infty(U)$ in $W^{k,p}(U)$.*
(d) *We write $H^k(U) = W^{k,2}(U)$, $H_0^k(U) = W_0^{k,2}(U)$.*

Theorem 5.1.8. *Let $1 \le p < \infty$, $k \in \mathbf{N}$ and $u \in W^{k,p}(\mathbf{R}^N)$. Then there exists a sequence $\{\phi_n\}$ in $C_c^\infty(\mathbf{R}^N)$ such that*

(i) $\phi_n \to u$ *in* $W^{k,p}(\mathbf{R}^N)$ *as* $n \to \infty$,
(ii) $\phi_n \to u$, *a.e.*

Proof. By definition, the assertion yields that $W^{k,p}(\mathbf{R}^N) = W_0^{k,p}(\mathbf{R}^N)$.

(1) We claim that

$$C_c(\mathbf{R}^N) \text{ is dense in } L^p(\mathbf{R}^N). \tag{5.1}$$

Let $p = 1$ and $f \in L^1(\mathbf{R}^N)$. Set $f^R = f 1_{B_R \cap \{|f| \le R\}} \in L^\infty(B_R)$ for any open ball $B_R = B_R(0)$ with radius R centered at zero. There is $R > 0$, for any $\varepsilon > 0$, such that $\|f^R - f\|_{L^1(\mathbf{R}^N)} < \varepsilon$. By Lusin's theorem [140], there exists $\bar{f} \in C_c(B_R)$ such that $\|\bar{f} - f^R\|_{L^1(B_R)} < \varepsilon$. Thus we obtain (5.1).

For $p > 1$, suppose that $\Lambda(f) = 0$ for every $f \in C_c(\mathbf{R}^N)$, where Λ is a continuous linear functional on $L^p(\mathbf{R}^N)$. By Riesz's theorem in $L^p(\mathbf{R}^N)$, we find $v \in L^q(\mathbf{R}^N)$, $\frac{1}{p} + \frac{1}{q} = 1$, such that

$$\Lambda(f) = \int f(x)v(x)dx, \quad \text{for all } f \in L^p(\mathbf{R}^N).$$

Hence

$$\int f(x)v(x)dx = 0 \quad \text{for all } f \in C_c(U), \quad U := B_R, \quad \forall R > 0.$$

Now, we take $v_\varepsilon \in C_c(U)$ such that $\|v_\varepsilon - v\|_{L^1(U)} = \|v_\varepsilon - v1_{B_R}\|_{L^1(B_R)} < \varepsilon$. Suppose that $V_\varepsilon = \{x \in U : v_\varepsilon(x) > \varepsilon\}$ is nonempty. We note $v_\varepsilon = 0$ on ∂U and set $g_n(x) = nd(x, \overline{V_\varepsilon^c})/(1 + nd(x, \overline{V_\varepsilon^c}))$. Then $g_n \nearrow 1_{V_\varepsilon}$ as $n \to \infty$. Also we have $g_n \in C_c(U)$, because $\bar{V}_\varepsilon$ is a compact subset of U. Hence

$$\begin{aligned}\int_U v_\varepsilon(x)1_{V_\varepsilon}(x)dx &= \lim_{n\to\infty} \int_U (v_\varepsilon(x) - v(x))g_n(x)dx \\ &\le \limsup_{n\to\infty} \int_U |v_\varepsilon(x) - v(x)|dx \|g_n\|_{C(\bar{U})} \le \varepsilon.\end{aligned}$$

Thus

$$\int_U v(x)^+dx \le \int_U |v(x) - v_\varepsilon(x)|dx + \int_U v_\varepsilon(x)^+1_{V_\varepsilon}(x)dx + \varepsilon\int_U dx \to 0 \quad \text{as } \varepsilon \to 0.$$

This implies that $v \le 0$ a.e., similarly $-v \le 0$ a.e. Thus $v = 0$ a.e. and $\Lambda = 0$ on $L^p(\mathbf{R}^N)$. Therefore, applying the Hahn–Banach theorem (cf. [139, Proposition 7, p. 190]), we deduce (5.1).

(2) Let $u \in L^p(\mathbf{R}^N)$ and $u_h \in C^\infty(\mathbf{R}^N)$ be the mollification of u in Definition 5.1.3. Let $p = 1$. Then

$$\begin{aligned}\|u_h\|_{L^1(\mathbf{R}^N)} &\le \int\int |\rho(z)||u(x - hz)|dzdx \\ &= \int |\rho(z)|(\int |u(x - hz)|dx)dz \le \|u\|_{L^1(\mathbf{R}^N)}.\end{aligned}$$

For $p > 1$, Hölder's inequality gives

$$|u_h(x)| \le \int \rho(z)^{1-1/p}\rho(z)^{1/p}|u(x - hz)|dz \le (\int \rho(z)|u(x - hz)|^p dz)^{1/p},$$

which implies

$$\begin{aligned}\|u_h\|_{L^p(\mathbf{R}^N)} &\le (\int(\int \rho(z)|u(x - hz)|^p dz)dx)^{1/p} \\ &= (\int \rho(z)(\int |u(x - hz)|^p dx)dz)^{1/p} = \|u\|_{L^p(\mathbf{R}^N)}.\end{aligned}$$

By (1), we take $\phi \in C_c(\mathbf{R}^N)$, for any $\varepsilon > 0$, such that $\|u - \phi\|_{L^p(\mathbf{R}^N)} < \varepsilon/3$. By Lemma 5.1.4 (i), we see that $\phi_h \in C_c^\infty(\mathbf{R}^N)$ and $\|\phi - \phi_h\|_{L^p(\mathbf{R}^N)} < \varepsilon/3$ for sufficiently small $h > 0$. Hence

$$\begin{aligned}\|u - u_h\|_{L^p(\mathbf{R}^N)} &\le \|u - \phi\|_{L^p(\mathbf{R}^N)} + \|\phi - \phi_h\|_{L^p(\mathbf{R}^N)} + \|(\phi - u)_h\|_{L^p(\mathbf{R}^N)} \\ &\le 2\|u - \phi\|_{L^p(\mathbf{R}^N)} + \|\phi - \phi_h\|_{L^p(\mathbf{R}^N)} < \varepsilon.\end{aligned}$$

Thus we have $\lim_{h\to 0} \|u - u_h\|_{L^p(\mathbf{R}^N)} = 0$.

(3) Let $u \in W^{1,p}(\mathbf{R}^N)$. By (2), we have

$$\|u_h - u\|_{L^p(\mathbf{R}^N)} + \|(Du)_h - Du\|_{L^p(\mathbf{R}^N)} \to 0 \quad \text{as } h \to 0.$$

It follows from the chain rule and integration by parts that

$$\begin{aligned}Du_h(x) &= h^{-N} \int D_x\rho(\frac{x-y}{h})u(y)dy \\ &= (-1)h^{-N} \int D_y\rho(\frac{x-y}{h})u(y)dy \\ &= h^{-N} \int \rho(\frac{x-y}{h})D_y u(y)dy \\ &= (Du)_h(x),\end{aligned}$$

where $D_x u_h(x) = Du_h(x)$. Hence $u_h \to u$ in $W^{1,p}(\mathbf{R}^N)$ as $h \to 0$.

(4) Let $u \in W^{1,p}(\mathbf{R}^N)$ and $u^n = u_h$ for $h = 1/n$. We take $\zeta \in C_c^\infty(\mathbf{R}^N)$ such that $0 \le \zeta \le 1$, $\zeta = 1$ on $B_1(0)$ and $supp(\zeta) \subset B_2(0)$. Multiplying u^n by the cutoff function $\zeta(x/n)$, we define $\phi_n \in C_c^\infty(\mathbf{R}^N)$ by

$$\phi_n(x) = \zeta(x/n)u^n(x).$$

Clearly, $\zeta(x/n) = 1$ for $x \in B_n(0)$ so that $\phi_n(x) = u^n(x)$ for $x \in B_n(0)$. Hence, by (2)

$$\begin{aligned}\|u^n - \phi_n\|_{L^p(\mathbf{R}^N)} &= (\int_{|x|\ge n} |u^n(x) - \phi_n(x)|^p dx)^{1/p} \\ &\le 2(\int_{|x|\ge n} |u^n(x)|^p dx)^{1/p} \\ &\le 2\{(\int_{|x|\ge n} |u(x)|^p dx)^{1/p} \\ &\quad + (\int_{\mathbf{R}^N} |u^n(x) - u(x)|^p dx)^{1/p}\} \to 0 \qquad \text{as } n \to \infty.\end{aligned}$$

Thus, by (2), $\phi_n \to u$ in $L^p(\mathbf{R}^N)$. Similarly, by definition

$$D_i\phi_n = D_i u^n \quad \text{on } B_n(0), \quad i = 1, 2, \ldots, N.$$

By (3) and the same argument as above, we have $D_i\phi_n \to D_i u$ in $L^p(\mathbf{R}^N)$. Therefore $\phi_n \to u$ in $W^{1,p}(\mathbf{R}^N)$.

(5) Let $u \in W^{k,p}(\mathbf{R}^N)$ and $k > 1$. Then we continue the arguments of (3) and (4) up to k to obtain (i).

(6) As is seen in (4), we have $\phi_n \to u$ in $L^p(\mathbf{R}^N)$, and then $\phi_n \to u$ in measure. Extracting a subsequence, we see that $\phi_n \to u$ a.e., which implies (ii). ∎

Theorem 5.1.9 (Friedrichs). *Let U be an open subset of $\mathbf{R}^N$ and $u \in W^{1,p}(U)$ for $1 \le p < \infty$. Then there exists a sequence $\{\varphi_n\}$ in $C_c^\infty(\mathbf{R}^N)$ such that*

(i) $\varphi_n \to u \quad in\ L^p(U) \quad as\ n \to \infty,$
(ii) $D_i\varphi_n|_V \to D_i u|_V \quad in\ L^p(V), \quad \forall V \subset\subset U : open,\ i = 1, \ldots, N.$

Proof. Let

$$\tilde{u}(x) = \begin{cases} u(x) & \text{if } x \in U, \\ 0 & \text{if } x \in \mathbf{R}^N \setminus U, \end{cases}$$

and $\tilde{u}_n$ be the mollification of $\tilde{u}$ for $h := 1/n > 0$. By Theorem 5.1.8 (2), $\tilde{u}_n \to \tilde{u}$ in $L^p(\mathbf{R}^N)$ and so $\tilde{u}_n \to u$ in $L^p(U)$. It is obvious that $\bar{V}$ is a compact subset of U and $d(V, \partial U) > 0$ for $V \subset\subset U$. Then, for $x \in V$ and sufficiently large $n \in \mathbf{N}$, we have

$$\tilde{u}_n(x) = \int_{B_1(0)} \rho(z) u(x - \frac{z}{n}) dz.$$

Since $D_i u \in L^p(U)$, we see that

$$D_i \tilde{u}_n = \int_{B_1(0)} \rho(z) D_i u(\cdot - \frac{z}{n}) dz \to D_i \tilde{u} = D_i u \quad \text{in } L^p(V).$$

Define $\varphi_n(x) = \zeta(x/n)\tilde{u}_n(x)$ for $\zeta \in C_c^\infty(\mathbf{R}^N)$. By the same line as Theorem 5.1.8 (4), we have

$$\|\tilde{u}_n - \varphi_n\|_{L^p(\mathbf{R}^N)} \to 0 \quad \text{as } n \to \infty,$$

which implies (i). Taking $B_n(0) \supset V$, we obtain $D_i\varphi_n = D_i\tilde{u}_n$ on V and thus (ii). ∎

Let us collect the basic properties of $W^{1,p}(\mathbf{R}^N)$ for the proof of the Sobolev imbedding theorem.

Theorem 5.1.10 (Gagliardo–Nirenberg–Sobolev). *Let $1 \le p < N$ and $p^* = \frac{Np}{N-p}$, that is, $\frac{1}{p^*} = \frac{1}{p} - \frac{1}{N}$. Then we have*

$$\|u\|_{L^{p^*}(\mathbf{R}^N)} \le C\|Du\|_{L^p(\mathbf{R}^N)},$$

for all $u \in W^{1,p}(\mathbf{R}^N)$, where $C = \frac{p(N-1)}{N-p}$.

Theorem 5.1.11. *Let $p = N$. Then, for any $s \ge N$, we have*

$$W^{1,p}(\mathbf{R}^N) \subset L^s(\mathbf{R}^N).$$

Theorem 5.1.12 (Morrey). *Let $N < p$. Then we have*

(i) $\|u\|_{C^{0,\gamma}(\mathbf{R}^N)} \le C_1\|u\|_{W^{1,p}(\mathbf{R}^N)}$,
(ii) $\|u\|_{L^\infty(\mathbf{R}^N)} \le C_2\|u\|_{W^{1,p}(\mathbf{R}^N)}$,

for all $u \in W^{1,p}(\mathbf{R}^N)$, *where* $\gamma = 1 - N/p$ *and the constants* $C_1, C_2 > 0$ *depend only on* N, p.

See H. Brezis [30, section IX.3, pp. 162–168] for the proofs of Theorems 5.1.10–5.1.12.

Theorem 5.1.13 (Sobolev). *Let* $k \in \mathbf{N}$, $p \geq 1$, $m \in \mathbf{Z}_+$ *and* $k > m + \frac{N}{p}$. *Then* $W^{k,p}(\mathbf{R}^N)$ *is continuously imbedded in* $C^m(\mathbf{R}^N)$, *that is, the following assertions are valid:*

(i) $W^{k,p}(\mathbf{R}^N) \subset C^m(\mathbf{R}^N)$.
(ii) *There exists a constant* $C > 0$ *such that*

$$\|u\|_{C^m(\mathbf{R}^N)} \leq C\|u\|_{W^{k,p}(\mathbf{R}^N)}, \quad u \in W^{k,p}(\mathbf{R}^N).$$

Proof.

(1) Let $u \in W^{k,p}(\mathbf{R}^N)$. Then $D^\beta u \in L^p(\mathbf{R}^N)$ for all $|\beta| \leq k$. By Theorem 5.1.10, we have

$$\|u\|_{L^{p^*}(\mathbf{R}^N)} \leq C(N,p)\|u\|_{W^{1,p}(\mathbf{R}^N)} \quad \text{if} \quad \frac{1}{p*} = \frac{1}{p} - \frac{1}{N} > 0,$$

where the constant $C(N,p) > 0$ depends only on N, p. Hence

$$\|D^\beta u\|_{L^{p^*}(\mathbf{R}^N)} \leq C(N,p)\|u\|_{W^{k,p}(\mathbf{R}^N)}$$
$$\text{if} \quad \frac{1}{p*} = \frac{1}{p} - \frac{1}{N} > 0, \quad |\beta| \leq k-1.$$

Similarly,

$$\|D^\beta u\|_{L^{p^{**}}(\mathbf{R}^N)} \leq C(N,p^*)\|u\|_{W^{k-1,p^*}(\mathbf{R}^N)}$$
$$\text{if} \quad \frac{1}{p^{**}} = \frac{1}{p^*} - \frac{1}{N} = \frac{1}{p} - \frac{2}{N} > 0, \quad |\beta| \leq k-2.$$

By continuation, we have

$$\|D^\beta u\|_{L^q(\mathbf{R}^N)} \leq C_{l,q}\|u\|_{W^{k,p}(\mathbf{R}^N)} \quad \text{if} \quad \frac{1}{q} = \frac{1}{p} - \frac{l}{N} > 0, \quad |\beta| \leq k-l,$$

where the constant $C_{l,q} > 0$ depends only on l, k, p, N.

(2) Let $k > m + \frac{N}{p}$, $\frac{N}{p} \notin \mathbf{N}$. We take $l = [\frac{N}{p}]$(integer part of $\frac{N}{p}$) and set $\frac{1}{r} = \frac{1}{p} - \frac{l}{N}$. Since $\frac{1}{r} > 0$, we see by (1), that

$$\|D^\beta u\|_{L^r(\mathbf{R}^N)} \leq C_{l,r}\|u\|_{W^{k,p}(\mathbf{R}^N)}, \quad \text{for} \quad |\beta| \leq k-l.$$

It is easy to see that

$$\frac{1}{r} < \frac{1}{p} - \frac{1}{N}(\frac{N}{p} - 1) = \frac{1}{N}.$$

By Theorem 5.1.12, there is $C_1 > 0$ such that

$$\|D^\beta u\|_{C^{0,\gamma}(\mathbf{R}^N)} \leq C_1\|D^\beta u\|_{W^{1,r}(\mathbf{R}^N)},$$

for $|\beta| \le k - l - 1$ and $\gamma = 1 - \frac{N}{r}$. We note that if $|\beta| \le m$, then $|\beta| + 1 \le m + 1 \le k - l$. Thus

$$\|D^\beta u\|_{W^{1,r}(\mathbf{R}^N)} \le C_{l,r}\|u\|_{W^{k,p}(\mathbf{R}^N)}.$$

Consequently, we have

$$\|D^\beta u\|_{C^{0,\gamma}(\mathbf{R}^N)} \le C\|u\|_{W^{k,p}(\mathbf{R}^N)}, \quad |\beta| \le m$$

for some constant $C > 0$.

(3) Let $\frac{N}{p} \in \mathbf{N}$. We take $l = [\frac{N}{p}] - 1 = \frac{N}{p} - 1$ and set $\frac{1}{r} = \frac{1}{p} - \frac{l}{N}$. By (1), we have

$$\|D^\beta u\|_{L^r(\mathbf{R}^N)} \le C_{l,r}\|u\|_{W^{k,p}(\mathbf{R}^N)}, \quad \text{for} \quad |\beta| \le k - l.$$

Clearly, $r = N$. By Theorem 5.1.11, there are $s > N$ and $C > 0$ such that

$$\|D^\beta u\|_{L^s(\mathbf{R}^N)} \le C\|D^\beta u\|_{W^{1,r}(\mathbf{R}^N)} \le CC_{l,r}\|u\|_{W^{k,p}(\mathbf{R}^N)},$$
$$|\beta| \le k - l - 1 = k - \frac{N}{p}.$$

By Theorem 5.1.12, we deduce

$$\|D^\beta u\|_{C^{0,\gamma}(\mathbf{R}^N)} \le C_1\|D^\beta u\|_{W^{1,s}(\mathbf{R}^N)} \le C_1 C C_{l,r}\|u\|_{W^{k,p}(\mathbf{R}^N)},$$
$$|\beta| \le m \le k - \frac{N}{p} - 1,$$

where $\gamma = 1 - \frac{N}{s}$.

(4) For $u \in W^{k,p}(\mathbf{R}^N)$, we take $\phi_n \in C_c^\infty(\mathbf{R}^N)$ of Theorem 5.1.8 (4). By (2) and (3), $D^\beta u \in C^{0,\gamma}(\mathbf{R}^N)$ for $|\beta| \le m$. Then we have

$$D^\beta \phi_n = \int_{B_1(0)} \rho(z) D^\beta u(\cdot - \frac{z}{n}) dz \ \rightarrow \ D^\beta u \quad \text{uniformly in } C(K)$$

for any compact subset K of $\mathbf{R}^N$. Hence $u \in C^m(\mathbf{R}^N)$. Thus, we obtain (i) and (ii). ∎

Corollary 5.1.14. *Let $\mathcal{O}$ be an open ball of $\mathbf{R}^N$, and $k \in \mathbf{N}$, $p \ge 1$, $m \in \mathbf{Z}_+$ with $k > m + \frac{N}{p}$. Then $W^{k,p}(\mathcal{O})$ is continuously imbedded in $C^m(\bar{\mathcal{O}})$.*

Proof. By using Theorem 2.1.15 below, for simplicity, we show only the assertion in the case $\frac{N}{p} \notin \mathbf{N}$. Let $u \in W^{k,p}(\mathcal{O})$. By Theorems 5.1.15 and 5.1.10, we have

$$\|u\|_{L^{p*}(\mathcal{O})} \le \|Eu\|_{L^{p*}(\mathbf{R}^N)} \le C(N,p)\|Eu\|_{W^{1,p}(\mathbf{R}^N)}$$
$$\le CC(N,p)\|u\|_{W^{1,p}(\mathcal{O})} \quad \text{if} \quad \frac{1}{p^*} = \frac{1}{p} - \frac{1}{N} > 0.$$

By the same line as Theorem 5.1.13 (1) and (2), we have $\|D^\beta u\|_{W^{1,r}(\mathcal{O})} \le C_{l,r}\|u\|_{W^{k,p}(\mathcal{O})}$ for $|\beta| \le m$. Hence, by Theorems 5.1.12 and 5.1.15,

$$\|D^\beta u\|_{C^{0,\gamma}(\mathcal{O})} \le \|E(D^\beta u)\|_{C^{0,\gamma}(\mathbf{R}^N)} \le C\|E(D^\beta u)\|_{W^{1,r}(\mathbf{R}^N)} \le C'\|D^\beta u\|_{W^{1,r}(\mathcal{O})}$$

for some constants $C, C' > 0$. Therefore, $D^\beta u \in C^{0,\gamma}(\mathcal{O})$. By extension, we may consider $D^\beta u \in C(\bar{\mathcal{O}})$.

Define $\tilde{u}(x) = u(x)$ if $x \in \mathcal{O}$, $= 0$ otherwise. Then $\tilde{u} \in W^{k,p}(\mathcal{O}')$ for any ball $\mathcal{O}' \supset\supset \mathcal{O}$ and hence $D^\beta \tilde{u} \in C^{0,\gamma}(\mathcal{O}')$. Let ψ_n be the mollification of $E\tilde{u} \in W^{1,r}(\mathbf{R}^N)$ for $h = 1/n$. Since $E\tilde{u} = \tilde{u}$ on $\mathcal{O}'$, we get

$$D^\beta \psi_n = \int_{B_1(0)} \rho(z) D^\beta \tilde{u}(\cdot - \frac{z}{n}) dz \to D^\beta \tilde{u} = D^\beta u \quad \text{uniformly in } C(\bar{\mathcal{O}}).$$

Thus we deduce $u \in C^m(\bar{\mathcal{O}})$. ■

Theorem 5.1.15. *Let $\mathcal{O}$ be an open ball of $\mathbf{R}^N$ and $p \geq 1$. Then there exists a bounded linear operator $E : W^{1,p}(\mathcal{O}) \to W^{1,p}(\mathbf{R}^N)$ such that for each $u \in W^{1,p}(\mathcal{O})$*

(i) $Eu = u$ *a.e. in* $\mathcal{O}$,
(ii) $\|Eu\|_{L^p(\mathbf{R}^N)} \leq C\|u\|_{L^p(\mathcal{O})}$,
(iii) $\|Eu\|_{W^{1,p}(\mathbf{R}^N)} \leq C\|u\|_{W^{1,p}(\mathcal{O})}$,

where the constant $C > 0$ depends only on $\mathcal{O}$.

Proof.

(1) Let $Q = \{y = (y', y_N) = (y_1, \ldots, y_{N-1}, y_N) \in \mathbf{R}^N : \ |y'| < 1, \ -1 < y_N < 1\}$, $Q_+ = \{y \in Q : 0 < y_N < 1\}$, and $v \in W^{1,p}(Q_+)$. By reflection, we define

$$v^*(y', y_N) = \begin{cases} v(y', y_N) & \text{if } y_N > 0, \\ v(y', -y_N) & \text{if } y_N < 0. \end{cases}$$

Let $\zeta \in C^\infty(\mathbf{R})$ be such that

$$\zeta(t) = \begin{cases} 0 & \text{if } t < \frac{1}{2}, \\ 1 & \text{if } t > 1. \end{cases}$$

We set $\zeta_k(t) = \zeta(kt)$ for $k \in \mathbf{N}$.

(2) We show that

$$① \ D_i v^* = (D_i v)^*, \quad \text{for } 1 \leq i \leq N-1,$$

$$② \ D_N v^* = (D_N v)^\square := \begin{cases} D_N v(y', y_N) & \text{if } y_N > 0, \\ -D_N v(y', -y_N) & \text{if } y_N < 0. \end{cases}$$

For ①, let $\phi \in C_c^\infty(Q)$ and set $\psi(y', y_N) = \phi(y', y_N) + \phi(y', -y_N)$ for $y_N > 0$. Then we see that

$$\int_Q v^*(y) D_i \phi(y) dy = \int_{Q_+} v(y) D_i \psi(y) dy.$$

By definition, $\zeta_k\psi(y) := \zeta_k(y_N)\psi(y', y_N)$ belongs to $C_c^\infty(Q_+)$ and $D_i\zeta_k(y_N) = 0$. Hence

$$\int_{Q_+} v(y)\zeta_k(y_N)D_i\psi(y)dy = \int_{Q_+} v(y)D_i(\zeta_k\psi)(y)dy$$
$$= -\int_{Q_+} D_i v(y)\zeta_k\psi(y)dy.$$

Letting $k \to \infty$, we get

$$\int_{Q_+} v(y)D_i\psi(y)dy = -\int_{Q_+} D_i v(y)\psi(y)dy.$$

Thus

$$\int_Q v^* D_i\phi(y)dy = -\int_{Q_+} D_i v(y)\psi(y)dy = -\int_Q (D_i v)^*(y)\phi(y)dy,$$

which implies ①.

Next, let $\phi \in C_c^\infty(Q)$ and set $\chi(y', y_N) = \phi(y', y_N) - \phi(y', -y_N)$ for $y_N > 0$. Then

$$\int_Q v^*(y)D_N\phi(y)dy = \int_{Q_+} v(y)D_N\chi(y)dy.$$

Note that $\zeta_k(y_N)\chi(y', y_N)$ belongs to $C_c^\infty(Q_+)$. Hence

$$\int_{Q_+} v(y)\{\zeta_k(y_N)D_N\chi(y) + k\zeta'(ky_N)\chi(y)\}dy = \int_{Q_+} v(y)D_N(\zeta_k\chi)(y)dy$$
$$= -\int_{Q_+} D_N v(y)\zeta_k\chi(y)dy.$$

Since

$$|\chi(y', y_N)| = |\chi(y', y_N) - \chi(y', 0)| \le C|y_N|,$$

for some constant $C > 0$, we have

$$\begin{aligned}\left|\int_{Q_+} v(y)k\zeta'(ky_N)\chi(y)dy\right| &\le kMC\int_{0<y_N<1/k} |v(y)|y_N\,dy \\ &\le MC\int_{0<y_N<1/k} |v(y)|dy \\ &\to 0 \quad \text{as } k \to \infty,\end{aligned}$$

where $M = \sup_{t\in[0,1]} |\zeta'(t)|$. Thus

$$\int_{Q_+} v(y)D_N\chi(y)dy = -\int_{Q_+} D_N v(y)\chi(y)dy.$$

By using the relation $D_N(v(-y)) = -D_N v(-y)$, we deduce

$$\int_Q v^*(y) D_N \phi(y) dy = -\int_{Q_+} D_N v(y)\chi(y) dy = -\int_Q (D_N v)^{\square}(y)\phi(y) dy,$$

which implies ②. By (1), ① and ②, we observe that

$$\|v^*\|_{L^p(Q)} \le 2\|v\|_{L^p(Q_+)}, \quad \|v^*\|_{W^{1,p}(Q)} \le 2\|v\|_{W^{1,p}(Q_+)}.$$

(3) Let $\mathcal{O} = B_r(x_0)$ and, for simplicity, we assume that $x_0 = 0$. We rewrite $\mathcal{O}$ as a finite union of open sets S of the form:

$$S : x_1^2 + x_2^2 + \cdots + x_{N-1}^2 < s^2,$$
$$-r + (r^2 - \sum_{i=1}^{N-1} x_i^2)^{1/2} < x_N < (r^2 - \sum_{i=1}^{N-1} x_i^2)^{1/2}, \quad 0 < s < r.$$

Changing variables of $x \in S$, we set

$$\begin{cases} y_i = x_i/s, & i = 1, \ldots, N-1, \\ y_N = \{(r^2 - \sum_{i=1}^{N-1} x_i^2)^{1/2} - x_N\}/r, \end{cases}$$

and write $y = \Phi(x)$. Similarly,

$$\begin{cases} x_i = s y_i, & i = 1, \ldots, N-1, \\ x_N = \{r^2 - \sum_{i=1}^{N-1} (s y_i)^2\}^{1/2} - r y_N, \end{cases}$$

and write $x = \Psi(y)$. Then $\Psi = \Phi^{-1}$ and $\Psi : Q_+ \to S$ is a bijective mapping such that

$$\Psi \in C^1(Q_+), \quad \Psi^{-1} \in C^1(S), \quad \Psi(\{y \in \bar{Q}_+ : y_N = 0\}) = \partial S \cap \partial \mathcal{O},$$
$$\|D\Psi\| \in L^\infty, \quad \|D\Psi^{-1}\| \in L^\infty, \quad \det D\Psi \neq 0.$$

For $u \in W^{1,p}(S)$, we define

$$v(y) = u(x) \quad \text{for} \quad x = \Psi(y), \quad y \in Q_+.$$

By Theorem 5.1.9, there exists $u_n \in C_c^\infty(\mathbf{R}^N)$ such that

$$u_n \to u \quad \text{in } L^p(S),$$
$$D_i u_n|_{S'} \to D_i u|_{S'} \quad \text{in } L^p(S'), \quad S' \subset\subset S: \text{ open}, \quad i = 1, \ldots, N.$$

Let $Q' \subset\subset Q_+$ be any open set, which satisfies $\Psi(Q') \subset\subset S$. Then, by the change of variables formula

$$u_n \circ \Psi \to v = u \circ \Psi \quad \text{in } L^p(Q_+),$$
$$\frac{\partial}{\partial y_j} u_n \circ \Psi = \sum_i (D_i u_n \circ \Psi) \frac{\partial \Psi_i}{\partial y_j} \to \sum_i (D_i u \circ \Psi) \frac{\partial \Psi_i}{\partial y_j} \quad \text{in } L^p(Q'),$$
$$j = 1, \ldots, N.$$

For $\phi \in C_c^\infty(Q_+)$, we have

$$\int_{Q_+} (u_n \circ \Psi)(y) \frac{\partial \phi}{\partial y_j}(y) dy = -\int_{Q_+} \sum_i (D_i u_n \circ \Psi)(y) \frac{\partial \Psi_i}{\partial y_j}(y) \phi(y) dy.$$

Letting $n \to \infty$, we deduce $\partial v / \partial y_j = \sum_i (D_i u \circ \Psi) \partial \Psi_i / \partial y_j$ on Q_+ and

$$\|v\|_{W^{1,p}(Q_+)} \leq C \|u\|_{W^{1,p}(S)}.$$

(4) By virtue of (3), there exist a collection $\{U_i, i = 1, \ldots, k\}$ of open sets and bijective mappings $\Psi_i : Q \to U_i$ such that

$$\begin{aligned}
&\partial \mathcal{O} \subset \bigcup_{i=1}^{k} U_i, \\
&\Psi_i \in C^1(\bar{Q}), \quad \Psi_i^{-1} \in C^1(\bar{U}_i), \quad \Psi_i(Q_+) = U_i \cap \mathcal{O}, \\
&\Psi_i(\{y \in \bar{Q}_+ : y_N = 0\}) = U_i \cap \partial \mathcal{O}.
\end{aligned}$$

Then there exists a partition of unity $\{\theta_i, i = 0, 1, \ldots, k\}$ in $C^\infty(\mathbf{R}^N)$ having the following properties:

① $0 \leq \theta_i \leq 1, \quad \sum_{i=0}^{k} \theta_i = 1$ on $\mathcal{O}$,

② $\operatorname{supp}(\theta_i)$: compact, $\quad \operatorname{supp}(\theta_i) \subset U_i, \quad i = 1, \ldots, k, \quad \operatorname{supp}(\theta_0) \subset \mathcal{O}$.

For any $x \in U_i$, there is a ball $B_{r_x}(x) \subset\subset U_i$. Hence $\partial \mathcal{O} \subset \cup_{x \in \partial \mathcal{O}} B_{r_x}(x)$. Since $\partial \mathcal{O}$ is compact, there are finitely many balls $B_{r_i}(x^i), i = 1, \ldots, k$, such that $\partial \mathcal{O} \subset \cup_{i=1}^k B_{r_i}(x^i)$ and $B_{r_i}(x^i) \subset\subset U_i$. Furthermore, we take a ball $B_{r_0}(x^0)$ such that $\mathcal{O} \setminus \cup_{i=1}^k B_{r_i}(x^i) \subset B_{r_0}(x^0) \subset\subset \mathcal{O}$. For each $i = 0, 1, \ldots, k$, as in the mollifier ρ of Definition 5.1.3 (a), we construct $\psi_i \in C^\infty(\mathbf{R}^N)$ such that

$$\psi_i(x) \begin{cases} > 0 & \text{if} \quad x \in B_{r_i}(x^i), \\ = 0 & \text{otherwise.} \end{cases}$$

Note that $\sum_{i=0}^k \psi_i(x) > 0$ for $x \in \mathcal{O} \subset \cup_{i=0}^k B_{r_i}(x^i)$. Thus the functions

$$\theta_i := \psi_i / \sum_{i=0}^{k} \psi_i$$

satisfy ①, ②.

(5) For $u \in W^{1,p}(\mathcal{O})$, we define

$$u_i = \theta_i u,$$

so that $\sum_{i=0}^k u_i = u$. We set

$$\bar{u}_0(x) = \begin{cases} u_0(x) & \text{if } x \in \mathcal{O}, \\ 0 & \text{otherwise.} \end{cases}$$

Let $\phi \in C_c^\infty(\mathbf{R}^N)$ be arbitrary. Since $\mathrm{supp}(u_0) \subset \mathcal{O}$, we have

$$\begin{aligned}\int_{\mathbf{R}^N} \bar{u}_0(x) D_j\phi(x)dx &= \int_{\mathcal{O}} u_0(x) D_j\phi(x)dx = \int_{\mathcal{O}} u(x)\theta_0(x) D_j\phi(x)dx \\ &= \int_{\mathcal{O}} u(x)\{D_j(\theta_0\phi)(x) - \phi(x)D_j\theta_0(x)\}dx \\ &= -\int_{\mathcal{O}} \{D_j u(x)\theta_0(x) + u(x)D_j\theta_0(x)\}\phi(x)dx \\ &= -\int_{\mathbf{R}^N} \{\overline{D_j u}(x)\theta_0(x) + \bar{u}(x)D_j\theta_0(x)\}\phi(x)dx.\end{aligned}$$

Thus

$$D_j\bar{u}_0 = \overline{D_j u}\,\theta_0 + \bar{u}\; D_j\theta_0 \in L^p(\mathbf{R}^N), \quad j = 1, 2, \ldots, N,$$

and

$$\|\bar{u}_0\|_{W^{1,p}(\mathbf{R}^N)} \leq C\|u\|_{W^{1,p}(\mathcal{O})}.$$

Next, for $i = 1, \ldots, k$, we define

$$v_i(y) = u(x) \quad \text{for} \quad x = \Psi_i(y), \quad y \in Q_+.$$

By (3), we get

$$\|v_i\|_{W^{1,p}(Q_+)} \leq C\|u\|_{W^{1,p}(U_i\cap\mathcal{O})}.$$

Furthermore, by (2)

$$\|v_i^*\|_{W^{1,p}(Q)} \leq 2\|v_i\|_{W^{1,p}(Q_+)}.$$

We set

$$w_i(x) = v_i^*(y), \quad \text{for} \quad y = \Psi_i^{-1}(x), \quad x \in U_i.$$

By the same line as (3), we get

$$\begin{aligned}&w_i \in W^{1,p}(U_i), \quad w_i = u \quad \text{on} \quad U_i \cap \mathcal{O}, \\ &\|w_i\|_{W^{1,p}(U_i)} \leq C\|u\|_{W^{1,p}(U_i\cap\mathcal{O})}.\end{aligned}$$

We set

$$\tilde{u}_i(x) = \begin{cases} \theta_i(x)w_i(x) & \text{if } x \in U_i, \\ 0 & \text{if } x \in \mathbf{R}^n \setminus U_i. \end{cases}$$

Since $\mathrm{supp}(\theta_i) \subset U_i$, we have by the product rule of derivatives that

$$\begin{aligned}&\tilde{u}_i \in W^{1,p}(\mathbf{R}^N), \quad \tilde{u}_i = u_i \quad \text{on} \quad \mathcal{O}, \\ &\|\tilde{u}_i\|_{W^{1,p}(\mathbf{R}^N)} \leq C\|u\|_{W^{1,p}(U_i\cap\mathcal{O})}.\end{aligned}$$

Define

$$Eu(x) = \bar{u}_0(x) + \sum_{i=1}^{k} \tilde{u}_i(x) \quad \text{for} \quad x \in \mathbf{R}^N.$$

Then (i), (ii), and (iii) are immediate. ∎

5.1.3 Existence and Uniqueness

We consider the Dirichlet problem for the linear elliptic equation

$$\begin{cases} Lu := -\sum_{i,j} D_i(a_{ij}D_j u) + \alpha u = f & \text{in } \mathcal{O}, \\ u = 0 & \text{on } \partial\mathcal{O}, \end{cases} \tag{5.2}$$

where $\mathcal{O}$ is an open ball $B_R(x_0)$ of $\mathbf{R}^N$ with radius R centered at x_0 and $\alpha \geq 0$. Here we are given

$$\begin{cases} f \in L^2(\mathcal{O}), \\ a_{ij} \in C^1(\bar{\mathcal{O}}), \quad i, j = 1, 2, \ldots, N, \\ (a_{ij}) : \text{ symmetric, that is, } a_{ij} = a_{ji}, \\ L : \text{ uniformly elliptic, that is, } \sum_{i,j} a_{ij}(x)\xi_i\xi_j \geq \theta|\xi|^2 \quad \forall \xi \in \mathbf{R}^N \\ \qquad \text{for some constant } \theta > 0. \end{cases} \tag{5.3}$$

Definition 5.1.16. *$u \in H_0^1(\mathcal{O})$ is a weak solution of (5.2), if u satisfies*

$$\sum_{i,j} \int_{\mathcal{O}} a_{ij} D_j u D_i v dx + \int_{\mathcal{O}} \alpha u v dx = \int_{\mathcal{O}} f v dx, \quad \textit{for all } v \in H_0^1(\mathcal{O}).$$

The existence of a unique weak solution is a consequence of the Lax–Milgram theorem.

Theorem 5.1.17 (Lax–Milgram). *Assume (5.3). Then there exists a unique weak solution $u \in H_0^1(\mathcal{O})$ of (5.2), and we have the estimate*

$$\|u\|_{H_0^1(\mathcal{O})} \leq C\|f\|_{L^2(\mathcal{O})} \tag{5.4}$$

for some constant $C > 0$, independent of u, f.

Proof. Let H denote the Hilbert space $H_0^1(\mathcal{O})$ with the inner product

$$(v, w) = \int_{\mathcal{O}} \{(Dv, Dw) + vw\} dx,$$

and we set

$$a(v, w) = \sum_{i,j} \int_{\mathcal{O}} a_{ij} D_j v D_i w dx + \int_{\mathcal{O}} \alpha v w dx$$

for $v, w \in H$.

(1) By Schwarz's inequality, we have

$$\begin{aligned} \left|\int_{\mathcal{O}} a_{ij} D_j v D_i w dx\right| &\leq C\left(\int_{\mathcal{O}} |D_j v|^2 dx\right)^{1/2}\left(\int_{\mathcal{O}} |D_i w|^2 dx\right)^{1/2} \\ &\leq C\|v\|_H\|w\|_H, \end{aligned}$$

and

$$\left|\int_{\mathcal{O}} \alpha v w dx\right| \leq \alpha\left(\int_{\mathcal{O}} |v|^2 dx\right)^{1/2}\left(\int_{\mathcal{O}} |w|^2 dx\right)^{1/2} \leq \alpha\|v\|_H\|w\|_H.$$

Thus, $a(v, w)$ is a bounded linear form on H, that is,

$$|a(v, w)| \leq C\|v\|_H\|w\|_H$$

for some constant $C > 0$.

(2) By (5.3), we have

$$\int_{\mathcal{O}} \sum_{i,j} a_{ij} D_j v D_i v dx \geq \int_{\mathcal{O}} \theta |Dv|^2 dx.$$

Hence, $a(v, w)$ is coercive if $\alpha > 0$, that is, for some $\theta' > 0$

$$a(v, v) \geq \theta'\|v\|_H^2, \quad v \in H.$$

Let $w \in C_c^\infty(\mathcal{O})$. Then

$$\begin{aligned}\int_{\mathcal{O}} |w(x)|^2 dx &= \int_{\mathcal{O}} \frac{\partial x_1}{\partial x_1}|w(x)|^2 dx = -\int_{\mathcal{O}} x_1 2w(x)\frac{\partial w(x)}{\partial x_1} dx \\ &\leq C\|w\|_{L^2(\mathcal{O})}\|Dw\|_{L^2(\mathcal{O})}.\end{aligned}$$

This gives Poincaré's inequality:

$$\|v\|_{L^2(\mathcal{O})} \leq C\|Dv\|_{L^2(\mathcal{O})}, \quad v \in H.$$

Thus we see that $a(v, w)$ is coercive for $\alpha = 0$.

(3) For each fixed $v \in H$, the map $w \to a(v, w)$ is a continuous linear functional on H. By the Riesz representation theorem on Hilbert spaces, there exists $v_0 \in H$, also denoted by $\mathcal{A}v$, such that

$$a(v, w) = (v_0, w), \quad \forall w \in H.$$

By (1), we see that $\mathcal{A} : H \to H$ is a linear operator and

$$\|\mathcal{A}v\|_H = \sup\{|(\mathcal{A}v, w)| : \|w\|_H \leq 1\} \leq C\|v\|_H, \quad \forall v \in H.$$

Furthermore, by (2),

$$\theta'\|v\|_H^2 \leq a(v, v) \leq \|\mathcal{A}v\|_H\|v\|_H,$$

and then

$$\theta'\|v\|_H \leq \|\mathcal{A}v\|_H.$$

This yields that $\mathcal{A}$ is a one-to-one mapping.

(4) We claim that

$$R(\mathcal{A}) := \{\mathcal{A}v : v \in H\} \text{ is closed in } H.$$

Let $\mathcal{A}v_n \to w$ in H. By (2), we see that

$$\theta'\|v_n - v_m\|_H \leq \|\mathcal{A}v_n - \mathcal{A}v_m\|_H \to 0 \quad \text{as } n, m \to \infty.$$

Hence $\{v_n\}$ is a Cauchy sequence, and then $v_n \to v \in H$. By (3), $\mathcal{A}v_n \to \mathcal{A}v$, and thus, $w = \mathcal{A}v \in R(\mathcal{A})$.

(5) We claim that

$$R(\mathcal{A}) = H.$$

Suppose $R(\mathcal{A}) \neq H$. By (4), the orthogonal decomposition theorem shows that there is $v' \in H$, $v' \neq 0$, such that $(v', w) = 0$, for all $w \in R(\mathcal{A})$. But, by coercivity,

$$0 < \theta' \|v'\|_H^2 \leq a(v', v') = (\mathcal{A}v', v') = 0,$$

which is a contradiction.

(6) By using the Riesz representation theorem again, we observe that

$$\int_{\mathcal{O}} f v dx = (\bar{f}, v), \quad \forall v \in H$$

for some element $\bar{f} \in H$. By (5), we find $u \in H$ such that $\mathcal{A}u = \bar{f}$. Then

$$a(u, v) = (\mathcal{A}u, v) = (\bar{f}, v) = \int_{\mathcal{O}} f v dx, \quad \forall v \in H,$$

which implies that u is a weak solution of (5.2). Furthermore, (5.4) is immediate.

(7) Let $\tilde{u} \in H$ be another weak solution of (5.2). Then

$$a(u - \tilde{u}, v) = (\mathcal{A}u, v) - (\mathcal{A}\tilde{u}, v) = 0, \quad \forall v \in H.$$

Taking $v = u - \tilde{u}$, we get

$$\theta' \|u - \tilde{u}\|_H \leq a(u - \tilde{u}, u - \tilde{u}) = 0,$$

which implies $u = \tilde{u}$. ■

The following results state the regularity of the weak solution u of (5.2).

Theorem 5.1.18 (Interior regularity). *We assume* (5.3). *Let $u \in H^1(\mathcal{O})$ be a weak solution of*

$$Lu = f \quad \text{in } \mathcal{O}.$$

Then, for any $\mathcal{O}' \subset\subset \mathcal{O}$, we have $u \in H^2(\mathcal{O}')$, and

$$\|u\|_{H^2(\mathcal{O}')} \leq C(\|u\|_{H^1(\mathcal{O})} + \|f\|_{L^2(\mathcal{O})}),$$

where the constant $C > 0$ depends only on $d(\mathcal{O}', \partial\mathcal{O}), \theta, N$ and the coefficients of L.

Theorem 5.1.19 (Global regularity). *We assume* (5.3). *Let $u \in H_0^1(\mathcal{O})$ be a weak solution of* (5.2). *Then we have $u \in H^2(\mathcal{O})$, and*

$$\|u\|_{H^2(\mathcal{O})} \leq C(\|u\|_{L^2(\mathcal{O})} + \|f\|_{L^2(\mathcal{O})}),$$

where the constant $C > 0$ depends only on $\mathcal{O}, \theta, N$ and the coefficients of L.

See Y. Z. Chen and L. C. Wu [32, chapter 1, pp. 13–16] or L. C. Evans [59, chapter 6, pp. 308–322] for the proofs of Theorems 5.1.18 and 5.1.19.

5.2 Linear Elliptic Equations: Classical Solutions

5.2.1 Schauder Estimates

We consider the $C^{2,\gamma}$-estimates for the solutions of

$$Lu = -\sum_{i,j} a_{ij} D_{ij} u + \sum_{j} b_j D_j u + \alpha u = f \quad \text{in } \mathcal{O}, \tag{5.5}$$

where $\mathcal{O}$ is an open ball $B_R(x_0)$ of $\mathbf{R}^N$, $\alpha \geq 0$ and

$$\begin{cases} f \in C^{0,\gamma}(\bar{\mathcal{O}}) \quad \text{for } 0 < \gamma < 1, \\ a_{ij} \in C^{0,\gamma}(\bar{\mathcal{O}}), \quad i, j = 1, 2, \dots, N, \quad (a_{ij}): \text{ symmetric}, \\ L: \text{uniformly elliptic}, \\ b_j \in C^{0,\gamma}(\bar{\mathcal{O}}), \quad j = 1, 2, \dots, N. \end{cases} \tag{5.6}$$

Let $\Theta > 0$ be a positive constant such that

$$\max_{i,j} \|a_{ij}\|_{C^{0,\gamma}(\bar{\mathcal{O}})} + \max_{j} \|b_j\|_{C^{0,\gamma}(\bar{\mathcal{O}})} + \alpha \leq \Theta. \tag{5.7}$$

In order to show the existence result, we need the interior and global estimates on classical solutions u of (5.5).

Theorem 5.2.1 (Interior Schauder estimates). *Assume* (5.6) *and* (5.7). *Let* $u \in C^{2,\gamma}(\bar{\mathcal{O}})$ *satisfy* (5.5) *for* $0 < \gamma < 1$. *Then, for* $\mathcal{O}' \subset\subset \mathcal{O}$, *we have*

$$\|u\|_{C^{2,\gamma}(\bar{\mathcal{O}}')} \leq C(\|u\|_{C(\bar{\mathcal{O}})} + \|f\|_{C^{0,\gamma}(\bar{\mathcal{O}})}),$$

where $C > 0$ *depends only on* $\theta, \Theta, d(\mathcal{O}', \partial\mathcal{O}), \gamma$, *and* N.

Theorem 5.2.2 (Global Schauder estimates). *Assume* (5.6) *and* (5.7). *Let* $u \in C^{2,\gamma}(\bar{\mathcal{O}})$ *satisfy* (5.5) *and the boundary condition* $u = 0$ *on* $\partial\mathcal{O}$ *for* $0 < \gamma < 1$. *Then we have*

$$\|u\|_{C^{2,\gamma}(\bar{\mathcal{O}})} \leq C(\|u\|_{C(\bar{\mathcal{O}})} + \|f\|_{C^{0,\gamma}(\bar{\mathcal{O}})}),$$

where $C > 0$ *depends only on* $\theta, \Theta, \mathcal{O}, \gamma$, *and* N.

See Y. Z. Chen and L. C. Wu [32, chapter 2, pp. 27–32] or D. Gilbarg and N. S. Trudinger [74, chapter 6, pp. 84–94] for the proofs of Theorems 5.2.1 and 5.2.2.

5.2.2 The Maximum Principle

Theorem 5.2.3. *Assume* (5.6) *and* (5.7). *Let* $u \in C^2(\mathcal{O}) \cap C(\bar{\mathcal{O}})$ *satisfy* (5.5). *Then we have*

$$\|u\|_{C(\bar{\mathcal{O}})} \leq \sup_{\partial\mathcal{O}} |u| + C\|f\|_{C(\bar{\mathcal{O}})}, \tag{5.8}$$

where $C > 0$ *depends only on* $\theta, \Theta, \mathcal{O}$, *and* N.

Proof.

(1) Let $\alpha > 0$ and $v = u - \sup_{\partial\mathcal{O}} |u|$. Then v satisfies

$$\begin{cases} -\sum_{i,j} a_{ij} D_{ij} v + \sum_j b_j D_j v + \alpha v = f - \alpha \sup_{\partial\mathcal{O}} |u| \le |f| & \text{in } \mathcal{O}, \\ v \le 0 & \text{on } \partial\mathcal{O}. \end{cases}$$

We may suppose that v attains its positive maximum at $\bar{x} \in \mathcal{O}$. By the maximum principle in $\mathbf{R}^N$, we have

$$Dv(\bar{x}) = 0, \quad D^2 v(\bar{x}) \le 0,$$

where the inequality means that the Hessian matrix $D^2 v$ is nonpositive definite at $\bar{x}$. Since $A := (a_{ij}(\bar{x}))$ is symmetric and positive definite, we have

$$\sum_{i,j} a_{ij}(\bar{x}) D_{ij} v(\bar{x}) = tr(A \cdot D^2 v(\bar{x})) \le 0.$$

Hence

$$-\sum_{i,j} a_{ij}(\bar{x}) D_{ij} v(\bar{x}) + \sum_j b_j(\bar{x}) D_j v(\bar{x}) \ge 0.$$

Thus

$$\alpha v(\bar{x}) \le \sup_{\bar{\mathcal{O}}} |f|,$$

that is,

$$\sup_{\mathcal{O}} v \le \frac{1}{\alpha} \sup_{\bar{\mathcal{O}}} |f|.$$

Therefore,

$$\sup_{\bar{\mathcal{O}}} u \le \sup_{\partial\mathcal{O}} |u| + \frac{1}{\alpha} \sup_{\bar{\mathcal{O}}} |f|.$$

Replacing u by $-u$, we obtain

$$\sup_{\bar{\mathcal{O}}} |u| \le \sup_{\partial\mathcal{O}} |u| + \frac{1}{\alpha} \sup_{\bar{\mathcal{O}}} |f| \quad \text{if} \quad \alpha > 0.$$

(2) Let $\alpha = 0$. Let $d_0 < 0 < d_1$ be such that $\mathcal{O} = B_R(x_0) \subset \{(x_1, x') : d_0 < x_1 < d_1, x' \in \mathbf{R}^{N-1}\}$. Define

$$z(x) = e^{2\sigma d_1} - e^{\sigma x_1} > 0, \quad \text{for } \sigma > 0.$$

We put $w = v/z$. Then, by the product rule for derivatives,

$$Lv = z\{-\sum_{i,j} a_{ij} D_{ij} w + \sum_j (b_j - \frac{2}{z} \sum_i a_{ij} D_i z) D_j w\}$$
$$+ (\sum_j b_j D_j z - \sum_{i,j} a_{ij} D_{ij} z) w = f.$$

By uniform ellipticity,

$$\begin{aligned}(\sum_j b_j D_j z - \sum_{i,j} a_{ij} D_{ij} z)/z &= (a_{11}\sigma^2 - b_1\sigma)e^{\sigma x_1}/z \\ &\geq (\theta\sigma^2 - \|b_1\|_{C(\bar{\mathcal{O}})}\,\sigma)e^{\sigma d_0}/\max_{\bar{\mathcal{O}}} z =: \hat{\alpha}, \\ &\qquad \text{for} \quad x \in \bar{\mathcal{O}},\end{aligned}$$

where we take σ sufficiently large such that $\hat{\alpha} > 0$. Since $w(x) \leq 0$, for $x \in \partial\mathcal{O}$, we apply (1) to obtain

$$\sup_{\bar{\mathcal{O}}} w \leq \frac{1}{\hat{\alpha}} \sup_{\bar{\mathcal{O}}} |f/z|,$$

that is,

$$\sup_{\bar{\mathcal{O}}} v \leq C \sup_{\bar{\mathcal{O}}} |f|.$$

Thus (5.8) follows. ∎

5.2.3 *Existence*

We consider the Dirichlet problem,

$$\begin{cases} Lu = -\sum_{i,j} D_i(a_{ij} D_j u) + \alpha u = f & \text{in } \mathcal{O} = B_R(x_0), \\ u = \varphi & \text{on } \partial\mathcal{O}, \end{cases} \tag{5.9}$$

where $\alpha \geq 0$,

$$\begin{cases} f \in C^{0,\gamma}(\bar{\mathcal{O}}), \quad \text{for } 0 < \gamma < 1, \\ a_{ij} \in C^{1,\gamma}(\bar{\mathcal{O}}), \quad i, j = 1, 2, \ldots, N, \quad (a_{ij}) : \text{ symmetric}, \\ (a_{ij}) \text{ fulfills (5.7)}, \\ L : \text{uniformly elliptic}, \end{cases} \tag{5.10}$$

and

$$\varphi \in C^{2,\gamma}(\bar{\mathcal{O}}). \tag{5.11}$$

Theorem 5.2.4. *We assume* (5.10)*, and* (5.11) *and that* $a_{ij}, f, \varphi \in C^\infty(\bar{\mathcal{O}})$*, for all* $1 \leq i, j \leq N$*. Then there exists a unique solution* $u \in C^{2,\gamma}(\bar{\mathcal{O}})$ *of* (5.9)*.*

Proof.

(1) Subtracting φ from u, we observe by Theorem 5.1.17 that there exists a unique solution u of (5.9) such that $u - \varphi \in H_0^1(\mathcal{O})$. By Theorem 5.1.19,

$u \in H^2(\mathcal{O})$, so that $w := D_k u \in H^1(\mathcal{O})$, for $k = 1, 2, \ldots, N$. By definition, we see that

$$D_i D_j u = D_j D_i u = D_{ij} u,$$
$$D_i(a_{ij} D_j u) = D_i a_{ij} \cdot D_j u + a_{ij} D_{ij} u.$$

Differentiating (5.9), we have

$$-\sum_{i,j} D_i(a_{ij} D_j w) - \sum_{i,j} D_i(D_k a_{ij} D_j u) + \alpha w = D_k f.$$

By Theorem 5.1.18

$$\|w\|_{H^2(\mathcal{O}_n)} \le C(\|w\|_{H^1(\mathcal{O})} + \|D^2 u\|_{L^2(\mathcal{O})} + \|Du\|_{L^2(\mathcal{O})} + \|Df\|_{L^2(\mathcal{O})}),$$

where $\mathcal{O}_n$ is a ball such that $\mathcal{O}_n \subset\subset \mathcal{O}$ and $\sup_{x \in \partial\mathcal{O}_n} d(x, \partial\mathcal{O}) \le 1/n$ for each $n \in \mathbf{N}$. Since the constant $C > 0$ is independent of $n \ge n_0$ for some $n_0 \in \mathbf{N}$, we get

$$\|Du\|_{H^2(\mathcal{O})} \le C(\|Du\|_{H^1(\mathcal{O})} + \|Df\|_{L^2(\mathcal{O})}),$$

which implies $u \in H^3(\mathcal{O})$. By induction,

$$u - \varphi \in H^k(\mathcal{O}) \cap H_0^1(\mathcal{O}) \quad \text{for } k \ge 3 + N/2.$$

By the imbedding result (Corollary 5.1.14), we obtain $u \in C^{2,\gamma}(\bar{\mathcal{O}})$.

(2) To check the boundary condition, we recall the bijective mapping $\Psi : Q \to S$ in Theorem 5.1.15 (3). Then it suffices to show that if $v \in C(\bar{Q}_+) \cap H_0^1(Q_+)$, then $v(y) = 0$ for $y \in \bar{Q}_+ \cap \{y_N = 0\}$. Let $v_n \in C_c^\infty(Q_+)$ and $v_n \to v$ in $H^1(Q_+)$. It is clear that

$$|v_n(y', y_N)| \le \int_0^{y_N} |\frac{\partial v_n}{\partial y_N}(y', t)| dt \quad \text{for } (y', y_N) \in Q_+.$$

Hence

$$\int_0^\varepsilon |v_n(y', y_N)| dy_N \le \int_0^\varepsilon \int_0^{y_N} |\frac{\partial v_n}{\partial y_N}(y', t)| dt dy_N$$
$$= \int_0^\varepsilon |\frac{\partial v_n}{\partial y_N}(y', t)|(\varepsilon - t) dt, \quad \varepsilon > 0.$$

Letting $n \to \infty$, we get

$$\frac{1}{\varepsilon} \int dy' \int_0^\varepsilon |v(y', y_N)| dy_N \le \int \int_0^\varepsilon |\frac{\partial v}{\partial y_N}(y', t)| dy' dt.$$

Letting $\varepsilon \to 0$, by continuity, we have

$$\int |v(y', 0)|dy' = 0,$$

which implies $v(y', 0) = 0$. The uniqueness is immediate from Theorem 5.2.3. ■

Theorem 5.2.5. *We assume* (5.10) *and* (5.11). *Then there exists a unique solution* $u \in C^{2,\gamma}(\bar{\mathcal{O}})$ *of* (5.9), *and we have the estimate*

$$\|u\|_{C^{2,\gamma}(\bar{\mathcal{O}})} \le C(\|f\|_{C^{0,\gamma}(\bar{\mathcal{O}})} + \|\varphi\|_{C^{2,\gamma}(\bar{\mathcal{O}})}), \tag{5.12}$$

where $C > 0$ *depends only on* $\theta, \Theta, \mathcal{O}, \gamma,$ *and* N.

Proof. Subtracting φ from u, we may consider $\varphi = 0$. We rewrite (5.9) as

$$\begin{cases} -\sum_{i,j} a_{ij} D_{ij} u + \sum_j d_j D_j u + \alpha u = f & \text{in } \mathcal{O}, \\ u = 0 & \text{on } \partial\mathcal{O}, \end{cases} \tag{5.13}$$

where $d_j = -\sum_i D_i a_{ij}$. By the same line as the proof of Theorem 5.1.15, we extend a_{ij}, f, d_j on $\mathbf{R}^N$. Then, by mollification, we have $a_{ij,n}, f_n, d_{j,n} \in C_c^\infty(\mathbf{R}^N)$, $n \in \mathbf{N}$, such that

$$a_{ij,n}, f_n, d_{j,n} \to a_{ij}, f, d_j \quad \text{uniformly in } C(\bar{\mathcal{O}}),$$

and

$$\|f_n\|_{C^{0,\gamma}(\bar{\mathcal{O}})} \le \bar{C}\|f\|_{C^{0,\gamma}(\bar{\mathcal{O}})},$$

where the constant $\bar{C} > 0$ is independent of n. We approximate (5.13) by

$$\begin{cases} -\sum_{i,,j} a_{ij,n} D_{ij} u_n + \sum_j d_{j,n} D_j u_n + \alpha u_n = f_n & \text{in } \mathcal{O}, \\ u_n = 0 & \text{on } \partial\mathcal{O}. \end{cases} \tag{5.14}$$

By Theorem 5.2.4, we find a unique solution $u_n \in C^{2,\gamma}(\bar{\mathcal{O}})$ of (5.14). By Theorems 5.2.2 and 5.2.3, we get

$$\begin{aligned} \|u_n\|_{C^{2,\gamma}(\bar{\mathcal{O}})} &\le C(\|f_n\|_{C^{0,\gamma}(\bar{\mathcal{O}})} + \|u_n\|_{C(\bar{\mathcal{O}})}) \\ &\le C(1 + C')\|f_n\|_{C^{0,\gamma}(\bar{\mathcal{O}})} \le \bar{C}C(1 + C')\|f\|_{C^{0,\gamma}(\bar{\mathcal{O}})}. \end{aligned}$$

We note that the constants $C, C' > 0$ are independent of n, since

$$\begin{aligned} &\sum_{i,j} a_{ij,n}(x)\, \xi_i \xi_j = \int \sum_{i,j} a_{ij}(x - \frac{z}{n}) \xi_i \xi_j \rho(z) dz \ge \theta |\xi|^2, \\ &\|a_{ij,n}\|_{C(\bar{\mathcal{O}})} \le \|a_{ij}\|_{C(\bar{\mathcal{O}})} \le \Theta. \end{aligned}$$

By the Ascoli–Arzelà theorem, extracting a subsequence, we have

$$u_n \to u \quad \text{in } C^2(\bar{\mathcal{O}}).$$

Letting $n \to \infty$ in (5.14), we see that u solves (5.13). It is easy to see that

$$[D^2u]_{C^{0,\gamma}(\bar{\mathcal{O}})} \le \bar{C}C(1+C')\|f\|_{C^{0,\gamma}(\bar{\mathcal{O}})}.$$

Thus we obtain (5.12). The uniqueness is immediate. ■

5.3 The Dirichlet Problem for HJB Equations

5.3.1 The Comparison Principle

We present the comparison principle for the HJB equation:

$$Ku := -\sum_{i,j} a_{ij}D_{ij}u + \alpha u - H(x,u,Du) = f \quad \text{in } \mathcal{O} = B_R(x_0), \tag{5.15}$$

where

$$\begin{cases} f \in C^{0,\gamma}(\bar{\mathcal{O}}) \quad \text{for } 0<\gamma<1, \\ a_{ij} \in C^{0,1}(\bar{\mathcal{O}}), \quad i,j=1,2,\ldots,N, \quad (a_{ij}): \text{ symmetric}, \\ (a_{ij}(x)) = \sigma(x)\sigma^{\top}(x) \text{ for a Lipschitz continuous matrix } \sigma(x) \text{ on } \mathbf{R}^N, \\ K: \text{uniformly elliptic, that is, } \sum_{i,j} a_{ij}(x)\xi_i\xi_j \ge \theta|\xi|^2, \quad \forall \xi \in \mathbf{R}^N \text{ for } \theta>0. \end{cases} \tag{5.16}$$

Usually, $H(x,u,Du)$ is given by

$$H(x,u,Du) = \min_{p\in\Gamma}\{g(x,p) + (b(x,p),Du)\} + \alpha_0 u(x),$$

for g, b in Section 5.4.3. Here, we make the following assumptions on H:

$$H(x,r,\lambda): \bar{\mathcal{O}} \times \mathbf{R} \times \mathbf{R}^N \to \mathbf{R}, \text{ continuous}, \tag{5.17}$$

$$|H(x_1,r_1,\lambda^1) - H(x_2,r_2,\lambda^2)| \le \alpha_0|r_1-r_2| + C_0\{|x_1-x_2|(1+|\lambda^1|) + |\lambda^1-\lambda^2|\},$$
$$(x_i,r_i,\lambda^i) \in \bar{\mathcal{O}} \times \mathbf{R} \times \mathbf{R}^N, \quad i=1,2, \tag{5.18}$$

for some constants $0 \le \alpha_0 < \alpha$ and $C_0 > 0$.

Theorem 5.3.1. *We assume* (5.16), (5.17), *and* (5.18). *Let $u \in C(\bar{\mathcal{O}})$ and $v \in C(\bar{\mathcal{O}})$ be a viscosity subsolution and supersolution of* (5.15), *respectively. If $u \le v$ on ∂O, then we have $u \le v$ in $\bar{\mathcal{O}}$.*

Proof.

(1) Suppose there exists $\bar{x} \in \mathcal{O}$ such that $u(\bar{x}) - v(\bar{x}) > 0$. Then $u(\bar{x}) - v(\bar{x}) \ge \delta$, for some $\delta > 0$. Define

$$\Psi_k(x,y) = u(x) - v(y) - \frac{k}{2}|x-y|^2,$$

for $k \in \mathbf{N}$. By continuity, there exists $(x_k, y_k) \in \bar{\mathcal{O}} \times \bar{\mathcal{O}}$ such that

$$\Psi_k(x_k,y_k) = \sup_{(x,y)\in\bar{\mathcal{O}}\times\bar{\mathcal{O}}} \Psi_k(x,y) \ge \Psi_k(\bar{x},\bar{x}) \ge \delta,$$

and

$$\Psi_k(x_k, y_k) \geq \Psi_k(x_k, x_k).$$

Hence

$$\frac{k}{2}|x_k - y_k|^2 \leq v(x_k) - v(y_k) \leq 2\|v\|_{C(\bar{\mathcal{O}})}.$$

Therefore, by uniform continuity,

$$\frac{k}{2}|x_k - y_k|^2 \leq \sup\{|v(x) - v(y)| : |x - y| \leq (4\|v\|_{C(\bar{\mathcal{O}})}/k)^{\frac{1}{2}}\} \to 0$$
$$\text{as } k \to \infty.$$

(2) If $(x_k, y_k) \in \partial(\mathcal{O} \times \mathcal{O})$, then we see by the boundary condition that

$$\begin{aligned}\delta &\leq \Psi_k(x_k, y_k) \leq u(x_k) - v(y_k)\\ &\leq \max(v(x_k) - v(y_k), u(x_k) - u(y_k)) \to 0 \quad \text{as } k \to \infty,\end{aligned}$$

which is a contradiction.

(3) Applying Ishii's lemma to $\Psi_k(x, y)$, we obtain $X = (X_{ij}),\ Y = (Y_{ij}) \in \mathbf{S}^N$ such that

$$\begin{aligned}&(\hat{\lambda}^1, X) \in \bar{J}^{2,+}u(x_k), \quad \hat{\lambda}^1 := k(x_k - y_k), \quad x_k \in \mathcal{O},\\ &(\hat{\lambda}^2, Y) \in \bar{J}^{2,-}v(y_k), \quad \hat{\lambda}^2 := k(x_k - y_k), \quad y_k \in \mathcal{O},\\ &\begin{pmatrix} X & 0 \\ 0 & -Y \end{pmatrix} \leq 3k \begin{pmatrix} I & -I \\ -I & I \end{pmatrix}.\end{aligned}$$

By the viscosity properties of u and v, we have

$$-\sum_{i,j} a_{ij}(x_k)X_{ij} + \alpha u(x_k) - H(x_k, u, \hat{\lambda}^1) - f(x_k) \leq 0,$$
$$-\sum_{i,j} a_{ij}(y_k)Y_{ij} + \alpha v(y_k) - H(y_k, v, \hat{\lambda}^2) - f(y_k) \geq 0.$$

Putting these inequalities together, by (5.18), we get

$$\begin{aligned}\alpha(u(x_k) - v(y_k)) \leq &\sum_{i,j}\{a_{ij}(x_k)X_{ij} - a_{ij}(y_k)Y_{ij}\} + \alpha_0|u(x_k) - v(y_k)|\\ &+ C_0\{|x_k - y_k|(1 + |\hat{\lambda}^1|) + |\hat{\lambda}^1 - \hat{\lambda}^2|\}\\ &+ f(x_k) - f(y_k).\end{aligned}$$

Since $\sigma(x)$ is Lipschitz, we see by Theorem 4.4.9 (5) that

$$\begin{aligned}\sum_{i,j}\{a_{ij}(x_k)X_{ij} - a_{ij}(y_k)Y_{ij}\} &= tr(\sigma(x_k)\sigma^{\top}(x_k)X) - tr(\sigma(y_k)\sigma^{\top}(y_k)Y)\\ &\leq 3k\|\sigma(x_k) - \sigma(y_k)\|^2 \ \to 0 \quad \text{as } k \to \infty.\end{aligned}$$

Thus

$$\begin{aligned}(\alpha-\alpha_0)\delta &\le (\alpha-\alpha_0)(u(x_k)-v(y_k))\\ &\le 3k\|\sigma(x_k)-\sigma(y_k)\|^2 + C_0|x_k-y_k|(1+k|x_k-y_k|)\\ &\quad + f(x_k)-f(y_k)\\ &\to 0 \quad \text{as } k\to\infty,\end{aligned}$$

which is a contradiction. ∎

Corollary 5.3.2. *Let $u_i \in C^2(\mathcal{O})\cap C(\bar{\mathcal{O}})$, $i=1,2$, be two solutions of* (5.15) *with $f_i \in C^{0,\gamma}(\bar{\mathcal{O}})$ replacing f. Then, under the assumptions of Theorem 5.3.1, we have*

$$\|u_1-u_2\|_{C(\bar{\mathcal{O}})} \le \sup_{\partial\mathcal{O}}|u_1-u_2| + \frac{1}{\alpha-\alpha_0}\|f_1-f_2\|_{C(\bar{\mathcal{O}})}.$$

Proof. Let $\tilde{v}$ be on the right-hand side and let $\tilde{u}=u_1-u_2$. Then, by (5.18), we have

$$-\sum_{i,j} a_{ij}D_{ij}\tilde{u}+\alpha\tilde{u} \le \alpha_0|\tilde{u}|+C_0|D\tilde{u}|+\|f_1-f_2\|_{C(\bar{\mathcal{O}})}+(\alpha-\alpha_0)\sup_{\partial\mathcal{O}}|\tilde{u}|,$$

$$-\sum_{i,j} a_{ij}D_{ij}\tilde{v}+\alpha\tilde{v} = \alpha_0|\tilde{v}|+C_0|D\tilde{v}|+\|f_1-f_2\|_{C(\bar{\mathcal{O}})}+(\alpha-\alpha_0)\sup_{\partial\mathcal{O}}|\tilde{u}| \ \text{ in } \mathcal{O}.$$

By Theorem 5.3.1, we have $\tilde{u}\le\tilde{v}$. Exchanging u_1 and u_2, we obtain $-\tilde{u}\le\tilde{v}$. ∎

We extend the maximum principle of Section 5.2 for the HJB equation (5.15).

Theorem 5.3.3. *Assume* (5.16), (5.17), *and* (5.18). *Let $u \in C^2(\mathcal{O})\cap C(\bar{\mathcal{O}})$ satisfy* (5.15). *Then we have*

$$\|u\|_{C(\bar{\mathcal{O}})} \le \sup_{\partial\mathcal{O}}|u| + \bar{C}(\|f\|_{C(\bar{\mathcal{O}})}+1), \tag{5.19}$$

where $\bar{C}=\max(1, \|H(\cdot,0,0)\|_{C(\bar{\mathcal{O}})})/(\alpha-\alpha_0)>0$.

Proof. Let $v=-u$, and let $\bar{u}$ be the right-hand side of (5.19). Clearly, by (5.15)

$$\tilde{K}v := -\sum_{i,j} a_{ij}D_{ij}v+\alpha v+\tilde{H}(x,v,Dv) = -f \le |f|,$$

where $\tilde{H}(x,r,\lambda)=H(x,-r,-\lambda)$. Also, by (5.18),

$$\begin{aligned}\tilde{K}\bar{u} &= \alpha\bar{u}+H(x,-\bar{u},0) \ge (\alpha-\alpha_0)\bar{u}+H(x,0,0)\\ &\ge \bar{C}(\alpha-\alpha_0)\|f\|_{C(\bar{\mathcal{O}})}+\bar{C}(\alpha-\alpha_0)-\|H(\cdot,0,0)\|_{C(\bar{\mathcal{O}})}\\ &\ge |f|.\end{aligned}$$

Thus v is a viscosity subsolution and $\bar{u}$ is a viscosity supersolution of $\tilde{K}(\cdot)=|f|$. It is obvious that $\tilde{H}$ satisfies (5.17), (5.18), and

$$v \le \bar{u} \quad \text{on } \partial\mathcal{O}.$$

Therefore, by Theorem 5.3.1, we get

$$v \le \bar{u} \quad \text{in } \bar{\mathcal{O}}.$$

Replacing $-u$ by u, we obtain (5.19). ■

5.3.2 Existence

The following is an existence result on classical solutions of the Dirichlet problem for HJB equations.

Theorem 5.3.4. *Let $\Gamma = \mathcal{O} \times \mathbf{R} \times \mathbf{R}^N \times \mathbf{S}^N$ and $\varphi \in C(\partial\mathcal{O})$. We assume that $F^{(i)} \in C^2(\Gamma)$, $i = 1, 2, \ldots$, satisfy the structure conditions:*

(a) $(z, p, q) \mapsto F^{(i)}(x, z, p, q)$ *: concave,*
(b) $z \mapsto F^{(i)}(x, z, p, q)$*: nonincreasing,*
(c) $\lambda_0 \le F_q^{(i)} \equiv (\frac{\partial F^{(i)}}{\partial q_{k\ell}}\ ; k, \ell = 1, \ldots, N) \le \lambda_1$ *for some constants* $0 < \lambda_0 < \lambda_1$,
(d) $|F_p^{(i)}|, |F_z^{(i)}|, |F_{qx}^{(i)}|, |F_{px}^{(i)}|, |F_{zx}^{(i)}| \le C$,
(e) $|F_x^{(i)}|, |F_{xx}^{(i)}| \le C(1 + |p| + |q|)$,

where the constant $C > 0$ is independent of (x, z, p, q) and i. Then there exists a unique solution $u \in C^2(\mathcal{O}) \cap C(\bar{\mathcal{O}})$ of

$$\begin{cases} \inf_{i\in\{1,2\ldots\}} F^{(i)}(x, w, Dw, D^2w) = 0 & \text{in } \mathcal{O}, \\ w = \varphi & \text{on } \partial\mathcal{O}. \end{cases}$$

For the proof, the theory of fully nonlinear elliptic equations is needed and see D. Gilbarg and N. S. Trudinger [74, Theorem 17.18, p. 466].

Instead, for the applications in Part II, we present the proof of one-dimensional case below. Let us consider the equation,

$$\begin{cases} Ku := -aD^2u + \alpha u - H(x, u, Du) = f & \text{in } \mathcal{O} = B_R(x_0), \\ u = \varphi & \text{on } \partial\mathcal{O}, \end{cases} \tag{5.20}$$

where $N = 1$, $a(x) := a_{ij}(x)$ and

$$\varphi \in C(\bar{\mathcal{O}}). \tag{5.21}$$

If

$$a \in C^2(\bar{\mathcal{O}}), \tag{5.22}$$

we can rewrite (5.20) as

$$\begin{cases} -D(aDu) + (\alpha + \mu)u - \hat{H}(x, u, Du) = f + \mu u & \text{in } \mathcal{O}, \\ u = \varphi & \text{on } \partial\mathcal{O}, \end{cases} \tag{5.23}$$

where $\hat{H}(x, r, \lambda) = H(x, r, \lambda) + d(x)\lambda$ for $\lambda \in \mathbf{R}$, $d = -Da \in C^1(\bar{\mathcal{O}})$ and $\mu > 0$ is sufficiently large.

Lemma 5.3.5. *We assume* (5.16), *and* (5.22), *and that* $\varphi = \hat{H} = 0$. *Let* $u \in C^2(\bar{\mathcal{O}})$ *satisfy* (5.23) *with* $f \in C^{0,\gamma}(\bar{\mathcal{O}})$ *replacing* $f + \mu u$. *Then*

$$\|u\|_{C^2(\bar{\mathcal{O}})} \leq C\|f\|_{C(\bar{\mathcal{O}})}, \tag{5.24}$$

where $C > 0$ *depends only on* $\theta, a, \alpha, \mathcal{O}$, *and* γ.

Proof.

(1) Suppose that $a = 1$ and $\alpha + \mu = 0$. Then u satisfies

$$-D^2 u = f \quad \text{in} \quad \mathcal{O}.$$

Clearly, $\|u\|_{C(\bar{\mathcal{O}})} \leq R\|Du\|_{C(\bar{\mathcal{O}})}$. Since $N = 1$, the interpolation inequality gives

$$\|Du\|_{C(\bar{\mathcal{O}})} \leq \varepsilon\|D^2u\|_{C(\bar{\mathcal{O}})} + C_\varepsilon\|u\|_{C(\bar{\mathcal{O}})} = \varepsilon\|f\|_{C(\bar{\mathcal{O}})} + C_\varepsilon\|u\|_{C(\bar{\mathcal{O}})}$$

for $\varepsilon = R/2$. Thus, by Theorem 5.2.3, we deduce (5.24).

(2) In case of $a \neq 1$, we have

$$-aD^2u = -dDu - (\alpha + \mu)u + f \quad \text{in} \quad \mathcal{O}.$$

By Theorem 5.2.3 (1), we note that

$$(\alpha + \mu)\|u\|_{C(\bar{\mathcal{O}})} \leq \|f\|_{C(\bar{\mathcal{O}})}.$$

We divide both sides by $a \geq \theta > 0$. Then, applying (1) and the interpolation inequality, we get

$$\begin{aligned}\|u\|_{C^2(\bar{\mathcal{O}})} &\leq C\{\|Du\|_{C(\bar{\mathcal{O}})} + (\alpha + \mu)\|u\|_{C(\bar{\mathcal{O}})} + \|f\|_{C(\bar{\mathcal{O}})}\} \\ &\leq C\{(\varepsilon\|D^2u\|_{C(\bar{\mathcal{O}})} + C_\varepsilon\|u\|_{C(\bar{\mathcal{O}})}) + 2\|f\|_{C(\bar{\mathcal{O}})}\}\end{aligned}$$

for sufficiently small $\varepsilon > 0$. Choosing $C\varepsilon \leq 1/2$, we obtain

$$\|u\|_{C^2(\bar{\mathcal{O}})} \leq C'(\|u\|_{C(\bar{\mathcal{O}})} + \|f\|_{C(\bar{\mathcal{O}})}) \leq C'(\frac{1}{\alpha} + 1)\|f\|_{C(\bar{\mathcal{O}})},$$

where the constant $C' > 0$ is independent of $\mu > 0$. This implies (5.24). ■

Theorem 5.3.6. *We assume* (5.11), (5.16), (5.17), (5.18), *and*(5.22). *Then there exists a unique solution* $u \in C^{2,\gamma}(\bar{\mathcal{O}})$ *of* (5.23).

Proof.

(1) Let $z \in C^2(\bar{\mathcal{O}})$ be arbitrary. We consider the equation,

$$\begin{cases} -D(aD\psi) + (\alpha + \mu)\psi - \hat{H}(x, z, Dz) = f & \text{in } \mathcal{O}, \\ \psi = \varphi & \text{on } \partial\mathcal{O}. \end{cases} \tag{5.25}$$

It is clear that $\hat{H}$ satisfies (5.17) and (5.18). By (5.18), $\hat{H}(x, z, Dz)$ belongs to $C^{0,\gamma}(\bar{\mathcal{O}})$. By Theorem 5.2.5, there exists a unique solution $\psi \in C^{2,\gamma}(\bar{\mathcal{O}})$ of (5.25).

(2) We define the mapping $\mathcal{T} : C^2(\bar{\mathcal{O}}) \to C^2(\bar{\mathcal{O}})$ by $\mathcal{T}z = \psi$. For $z_1, z_2 \in C^2(\bar{\mathcal{O}})$, we set $\psi_1 = \mathcal{T}z_1$, $\psi_2 = \mathcal{T}z_2$. Then

$$\begin{cases} -D(aD(\psi_1 - \psi_2)) + (\alpha + \mu)(\psi_1 - \psi_2) \\ \quad = \hat{H}(x, z_1, Dz_1) - \hat{H}(x, z_2, Dz_2) & \text{in } \mathcal{O}, \\ \psi_1 - \psi_2 = 0 & \text{on } \partial\mathcal{O}. \end{cases}$$

By (5.18), we can easily see that

$$\|\hat{H}(\cdot, z_1, Dz_1) - \hat{H}(\cdot, z_2, Dz_2)\|_{C(\bar{\mathcal{O}})} \le C_1 \|z_1 - z_2\|_{C^1(\bar{\mathcal{O}})},$$

for some $C_1 > 0$. By Lemma 5.3.5, we get

$$\|\psi_1 - \psi_2\|_{C^2(\bar{\mathcal{O}})} \le C \|\hat{H}(\cdot, z_1, Dz_1) - \hat{H}(\cdot, z_2, Dz_2)\|_{C(\bar{\mathcal{O}})}.$$

Hence

$$\|\psi_1 - \psi_2\|_{C^2(\bar{\mathcal{O}})} \le C_2 \|z_1 - z_2\|_{C^1(\bar{\mathcal{O}})}, \quad C_2 := CC_1 > 0.$$

Then

$$\begin{aligned} \mu\|\psi_1 - \psi_2\|_{C(\bar{\mathcal{O}})} &\le (\alpha + \mu)\|\psi_1 - \psi_2\|_{C(\bar{\mathcal{O}})} \\ &\le \|D(aD(\psi_1 - \psi_2))\|_{C(\bar{\mathcal{O}})} + \|\hat{H}(\cdot, z_1, Dz_1) \\ &\quad - \hat{H}(\cdot, z_2, Dz_2)\|_{C(\bar{\mathcal{O}})} \\ &\le \|aD^2(\psi_1 - \psi_2) - dD(\psi_1 - \psi_2)\|_{C(\bar{\mathcal{O}})} \\ &\quad + C_1\|z_1 - z_2\|_{C^1(\bar{\mathcal{O}})} \\ &\le \nu\|\psi_1 - \psi_2\|_{C^2(\bar{\mathcal{O}})} + C_1\|z_1 - z_2\|_{C^1(\bar{\mathcal{O}})} \\ &\le (\nu C_2 + C_1)\|z_1 - z_2\|_{C^1(\bar{\mathcal{O}})}, \end{aligned}$$

and thus

$$\|\psi_1 - \psi_2\|_{(\mu)} := \|\psi_1 - \psi_2\|_{C^2(\bar{\mathcal{O}})} + \mu\|\psi_1 - \psi_2\|_{C(\bar{\mathcal{O}})} \le C_3\|z_1 - z_2\|_{C^1(\bar{\mathcal{O}})},$$

where $\nu = \|a\|_{C(\bar{\mathcal{O}})} + \|d\|_{C(\bar{\mathcal{O}})}$ and $C_3 = (\nu + 1)C_2 + C_1$. By Theorem 5.1.2 (1) ②, there is $C_\varepsilon > 0$, for any $\varepsilon \in (0, R]$ such that

$$\|z_1 - z_2\|_{C^1(\bar{\mathcal{O}})} \le \varepsilon\|z_1 - z_2\|_{C^2(\bar{\mathcal{O}})} + C_\varepsilon\|z_1 - z_2\|_{C(\bar{\mathcal{O}})}.$$

Taking $\varepsilon \le 1/(2C_3)$ and $\mu > 2C_3C_\varepsilon$, we have

$$\|\mathcal{T}z_1 - \mathcal{T}z_2\|_{(\mu)} = \|\psi_1 - \psi_2\|_{(\mu)} \le \frac{1}{2}\|z_1 - z_2\|_{(\mu)}.$$

Since $\mathcal{T}$ is a contraction mapping under the norm $\|\cdot\|_{(\mu)}$ equivalent to $\|\cdot\|_{C^2(\bar{\mathcal{O}})}$, it has a fixed point $u^0 \in C^2(\bar{\mathcal{O}})$, which solves

$$\begin{cases} -D(aDu^0) + (\alpha + \mu)u^0 - \hat{H}(x, u^0, Du^0) = f & \text{in } \mathcal{O}, \\ u^0 = \varphi & \text{on } \partial\mathcal{O}. \end{cases}$$

(3) Define $u^n \in C^2(\bar{\mathcal{O}})$ by

$$\begin{cases} -D(aDu^{n+1}) + (\alpha+\mu)u^{n+1} - \hat{H}(x, u^{n+1}, Du^{n+1}) = f + \mu u^n & \text{in } \mathcal{O}, \\ u^{n+1} = \varphi \quad \text{on } \partial\mathcal{O}, \quad n = 0, 1, \ldots. \end{cases} \tag{5.26}$$

By Corollary 5.3.2, we note that

$$\begin{aligned} \|u^{n+1} - u^n\|_{C(\bar{\mathcal{O}})} &\le \frac{1}{\alpha+\mu-\alpha_0} \|(f+\mu u^n) - (f + \mu u^{n-1})\|_{C(\bar{\mathcal{O}})} \\ &= \delta \|u^n - u^{n-1}\|_{C(\bar{\mathcal{O}})}, \end{aligned}$$

where $0 < \delta := \mu/(\alpha+\mu-\alpha_0) < 1$. By (5.26), it is clear that

$$\begin{aligned} &-D(aD(u^{n+1}-u^n)) + (\alpha+\mu)(u^{n+1}-u^n) \\ &\quad = \hat{H}(x, u^{n+1}, Du^{n+1}) - \hat{H}(x, u^n, Du^n) + \mu(u^n - u^{n-1}) \quad \text{in } \mathcal{O}. \end{aligned}$$

Applying Lemma 5.3.5, by (5.18), we see that

$$\begin{aligned} \|u^{n+1} - u^n\|_{C^2(\bar{\mathcal{O}})} &\le C\|\hat{H}(\cdot, u^{n+1}, Du^{n+1}) - \hat{H}(\cdot, u^n, Du^n) \\ &\qquad + \mu(u^n - u^{n-1})\|_{C(\bar{\mathcal{O}})} \\ &\le C\{C_1\|u^{n+1} - u^n\|_{C^1(\bar{\mathcal{O}})} + \mu\|u^n - u^{n-1}\|_{C^1(\bar{\mathcal{O}})}\}. \end{aligned}$$

By Theorem 5.1.2 (1) ②, for any $0 < \varepsilon \le R$

$$\begin{aligned} \|u^{n+1} - u^n\|_{C^1(\bar{\mathcal{O}})} &\le \varepsilon\|u^{n+1} - u^n\|_{C^2(\bar{\mathcal{O}})} + C_\varepsilon\|u^{n+1} - u^n\|_{C(\bar{\mathcal{O}})}, \\ \|u^n - u^{n-1}\|_{C^1(\bar{\mathcal{O}})} &\le \varepsilon\|u^n - u^{n-1}\|_{C^2(\bar{\mathcal{O}})} + C_\varepsilon\|u^n - u^{n-1}\|_{C(\bar{\mathcal{O}})}. \end{aligned}$$

Thus

$$\|u^{n+1} - u^n\|_{C^2(\bar{\mathcal{O}})} \le \tilde{\varepsilon}\|u^n - u^{n-1}\|_{C^2(\bar{\mathcal{O}})} + \tilde{C}\|u^n - u^{n-1}\|_{C(\bar{\mathcal{O}})},$$

where $\tilde{\varepsilon} := \mu C\varepsilon/(1 - CC_1\varepsilon) > 0$ and $\tilde{C} := CC_\varepsilon(\delta C_1 + \mu)/(1 - CC_1\varepsilon)$ for sufficiently small $\varepsilon > 0$. Hence

$$\|u^{n+1} - u^n\|_{C^2(\bar{\mathcal{O}})} \le \tilde{\varepsilon}\|u^n - u^{n-1}\|_{C^2(\bar{\mathcal{O}})} + \delta^{n-1}\tilde{C}\|u^1 - u^0\|_{C(\bar{\mathcal{O}})}.$$

By induction, we get

$$\begin{aligned} \|u^{n+1} - u^n\|_{C^2(\bar{\mathcal{O}})} &\le \tilde{\varepsilon}^n\|u^1 - u^0\|_{C^2(\bar{\mathcal{O}})} \\ &\qquad + (\tilde{\varepsilon} + \delta)^{n-1}\tilde{C}\|u^1 - u^0\|_{C(\bar{\mathcal{O}})}, \quad n \in \mathbf{N}. \end{aligned}$$

Choosing $\varepsilon > 0$ such that $\tilde{\varepsilon} > 0$ and $0 < \tilde{\varepsilon} + \delta < 1$, we deduce

$$\sum_{n=1}^{\infty} \|u^{n+1} - u^n\|_{C^2(\bar{\mathcal{O}})} < \infty.$$

This yields that $\{u^n\}$ is a Cauchy sequence in $C^2(\bar{\mathcal{O}})$. Therefore, we have

$$u^n \to u \quad \text{in } C^2(\bar{\mathcal{O}}) \quad \text{as } n \to \infty.$$

Passing to the limit in (5.26), we observe that u solves (5.23). By Theorem 5.2.5 again, we obtain $u \in C^{2,\gamma}(\bar{\mathcal{O}})$.

(4) Let $u_i \in C^{2,\gamma}(\bar{\mathcal{O}})$, $i = 1, 2$, be two solutions of (5.23). Then

$$-D(aDu_i) + \alpha u_i - \hat{H}(x, u_i, Du_i) = f \quad \text{in } \mathcal{O},$$

and $u_1 - u_2 = 0$ on $\partial\mathcal{O}$. By Corollary 5.3.2,

$$\|u_1 - u_2\|_{C(\bar{\mathcal{O}})} \le \frac{1}{\alpha - \alpha_0}\|f - f\|_{C(\bar{\mathcal{O}})} = 0,$$

which implies uniqueness. ∎

Theorem 5.3.7. *We assume* (5.16), (5.17), (5.18), *and* (5.21). *Then there exists a unique solution $u \in C^2(\mathcal{O}) \cap C(\bar{\mathcal{O}})$ of* (5.20).

Proof.

(1) Let $\varphi \in C^{2,\gamma}(\bar{\mathcal{O}})$. For σ of (5.16), by mollification, we find σ_n in $C^2(\mathbf{R})$ such that $[\sigma_n]_{C^{0,1}(\bar{\mathcal{O}})} \le [\sigma]_{C^{0,1}(\bar{\mathcal{O}})}$ and $\|\sigma_n - \sigma\|_{C(\bar{\mathcal{O}})} \to 0$ as $n \to \infty$. Define $a_n \in C^2(\bar{\mathcal{O}})$ by $a_n(x) = \sigma_n(x)^2$. It is easy to check that

$$\begin{aligned}
\|a_n\|_{C^{0,1}(\bar{\mathcal{O}})} &\le \|\sigma_n\|^2_{C(\bar{\mathcal{O}})} + 2\|\sigma_n\|_{C(\bar{\mathcal{O}})}[\sigma_n]_{C^{0,1}(\bar{\mathcal{O}})} \le 3\|\sigma_n\|^2_{C^{0,1}(\bar{\mathcal{O}})} \\
&\le 3(\|\sigma\|_{C^{0,1}(\bar{\mathcal{O}})} + \sup_n \|\sigma_n - \sigma\|_{C(\bar{\mathcal{O}})})^2, \\
\|a_n - a\|_{C(\bar{\mathcal{O}})} &\to 0 \text{ as } n \to \infty.
\end{aligned}$$

Furthermore, we take $n_0 \in \mathbf{N}$ such that $\|\sigma_n - \sigma\|_{C(\bar{\mathcal{O}})} < \sqrt{\theta}/2$ for $n \ge n_0$. Then, by uniform ellipticity,

$$\begin{aligned}
a_n(x)\xi^2 = |\sigma_n\xi|^2 &\ge (|\sigma\xi| - \|\sigma_n - \sigma\|_{C(\bar{\mathcal{O}})}|\xi|)^2 \\
&\ge \frac{\theta}{4}|\xi|^2, \quad \forall n \ge n_0,\ \xi \in \mathbf{R}.
\end{aligned}$$

By Theorem 5.3.6, there exists a unique solution $u_n \in C^{2,\gamma}(\bar{\mathcal{O}})$ of

$$\begin{cases} -a_n D^2 u_n + \alpha u_n - H(x, u_n, Du_n) = f & \text{in } \mathcal{O}, \\ u_n = \varphi & \text{on } \partial\mathcal{O}. \end{cases}$$

Theorem 5.2.2 yields that

$$\begin{aligned}
\|u_n\|_{C^{2,\gamma}(\bar{\mathcal{O}})} \le C(&\|H(\cdot, u_n, Du_n)\|_{C^{0,\gamma}(\bar{\mathcal{O}})} + \|f\|_{C^{0,\gamma}(\bar{\mathcal{O}})} \\
&+ \|u_n\|_{C(\bar{\mathcal{O}})} + \|\varphi\|_{C^{2,\gamma}(\bar{\mathcal{O}})})
\end{aligned}$$

for some constant $C > 0$, independent of $n \ge n_0$. By Theorem 5.3.3, (5.18), and the interpolation inequality, we note that

$$\begin{aligned}
&\|u_n\|_{C(\bar{\mathcal{O}})} \le \|\varphi\|_{C(\bar{\mathcal{O}})} + C(\|f\|_{C(\bar{\mathcal{O}})} + 1), \\
&\|H(\cdot, u_n, Du_n)\|_{C^{0,\gamma}(\bar{\mathcal{O}})} \le C(1 + \|u_n\|_{C^{1,\gamma}(\bar{\mathcal{O}})}), \\
&\|u_n\|_{C^{1,\gamma}(\bar{\mathcal{O}})} \le \varepsilon\|u_n\|_{C^{2,\gamma}(\bar{\mathcal{O}})} + C_\varepsilon\|u_n\|_{C(\bar{\mathcal{O}})}, \quad C_\varepsilon > 0.
\end{aligned}$$

Hence, for sufficiently small $\varepsilon > 0$,

$$\|u_n\|_{C^{2,\gamma}(\bar{\mathcal{O}})} \leq \varepsilon \|u_n\|_{C^{2,\gamma}(\bar{\mathcal{O}})} + C_\varepsilon(\|f\|_{C^{0,\gamma}(\bar{\mathcal{O}})} + \|\varphi\|_{C^{2,\gamma}(\bar{\mathcal{O}})} + 1).$$

Taking $\varepsilon \leq 1/2$, we get

$$\|u_n\|_{C^{2,\gamma}(\bar{\mathcal{O}})} \leq C(\|f\|_{C^{0,\gamma}(\bar{\mathcal{O}})} + \|\varphi\|_{C^{2,\gamma}(\bar{\mathcal{O}})} + 1).$$

By the Ascoli–Arzelà theorem, extracting a subsequence, we have

$$u_n \to u \in C^2(\bar{\mathcal{O}}), \quad u \in C^{2,\gamma}(\bar{\mathcal{O}}).$$

Passing to the limit, we see that u solves (5.20).

(2) Let φ satisfy (5.21). By extension and mollification, we take $\varphi_n \in C^{2,\gamma}(\bar{\mathcal{O}})$ such that $\|\varphi_n - \varphi\|_{C(\bar{\mathcal{O}})} \leq 1/n$ for each $n \in \mathbf{N}$. Let $\mathcal{O}_n$ be a ball (interval) such that $\mathcal{O}_n \subset\subset \mathcal{O}$ and $\sup_{x\in\partial\mathcal{O}_n} d(x, \partial\mathcal{O}) \leq 1/n$. By (1), there exists a unique solution $v_n \in C^{2,\gamma}(\bar{\mathcal{O}}_n)$ of

$$\begin{cases} K(v_n) = -aD^2 v_n + \alpha v_n - H(x, v_n, Dv_n) = f & \text{in } \mathcal{O}_n, \\ v_n = \varphi_n & \text{on } \partial\mathcal{O}_n. \end{cases} \tag{5.27}$$

For any ball $\mathcal{O}' \subset\subset \mathcal{O}$, we choose $\kappa \in \mathbf{N}$ such that $\mathcal{O}' \subset\subset \mathcal{O}_\kappa \subset \mathcal{O}$. By Theorem 5.2.1, for $n > \kappa$, we have

$$\|v_n\|_{C^{2,\gamma}(\bar{\mathcal{O}}')} \leq C(\|H(\cdot, v_n, Dv_n)\|_{C^{0,\gamma}(\bar{\mathcal{O}}_\kappa)} + \|f\|_{C^{0,\gamma}(\bar{\mathcal{O}}_\kappa)} + \|v_n\|_{C(\bar{\mathcal{O}}_\kappa)}),$$

and by (5.18)

$$\|v_n\|_{C^{2,\gamma}(\bar{\mathcal{O}}')} \leq C(\|v_n\|_{C^2(\bar{\mathcal{O}}_\kappa)} + \|f\|_{C^{0,\gamma}(\bar{\mathcal{O}}_\kappa)} + \|v_n\|_{C(\bar{\mathcal{O}}_\kappa)} + 1),$$

where the constant $C > 0$ depends only on $d(\mathcal{O}', \partial\mathcal{O}_\kappa), \theta, \Theta, \gamma, N$. Furthermore, by Theorem 5.3.3

$$\begin{aligned} \|v_n\|_{C(\bar{\mathcal{O}}_\kappa)} \leq \|v_n\|_{C(\bar{\mathcal{O}}_n)} &\leq \sup\nolimits_{\partial\mathcal{O}_n} |v_n| + C_n(\|f\|_{C(\bar{\mathcal{O}}_n)} + 1) \\ &\leq \|\varphi_n\|_{C(\bar{\mathcal{O}})} + C_n(\|f\|_{C(\bar{\mathcal{O}})} + 1) \\ &\leq C'(\|\varphi_n\|_{C(\bar{\mathcal{O}})} + \|f\|_{C(\bar{\mathcal{O}})} + 1) \\ &\leq C'(\|\varphi\|_{C(\bar{\mathcal{O}})} + \|f\|_{C(\bar{\mathcal{O}})} + 1 + 1/n), \end{aligned} \tag{5.28}$$

where the constant $C' \geq 1$ is taken as $C' \geq C_n := \max(1, \|H(\cdot, 0, 0)\|_{C(\bar{\mathcal{O}}_n)})/(\alpha - \alpha_0)) > 0$. By (5.27) and (5.18), we have

$$\begin{aligned} \|D^2 v_n\|_{C(\bar{\mathcal{O}}_\kappa)} &\leq C(\|H(\cdot, v_n, Dv_n)\|_{C(\bar{\mathcal{O}}_\kappa)} + \|f\|_{C(\bar{\mathcal{O}}_\kappa)} + \|v_n\|_{C(\bar{\mathcal{O}}_\kappa)}) \\ &\leq \bar{C}(\|Dv_n\|_{C(\bar{\mathcal{O}}_\kappa)} + \|f\|_{C(\bar{\mathcal{O}}_\kappa)} + \|v_n\|_{C(\bar{\mathcal{O}}_\kappa)} + 1), \end{aligned}$$

for some constants $C, \bar{C} > 0$, independent of n. Also, by Theorem 5.1.2 (ii)

$$\|Dv_n\|_{C(\bar{\mathcal{O}}_\kappa)} \leq \varepsilon \|D^2 v_n\|_{C(\bar{\mathcal{O}}_\kappa)} + C_\varepsilon \|v_n\|_{C(\bar{\mathcal{O}}_\kappa)},$$

where the constant $C_\varepsilon > 0$ depends only on $\mathcal{O}_\kappa$ and sufficiently small $\varepsilon > 0$. Thus, choosing $\bar{C}\varepsilon \le 1/2$, we get

$$\|D^2 v_n\|_{C(\bar{\mathcal{O}}_\kappa)} \le 2\bar{C}\{(C_\varepsilon + 1)\|v_n\|_{C(\bar{\mathcal{O}}_\kappa)} + \|f\|_{C(\bar{\mathcal{O}})} + 1\}.$$

Hence, by (5.28) and the interpolation inequality,

$$\sup_{n>\kappa} \|v_n\|_{C^2(\bar{\mathcal{O}}_\kappa)} < \infty.$$

Therefore,

$$\sup_{n>\kappa} \|v_n\|_{C^{2,\gamma}(\bar{\mathcal{O}}')} < \infty.$$

By the Ascoli–Arzelà theorem, extracting a subsequence, we obtain $u \in C^2(\mathcal{O})$ such that

$$v_n \to u \in C^2(\bar{\mathcal{O}}'), \quad u \in C^{2,\gamma}(\bar{\mathcal{O}}').$$

Passing to the limit, we deduce that u solves the HJB equation of (5.20) in $\mathcal{O}$.

(3) To prove that $u \in C(\bar{\mathcal{O}})$ and $u = \varphi$ on $\partial\mathcal{O}$, we consider the barrier function $w(x)$ at $b \in \partial\mathcal{O}$, which has the following properties:

① $w(b) = 0, \quad w(x) > 0, \quad \text{for } x \in \bar{\mathcal{O}} \setminus \{b\},$
② $w \in C^2(\bar{\mathcal{O}}), \quad Kw - K0 \ge \hat{w}, \quad$ for some constant $\hat{w} > 0$.

Since $\mathcal{O} = B_R(x_0)$, we take a ball $B_r(z) \subset \mathbf{R} \setminus \bar{\mathcal{O}}$ such that $\overline{B_r(z)} \cap \bar{\mathcal{O}} = \{b\}$. Define

$$w(x) = e^{-cr^2} - e^{-c|x-z|^2}, \quad \text{for } c > 0.$$

It is easy to see that $r = |b - z|$ and w satisfies ①. By (5.18) and uniform ellipticity, we have

$$\begin{aligned} Kw - K0 &= -aD^2 w + \alpha w - [H(x, w, Dw) - H(x, 0, 0)] \\ &\ge -aD^2 w + \alpha w - \alpha_0|w| - C_0|Dw| \\ &= [4ac^2|x-z|^2 - 2ac]e^{-c|x-z|^2} + \alpha w - \alpha_0|w| - C_0|Dw| \\ &\ge [4\theta c^2|x-z|^2 - 2\hat{a}c - 2c|x-z|C_0]e^{-c|x-z|^2} \\ &\ge [4\theta c^2 r^2 - 2\hat{a}c - 2c(2R+r)C_0]e^{-c(2R+r)^2} =: \hat{w}, \end{aligned}$$

where $\hat{a} = \|a\|_{C(\bar{\mathcal{O}})}$. Here, we choose the constant $c > 0$ such that $\hat{w} > 0$. Therefore, we obtain ②.

(4) For any $\varepsilon > 0$, there is a neighborhood $V(b)$ of $b \in \partial\mathcal{O}$ such that for $x \in V(b) \cap \mathcal{O}$,

$$|\varphi(x) - \varphi(b)| < \varepsilon.$$

Since $\inf\{w(x) : x \in \mathcal{O} \setminus V(b)\} > 0$, we find $C > 0$ (depending on ε) such that

$$\begin{aligned} w_1(x) &:= -Cw(x) + \varphi(b) - \varepsilon < \varphi(x) < Cw(x) + \varphi(b) + \varepsilon \\ &=: w_2(x), \quad x \in \bar{\mathcal{O}}, \end{aligned}$$

and then

$$w_1(x) - \frac{1}{n} < \varphi_n(x) < w_2(x) + \frac{1}{n}, \quad x \in \bar{\mathcal{O}}.$$

We choose $C' > 0$ such that

$$-C'\hat{w} \leq f - K0 \leq C'\hat{w} \quad \text{in } \bar{\mathcal{O}}.$$

By (5.18) and (3), we have for $\lambda \in \mathbf{R}$

$$\begin{aligned} K(Cw + \lambda) - K0 &= -aD^2(Cw) + \alpha(Cw + \lambda) \\ &\quad - [H(x, Cw + \lambda, D(Cw)) - H(x, 0, 0)] \\ &\geq C\{-aD^2 w + \alpha w - \alpha_0|w| - C_0|Dw|\} \\ &\quad + (\alpha\lambda - \alpha_0|\lambda|) \\ &\geq C\hat{w} - (\alpha + \alpha_0)|\lambda| \\ &\geq C'\hat{w}, \end{aligned}$$

taking sufficiently large $C > 0$ if necessary. Hence

$$C'\hat{w} \leq K(w_2 + \frac{1}{n}) - K0.$$

Similarly,

$$K(w_1 - \frac{1}{n}) - K0 \leq -C'\hat{w}.$$

Thus

$$K(w_1 - \frac{1}{n}) \leq f \leq K(w_2 + \frac{1}{n}) \quad \text{in } \mathcal{O}_n.$$

By (2) and Theorem 5.3.1, we get

$$w_1(x) - \frac{1}{n} \leq v_n(x) \leq w_2(x) + \frac{1}{n}, \quad x \in \mathcal{O}_n.$$

Letting $n \to \infty$ and then $x \to b$, we deduce

$$\varphi(b) - \varepsilon \leq \liminf_{x \to b} u(x) \leq \limsup_{x \to b} u(x) \leq \varphi(b) + \varepsilon,$$

and hence

$$\lim_{x \to b} u(x) = \varphi(b).$$

Therefore we conclude that $u(x)$ can be extended continuously to $\partial\mathcal{O}$ and $u = \varphi$ on $\partial\mathcal{O}$. The uniqueness is immediate from Corollary 5.3.2. ■

5.4 Stochastic LQ Problems with Constraints

5.4.1 One-Dimensional Case

We study the minimization problem of the cost functional with discount rate $\alpha > 0$:

$$J(c) = E[\int_0^\infty e^{-\alpha t} X_t^2 dt]$$

over $\mathcal{A}$ subject to a one-dimensional stochastic differential system

$$dX_t = c_t dt + dB_t, \quad X_0 = x \in \mathbf{R}, \tag{5.29}$$

where $\mathcal{A}$ is the class of $\{\mathcal{F}_t^B\}$-progressively measurable processes $c = \{c_t\}$ with hard constraint

$$|c_t| \leq 1, \quad \text{a.s.,} \quad \text{for all } t \geq 0.$$

The control $c^* = \{c_t^*\} \in \mathcal{A}$ is said to be optimal if $J(c^*) \leq J(c)$ for all $c \in \mathcal{A}$. The HJB equation associated with this problem is given by

$$-\alpha v + \frac{1}{2} v'' + \min_{|c| \leq 1} cv' + x^2 = 0 \quad \text{in } \mathbf{R}. \tag{5.30}$$

Theorem 5.4.1. *Under $\alpha > 0$, there exists a convex solution $v \in C^2(\mathbf{R})$ of (5.30) uniquely.*

Proof. Define v by the value function

$$v(x) = \inf_{c \in \mathcal{A}} E[\int_0^\infty e^{-\alpha t} X_t^2 dt], \tag{5.31}$$

where the infimum is taken over all systems $(\Omega, \mathcal{F}, P, \{\mathcal{F}_t\}; \{B_t\}, \{c_t\})$.

(1) We claim that v is convex. Let $x_i \in \mathbf{R}$, $i = 1, 2$ and $0 \leq \xi \leq 1$. For any $\varepsilon > 0$, there is $c^{(i)} \in \mathcal{A}$ such that

$$E[\int_0^\infty e^{-\alpha t} (X_t^{(i)})^2 dt] < v(x_i) + \varepsilon,$$

where $\{X_t^{(i)}\}$ is the solution of (5.29) corresponding to $c^{(i)}$. We may consider that

$$dX_t^{(i)} = c_t^{(i)} dt + dB_t, \quad X_0^{(i)} = x_i \in \mathbf{R},$$

on the same probability space. We set $X_t^{(\xi)} = \xi X_t^{(1)} + (1 - \xi) X_t^{(2)}$ and $c_t^{(\xi)} = \xi c_t^{(1)} + (1 - \xi) c_t^{(2)}$, which belongs to $\mathcal{A}$. Clearly,

$$dX_t^{(\xi)} = c_t^{(\xi)} dt + dB_t, \quad X_0^{(\xi)} = \xi x_1 + (1 - \xi) x_2.$$

Hence

$$\begin{aligned}v(\xi x_1 + (1-\xi)x_2) &\le E[\int_0^\infty e^{-\alpha t}(X_t^{(\xi)})^2 dt] \\ &\le \xi E[\int_0^\infty e^{-\alpha t}(X_t^{(1)})^2 dt] \\ &\quad + (1-\xi)E[\int_0^\infty e^{-\alpha t}(X_t^{(2)})^2 dt] \\ &< \xi v(x_1) + (1-\xi)v(x_2) + \varepsilon.\end{aligned}$$

Thus, letting $\varepsilon \to 0$, we obtain the convexity of v.

(2) By (5.31), we have

$$0 \le v(x) \le E[\int_0^\infty e^{-\alpha t}(x + B_t)^2 dt] \le C(1 + |x|^2)$$

for some constant $C > 0$. By the same line as Proposition 3.3.1, taking $k = 2$ and sufficiently small $L > 0$, we see that v satisfies the uniformly continuity with 2 weight, that is,

$$\begin{aligned}\forall \rho > 0,\ \exists C_\rho > 0 :\ &|v(x) - v(y)| \le C_\rho |x - y|^2 \\ &+ \rho(1 + |x|^2 + |y|^2), \quad x, y \in \mathbf{R}.\end{aligned}$$

Hence, by Theorem 3.3.2, the DPP holds for v, that is,

$$v(x) = \inf_{c \in \mathcal{A}} E[\int_0^\tau e^{-\alpha t} X_t^2 dt + e^{-\alpha \tau} v(X_\tau)], \quad \tau \in \mathcal{S}_b.$$

We note that

$$\begin{aligned}E[\sup_{0 \le t \le h} |X_t - X_0|^2] &\le 2E[(\int_0^h |c_t| dt)^2 \\ &+ \sup_{0 \le t \le h} |B_t|^2] \le 2(h^2 + 4h) \quad \text{as } h \to 0,\end{aligned}$$

and

$$\begin{aligned}\sup_{0 \le t \le h} E[|X_t^2 - X_0^2|] \le &\sup_{0 \le t \le h} (E[|X_t - X_0|^2])^{1/2} (E[|X_t + X_0|^2])^{1/2} \\ &\to 0 \quad \text{as } h \to 0.\end{aligned}$$

By the same line as Theorem 4.3.1, v is a unique viscosity solution of (5.30).

(3) Along (a)–(c), we here show the smoothness of v. Let $a < b$ be arbitrary. We consider the boundary value problem:

$$\begin{cases} -\alpha w + \dfrac{1}{2} w'' + \min\limits_{|c| \le 1} c w' + x^2 = 0 \quad \text{in } (a, b), \\ w(a) = v(a), \quad w(b) = v(b), \end{cases} \tag{5.32}$$

where the boundary condition is given by the viscosity solution v of (5.30). It is clear that v is a viscosity solution of (5.32) in (a, b). By Theorem 5.3.7,

(5.32) admits a solution $w \in C^2(a,b) \cap C[a,b]$, which is also a viscosity solution of (5.32). By the comparison principle in Theorem 5.3.1, we have $w = v$ in $[a,b]$ and thus $v \in C^2(\mathbf{R})$. ∎

Let us consider

$$dX_t^* = F(v'(X_t^*))dt + dB_t. \quad X_0^* = x. \tag{5.33}$$

Here $F(x)$ denotes the minimizer of $\min_{|c|\leq 1} cx$, that is, $F(x) = -\text{sgn}(x)$.

Theorem 5.4.2. *We make the assumption of Theorem 5.4.1. Then the optimal control* $\{c_t^*\}$ *is given by the feedback form:*

$$c_t^* = F(v'(X_t^*)).$$

Proof.

(1) We claim that there exists a unique strong solution $\{X_t^*\}$ to (5.33). By Theorem 2.4.6, (5.33) admits a weak solution. Let $\{X_t^{(i)}\}$, $i = 1, 2$, be two solutions of (5.33). Then

$$d(X_t^{(1)} - X_t^{(2)}) = \{-\text{sgn}(v'(X_t^{(1)})) + \text{sgn}(v'(X_t^{(2)}))\}dt, \quad X_0^{(1)} - X_0^{(2)} = 0.$$

By convexity, $-\text{sgn}(v'(x))$ is nonincreasing and hence

$$(x - \bar{x})\{-\text{sgn}(v'(x)) + \text{sgn}(v'(\bar{x}))\} \leq 0, \quad x, \bar{x} \in \mathbf{R}.$$

Thus

$$(X_t^{(1)} - X_t^{(2)})^2 \leq 0, \quad \text{for all } t \geq 0 \quad \text{a.s.}$$

This implies that the pathwise uniqueness holds for (5.33). By Theorem 2.5.3, we obtain a strong solution $\{X_t^*\}$ to (5.33).

(2) For optimality, we apply Itô's formula to obtain

$$e^{-\alpha t}v(X_t^*) = v(x) + \int_0^t e^{-\alpha s}\{-\alpha v(X_t^*) + v'(X_s^*)c_s^* + \frac{1}{2}v''(X_s^*)\}ds$$
$$+ \int_0^t e^{-\alpha s}v'(X_s^*)dB_s, \quad \text{a.s.}$$

By (5.30), we have

$$E[e^{-\alpha(t\wedge\tau_n)}v(X_{t\wedge\tau_n}^*)] + E[\int_0^{t\wedge\tau_n} e^{-\alpha s}(X_s^*)^2 ds] = v(x),$$

where $\{\tau_n\}$ is a sequence of localizing stopping times $\tau_n \nearrow \infty$ of the local martingale. Note that $\{c_t^*\}$ belongs to $\mathcal{A}$ and

$$E[\sup_n v(X_{t\wedge\tau_n}^*)] \leq C(1 + |x|^2 + t^2 + t).$$

Letting $n \to \infty$ and then $t \to \infty$, we get

$$J(c^*) = E[\int_0^\infty e^{-\alpha t}(X_t^*)^2 dt] = v(x).$$

For any $c \in \mathcal{A}$, by the same calculation as above, we get

$$J(c) = E[\int_0^\infty e^{-\alpha t} X_t^2 dt] \geq v(x),$$

which implies the optimality of $\{c_t^*\}$. ■

5.4.2 Multi-Dimensional Case

We study the minimization problem of the cost functional with discount rate $\alpha > 0$:

$$J(c) = E[\int_0^\infty e^{-\alpha t}\{|X_t|^2 + |c_t|^2\}dt]$$

over $\mathcal{A}$ subject to an N-dimensional stochastic differential system

$$dX_t = c_t dt + dB_t, \quad X_0 = x \in \mathbf{R}^N, \tag{5.34}$$

where $\{B_t\}$ is an N-dimensional Brownian motion and $\mathcal{A}$ is the class of $\{\mathcal{F}_t^B\}$-progressively measurable processes $c = \{c_t\}$ taking values in $\mathbf{R}^N$ with hard constraint

$$|c_t| \leq 1, \quad \text{a.s.,} \quad \text{for all } t \geq 0.$$

The HJB equation associated with this optimization problem is given by

$$-\alpha v + \frac{1}{2}\Delta v + \min_{|c|\leq 1}\{|c|^2 + (c, Dv)\} + |x|^2 = 0 \quad \text{in } \mathbf{R}^N. \tag{5.35}$$

Theorem 5.4.3. *Under* $\alpha > 0$, *there exists a convex solution* $v \in C^2(\mathbf{R}^N)$ *of* (5.35) *uniquely.*

Proof. Define v by the value function

$$v(x) = \inf_{c\in\mathcal{A}} E[\int_0^\infty e^{-\alpha t}\{|X_t|^2 + |c_t|^2\}dt], \tag{5.36}$$

where the infimum is taken over all systems $(\Omega, \mathcal{F}, P, \{\mathcal{F}_t\}; \{B_t\}, \{c_t\})$.

(1) We claim that v is convex. Let $x_i \in \mathbf{R}^N$, $i = 1, 2$ and $0 \leq \xi \leq 1$. For any $\varepsilon > 0$, there is $c^{(i)} \in \mathcal{A}$ such that

$$E[\int_0^\infty e^{-\alpha t}\{|X_t^{(i)}|^2 + |c_t^{(i)}|^2\}dt] < v(x_i) + \varepsilon,$$

where $\{X_t^{(i)}\}$ is the solution of (5.34) corresponding to $c^{(i)}$. We may consider that

$$dX_t^{(i)} = c_t^{(i)} dt + dB_t, \quad X_0^{(i)} = x_i \in \mathbf{R}^N,$$

on the same probability space. We set $X_t^{(\xi)} = \xi X_t^{(1)} + (1-\xi)X_t^{(2)}$ and $c_t^{(\xi)} = \xi c_t^{(1)} + (1-\xi)c_t^{(2)}$, which belongs to $\mathcal{A}$. Clearly,

$$dX_t^{(\xi)} = c_t^{(\xi)} dt + dB_t, \quad X_0^{(\xi)} = \xi x_1 + (1-\xi)x_2.$$

Hence

$$\begin{aligned}
v(\xi x_1 + (1-\xi)x_2) &\le E[\int_0^\infty e^{-\alpha t}\{|X_t^{(\xi)}|^2 + |c_t^{(\xi)}|^2\}dt] \\
&\le \xi E[\int_0^\infty e^{-\alpha t}\{|X_t^{(1)}|^2 + |c_t^{(1)}|^2\}dt] \\
&\quad + (1-\xi)E[\int_0^\infty e^{-\alpha t}\{|X_t^{(2)}|^2 + |c_t^{(2)}|^2\}dt] \\
&< \xi v(x_1) + (1-\xi)v(x_2) + \varepsilon.
\end{aligned}$$

Thus, letting $\varepsilon \to 0$, we obtain the convexity of v.

(2) By (5.36), we have

$$0 \le v(x) \le E[\int_0^\infty e^{-\alpha t}\{(x + B_t)^2 + 0\}dt] \le C(1 + |x|^2),$$

for some constant $C > 0$. By the same line as Proposition 3.3.1, taking $k = 2$ and sufficiently small $L > 0$, we see that v satisfies the uniformly continuity with 2-weight, that is,

$$\begin{aligned}
&\forall \rho > 0,\ \exists C_\rho > 0 : |v(x) - v(y)| \le C_\rho |x - y|^2 \\
&\quad + \rho(1 + |x|^2 + |y|^2), \quad x, y \in \mathbf{R}^N.
\end{aligned}$$

Hence, by Theorem 3.3.2, the DPP holds for v, that is,

$$v(x) = \inf_{c \in \mathcal{A}} E[\int_0^\tau e^{-\alpha t}\{|X_t|^2 + |c_t|^2)dt + e^{-\alpha\tau} v(X_\tau)], \quad \tau \in \mathcal{S}_b.$$

By the same line as Theorem 4.3.1, v is a viscosity solution of (5.35).

(3) Let $\mathcal{O}$ be any ball in $\mathbf{R}^N$. We consider the boundary value problem:

$$\begin{cases} -\alpha w + \dfrac{1}{2}\Delta w + \min\limits_{|c| \le 1}\{|c|^2 + (c, Dw)\} + |x|^2 = 0 & \text{in } \mathcal{O}, \\ w = v & \text{on } \partial\mathcal{O}. \end{cases} \tag{5.37}$$

Now, we set

$$F^{(i)}(x, z, p, q) = -\alpha z + \frac{1}{2}tr(q) + |i|^2 + (i, p) + |x|^2$$

for a countable dense subset $\{i\}$ of the unit ball $B_1(0)$. Then, by Theorem 5.3.4, there exists a solution $w \in C^2(\mathcal{O}) \cap C(\bar{\mathcal{O}})$ of

$$\begin{cases} \inf\limits_i F^{(i)}(x, w, Dw, D^2 w) = 0 & \text{in } \mathcal{O}, \\ w = v & \text{on } \partial\mathcal{O}, \end{cases}$$

which is equivalent to (5.37). It is clear that v and w are two viscosity solutions of (5.37) in $\mathcal{O}$. Clearly, (5.17) and (5.18) are satisfied for (5.37). By Theorem 5.3.1, we have $w = v$ in $\bar{\mathcal{O}}$ and thus $v \in C^2(\mathbf{R}^N)$. ■

Let us consider

$$dX_t^* = G(Dv(X_t^*))dt + dB_t. \quad X_0^* = x. \tag{5.38}$$

Here $G(x)$ denotes the minimizer of $\min_{|c|\le 1}\{|c|^2 + (c, x)\}$, that is, $G(x) = \xi_1(-\frac{x}{2})$, for $\xi_1(x) = x$ if $|x| \le 1$, $\xi_1(x) = x/|x|$ otherwise.

Theorem 5.4.4. *We make the assumption of Theorem 5.4.3. Then the optimal control $\{c_t^*\}$ is given by the feedback form:*

$$c_t^* = G(Dv(X_t^*)).$$

Proof.

(1) Clearly, $G(Dv(x))$ is bounded and locally Lipschitz. By Theorem 1.9.1, there exists a unique strong solution $\{X_t^*\}$ of (5.38).

(2) For optimality, we apply Itô's formula to obtain

$$\begin{aligned} e^{-\alpha t} v(X_t^*) = v(x) + \int_0^t e^{-\alpha s}\{-\alpha v(X_t^*) + (Dv(X_s^*), c_s^*) + \frac{1}{2}\Delta v(X_s^*)\}ds \\ + \int_0^t e^{-\alpha s} Dv(X_s^*) dB_s, \quad \text{a.s.} \end{aligned}$$

By (5.35), we have

$$E[e^{-\alpha(t\wedge\tau_n)} v(X_{t\wedge\tau_n}^*)] + E[\int_0^{t\wedge\tau_n} e^{-\alpha s}\{|X_s^*|^2 + |c_s^*|^2\}ds] = v(x),$$

where $\{\tau_n\}$ is a sequence of localizing stopping times $\tau_n \nearrow \infty$ of the local martingale. Note that $\{c_t^*\}$ belongs to $\mathcal{A}$ and

$$E[\sup_n v(X_{t\wedge\tau_n}^*)] \le C(1 + |x|^2 + t^2 + t).$$

Letting $n \to \infty$ and then $t \to \infty$, we get

$$J(c^*) = E[\int_0^\infty e^{-\alpha t}\{|X_t^*|^2 + |c_t^*|^2\}dt] = v(x).$$

For any $c \in \mathcal{A}$, by the same calculation as above, we get

$$J(c) = E[\int_0^\infty e^{-\alpha t}\{|X_t|^2 + |c_t|^2)\}dt] \ge v(x),$$

which implies the optimality of $\{o_t^*\}$. ∎

Part II

Applications to Mathematical Models in Economics

6

Production Planning and Inventory

In this chapter, we investigate a firm's decision problem regarding production and inventory. Let the demand ξ for the firm's product a constant. To meet this demand, the firm can produce the demanded quantity $p(t)$ as current input and draw down its inventory $x(t)$. In the deterministic case, these variables are related to each other by

$$\dot{x}(t) = p(t) - \xi.$$

Production cost $p(t)^2$ and storage cost $x(t)^2$ are needed for holding costs. The firm's objective is to minimize the total cost $\int_0^\infty e^{-\alpha t}\{x(t)^2 + p(t)^2\}dt$ with the discount rate $\alpha > 0$ over the class of productions. We analyze a stochastic extension of this production planning problem.

6.1 The Model

Consider a mathematical problem of the producton planning with random disturbances. Define the following quantities:

$X(t) =$ inventory level at time t.
$f(X(t)) =$ storage cost of inventory level $X(t)$ at time t.
$p(t) =$ production rate at time t; $p(t) \geq 0$ a.s.
$|p(t)|^k =$ production cost at time t; $k \geq 2$.
$\xi =$ the constant demand at time t; $\xi > 0$.
$B(t) =$ the standard Brownian motion.
$\sigma =$ the nonzero diffusion constant.

We assume that the inventory level $X(t)$ evolves according to the stochastic differential equation,

$$dX(t) = (p(t) - \xi)dt + \sigma dB(t), \qquad X(0) = x \in \mathbf{R}, \tag{6.1}$$

on a complete probability space $(\Omega, \mathcal{F}, P)$ endowed with the natural filtration $\mathcal{F}_t$ generated by $\sigma(B(s), s \leq t)$. We note that the formal derivative $dB(t)/dt$, called the *white noise process*, is interpreted as sales returns or inventory spoilage. The policy $p = \{p(t)\}$ is admissible if

$$p(t) \text{ is a nonnegative } \{\mathcal{F}_t\}\text{-progressively measurable process,}$$
$$E[\int_0^\infty e^{-\alpha t}|p(t)|^k dt] < \infty, \quad \text{for a fixed integer } k \geq 2.$$

We denote by $\mathcal{A}$ the set of admissible policies.

The objective is to find an optimal production $\{p^*(t)\}$ so as to minimize the cost

$$J_x(p) = E[\int_0^\infty e^{-\alpha t}\{f(X(t)) + |p(t)|^k\}\, dt],$$

over all policies $p \in \mathcal{A}$, where $\alpha > 0$ is a discount rate. We assume that f satisfies the polynomial growth conditions:

$$\begin{aligned} &0 \leq f(x) \leq c_0(1 + |x|^k), \quad x \in \mathbf{R}, \\ &c_1|x|^k - c_2 \leq f(x), \end{aligned} \tag{6.2}$$

for some constants $c_0, c_1, c_2 > 0$. The HJB equation associated with the optimization problem is given by

$$\alpha v(x) - \frac{1}{2}\sigma^2 v''(x) + \xi v'(x) - \min_{p\geq 0}(v'(x)p + |p|^k) = f(x) \quad \text{in } \mathbf{R}. \tag{6.3}$$

To show the existence of a classical solution v of (6.3), we make the the following assumption on f:

There exists a constant $C > 0$ such that

$$|f(x) - f(y)| \leq \frac{C}{\varepsilon}|x - y|^k + \varepsilon(1 + |x|^k + |y|^k), \quad x, y \in \mathbf{R}, \tag{6.4}$$

for any $\varepsilon > 0$.

6.2 Viscosity Solutions of the HJB Equations

We consider a viscosity solution of the HJB equation:

$$\begin{aligned} F(x, u, u', u'') &:= \alpha u(x) - \frac{1}{2}\sigma^2 u''(x) + \xi u'(x) - \min_{0\leq p\leq L}(u'(x)p + |p|^k) \\ &= f(x) \quad \text{in } \mathbf{R}, \end{aligned} \tag{6.5}$$

for each $L \geq 1$. Let u_L be the value function for the optimization problem

$$u_L(x) = \inf_{p\in\mathcal{A}_L} E[\int_0^\infty e^{-\alpha t}\{f(X(t)) + |p(t)|^k\}dt], \tag{6.6}$$

where the infimum is taken as in Chapter 3 Section 2 and $\mathcal{A}_L = \{p \in \mathcal{A} : 0 \leq p(t) \leq L \text{ for all } t \geq 0\}$.

Lemma 6.2.1. *We assume* (6.2) *and* (6.4). *Then we have*

(i) $0 \le u_{\bar{L}} \le u_L$ *for* $L \le \bar{L}$,
(ii) $u_L(x) \le \bar{C}(1+|x|^k)$ *for some constant* $\bar{C} > 0$,
(iii) u_L *has the property* (6.4) *with some constant* $C > 0$, *independent of* L, ε.

Proof.

(1) Since $\mathcal{A}_L \subset \mathcal{A}_{\bar{L}}$, (i) is immediate from (6.6).
(2) Let $\{Z(t)\}$ be the unique solution of

$$dZ(t) = -\xi dt + \sigma dB(t), \qquad Z(0) = x.$$

Then, by (6.2),

$$\begin{aligned} u_L(x) &\le E[\int_0^\infty e^{-\alpha t} f(Z(t))dt] \le c_0 E[\int_0^\infty e^{-\alpha t}(1+|Z(t)|^k)dt] \\ &\le c_0 E[\int_0^\infty e^{-\alpha t}(1+3^k(|x|^k + |\xi t|^k + |\sigma B(t)|^k))dt] \\ &\le \bar{C}(1+|x|^k), \end{aligned}$$

for some constant $\bar{C} > 0$. Thus we obtain (ii).
(3) We set $\nu = (\bar{C} + 1 + c_2/\alpha)/c_1$. Let $\bar{\mathcal{A}}_L$ be the class of all $p \in \mathcal{A}_L$ such that

$$E[\int_0^\infty e^{-\alpha t}|X(t)|^k dt] \le \nu(1+|x|^k).$$

There exists $p \in \mathcal{A}_L$, for any $n \in \mathbf{N}$, such that $J_x(p) \le u_L(x) + 1/n$. Hence, by (6.2) and (ii), we have

$$\begin{aligned} c_1 E[\int_0^\infty e^{-\alpha t}|X(t)|^k dt] - c_2/\alpha &\le E[\int_0^\infty e^{-\alpha t} f(X(t))dt] \le J_x(p) \\ &\le (\bar{C} + 1/n)(1+|x|^k). \end{aligned}$$

This shows that such p belongs to $\bar{\mathcal{A}}_L$, and thus

$$u_L(x) = \inf_{p \in \bar{\mathcal{A}}_L} J_x(p).$$

Let $\{Y_t\}$ be the solution to (6.1) with $Y_0 = y$. Then, by (6.4),

$$\begin{aligned} |u_L(x) - u_L(y)| &\le \sup_{p \in \bar{\mathcal{A}}_L} |J_x(p) - J_x(p)| \\ &\le \sup_{p \in \bar{\mathcal{A}}_L} E[\int_0^\infty e^{-\alpha t}|f(X(t)) - f(Y(t))|dt] \\ &\le \sup_{p \in \bar{\mathcal{A}}_L} E[\int_0^\infty e^{-\alpha t}\{\frac{C}{\varepsilon}|X(t) - Y(t)|^k \\ &\quad + \varepsilon(1 + |X(t)|^k + |Y(t)|^k)\}dt] \\ &\le \frac{C}{\alpha\varepsilon}|x-y|^k + \varepsilon(\frac{1}{\alpha} + \nu(2 + |x|^k + |y|^k)), \end{aligned}$$

which implies (iii). ■

Theorem 6.2.2. *We assume* (6.2) *and* (6.4). *Then* u_L *of* (6.6) *is a viscosity solution of* (6.5).

Proof.

(1) Under (6.2) and (6.4), we note by Lemma 6.2.1 (ii), (iii) that the properties of Proposition 3.3.1 are satisfied. Thus, by Theorem 3.3.2, we observe that the dynamic programming principle holds for $u := u_L$, that is,

$$u(x) = \inf_{p \in \mathcal{A}_L} E[\int_0^\tau e^{-\alpha t}\{f(X(t)) + |p(t)|^k\}dt + e^{-\alpha\tau}u(X(\tau))], \quad \tau \in \mathcal{S}_b,$$

where the infimum is taken over all systems $(\Omega, \mathcal{F}, P, \{\mathcal{F}_t\}; \{B(t)\}, \{p(t)\})$.

(2) We will review the proof of Theorem 4.3.1 to see the viscosity property of u_L for f with polynomial growth. Since

$$\begin{aligned} E[\sup_{0\le s\le h} |X(s) - X(0)|^k] &\le 2^{k-1}E[\sup_{0\le s\le h}\{(\int_0^s (p(t)-\xi)dt)^k + |\sigma B(s)|^k\}] \\ &\le 2^{k-1}\{(L+\xi)^k h^k + (\frac{k}{k-1})^k|\sigma|^k h^{k/2}\} \to 0, \end{aligned}$$

we have by Theorem 4.3.1 (2)

$$\begin{aligned} \frac{1}{h}E[h - h\wedge\rho] &\le \frac{1}{h}E[(h-\rho)1_{\{\rho\le h\}}] \le P(\rho \le h) \\ &\le P(\sup_{0\le s\le h}|X(s) - x| \ge d(x, \partial\mathcal{O})) \to 0 \text{ as } h \to 0, \end{aligned}$$

where $\rho = \rho_x$ denotes the first exit time of $\{X(t)\}$ from a bounded neighborhood $\mathcal{O}$ of $x \in \mathbf{R}$.

(3) Subsolution property. Let $\varphi \in C^2$ and z be the local maximum point of $u - \varphi$. Let $\{X(t)\}$ be the solution to (6.1) with $X(0) = z$ for any constant control $p(t) = p$ in $\mathcal{A}_L$. For $h > 0$ and $\rho = \rho_z$, the DPP gives

$$\begin{aligned} \varphi(z) = u(z) &\le E[\int_0^{h\wedge\rho} e^{-\alpha t}\{f(X(t)) + |p|^k\}dt + e^{-\alpha(h\wedge\rho)}u(X(h\wedge\rho))] \\ &\le E[\int_0^{h\wedge\rho} e^{-\alpha t}\{f(X(t)) + |p|^k\}dt + e^{-\alpha(h\wedge\rho)}\varphi(X(h\wedge\rho))]. \end{aligned}$$

By the same calculations as Theorem 4.3.1 (3), applying, Itô's formula to φ, we have

$$E[\int_0^{h\wedge\rho} e^{-\alpha t}\{f(x) + |p|^k - \alpha\varphi + \frac{1}{2}\sigma^2\varphi'' + (p-\xi)\varphi'\}\Big|_{x=X(t)} dt] \ge 0.$$

Dividing by h and letting $h \to 0$, by (6.4) and (2), we deduce

$$f(z) - \alpha\varphi(z) + \frac{1}{2}\sigma^2\varphi''(z) - \xi\varphi'(z) + (\varphi'(z)p + |p|^k) \ge 0.$$

Therefore,

$$F(z, u(z), \varphi'(z), \varphi''(z)) \le f(z).$$

(4) Supersolution property. Let $\varphi \in C^2$ and z be the local minimum point of $u - \varphi$. We take $\tau = h \wedge \rho$. By the DPP, there exists a weak solution $\{X^h(t)\}$ to (6.1) with $X^h(0) = z$ corresponding to $p^h \in \mathcal{A}_L$ on some probability space $(\Omega^h, \mathcal{F}^h, P^h, \{\mathcal{F}_t^h\})$ such that

$$\begin{aligned}\varphi(z) = u(z) &\geq E^h\Big[\int_0^\tau e^{-\alpha t}\{f(X^h(t)) + |p^h(t)|^k\}dt + e^{-\alpha\tau}u(X^h(\tau))\Big] - h^2\\ &\geq E^h\Big[\int_0^\tau e^{-\alpha t}\{f(X^h(t)) + |p^h(t)|^k\}dt + e^{-\alpha\tau}\varphi(X^h(\tau))\Big] - h^2.\end{aligned}$$

We subtract $\varphi(z)$ from both sides and apply Itô's formula to obtain

$$\begin{aligned}0 \geq \frac{1}{h}E^h\Big[&\int_0^\tau e^{-\alpha t}\{f(X^h(t)) + |p^h(t)|^k\}dt\\ &+ \int_0^\tau e^{-\alpha t}\{-\alpha\varphi(x) + \frac{1}{2}\sigma^2\varphi'' + (p^h(t) - \xi)\varphi'\Big|_{x=X^h(t)}\}dt\Big] - h.\end{aligned}$$

By (6.4) and (2), we have

$$\begin{aligned}\sup_{0\leq t\leq h} E^h[|f(X^h(t)) - f(z)|] &\leq E^h\Big[\sup_{0\leq t\leq h}\{\frac{C}{\varepsilon}|X^h(t) - z|^k\\ &\qquad + \varepsilon(1 + |X^h(t)|^k + |z|^k)\}\Big]\\ &\to 0 \quad \text{as} \quad h \to 0 \quad \text{and} \quad \varepsilon \to 0.\end{aligned}$$

By (2) and the uniform continuity of $\varphi, \varphi', \varphi''$ on $\bar{\mathcal{O}}$, we see that

$$\begin{aligned}0 \geq \liminf_{h\to 0}\frac{1}{h}E^h\Big[&\int_0^\tau e^{-\alpha t}\{f(z) + |p^h(t)|^k\}dt\\ &+ \int_0^\tau e^{-\alpha t}\{-\alpha\varphi(z) + \frac{1}{2}\sigma^2\varphi''(z) + (p^h(t) - \xi)\varphi'(z)\}dt\Big].\end{aligned}$$

Thus

$$\begin{aligned}0 &\geq \liminf_{h\to 0}\frac{1}{h}E^h\Big[\int_0^h e^{-\alpha t}\{f(z) + |p^h(t)|^k\}dt\\ &\qquad + \int_0^h e^{-\alpha t}\{-\alpha\varphi(z) + \frac{1}{2}\sigma^2\varphi''(z) + (p^h(t) - \xi)\varphi'(z))dt\Big]\\ &\geq \lim_{h\to 0}\frac{1}{h}\int_0^h e^{-\alpha t}\{-\alpha\varphi(z) + \frac{1}{2}\sigma^2\varphi''(z) - \xi\varphi'(z)\\ &\qquad + \min_{0\leq p\leq L}(p\varphi'(z) + |p|^k) + f(z)\}dt\\ &= -F(z, u(z), \varphi'(z), \varphi''(z)) + f(z).\end{aligned}$$

■

Theorem 6.2.3. *We assume* (6.2) *and* (6.4). *Then there exists a viscosity solution* v *of* (6.3).

Proof. Since $u_L \geq 0$ is nonincreasing, we can define v by $v(x) = \lim_{L\to\infty} u_L(x)$. By Lemma 6.2.1 (iii)

$$|v(x) - v(y)| \leq \frac{C}{\varepsilon}|x-y|^k + \varepsilon(1 + |x|^k + |y|^k), \quad x, y \in \mathbf{R},$$

and hence $v \in C(\mathbf{R})$. Thus, by Dini's theorem, we observe the locally uniform convergence of u_L. Since

$$\Psi(x) := \min_{p\geq 0}(xp + |p|^k) = (1-k)(\frac{x^-}{k})^{k/(k-1)},$$

and the minimum is attained by

$$\psi(x) = \left(\frac{x^-}{k}\right)^{1/(k-1)}, \qquad \text{for } x \in \mathbf{R},$$

we have

$$\min_{0\leq p\leq L}(\lambda p + |p|^k) \to \min_{p\geq 0}(\lambda p + |p|^k) \quad \text{locally uniformly in } \mathbf{R}.$$

Thus we deduce by Theorem 4.5.1 that v is a viscosity solution of (6.3). ■

6.3 Classical Solutions

We study the smoothness of a viscosity solution $u_L \in C(\mathbf{R})$ of (6.5) to obtain a classical solution v of (6.3). Furthermore, make the following assumption on f:

$$f : \mathbf{R} \to \mathbf{R}, \text{ convex.} \tag{6.7}$$

Theorem 6.3.1. *We assume* (6.2), (6.4), *and* (6.7). *Then we have* $u_L \in C^2(\mathbf{R})$.

Proof. Let $a < b$ be arbitrary. Consider the boundary value problem:

$$\begin{cases} F(x, w, w', w'') := -\frac{1}{2}\sigma^2 w'' + H(w') + \alpha w = f & \text{in } (a,b), \\ w(a) = u_L(a), \quad w(b) = u_L(b), \end{cases} \tag{6.8}$$

where $H(\lambda) = \xi\lambda - \min_{0\leq p\leq L}(\lambda p + |p|^k)$ for $\lambda \in \mathbf{R}$. Clearly, u_L is a viscosity solution of (6.8) in (a,b). Since f is Lipschitz on $[a,b]$, f satisfies the sufficient conditions of Theorem 5.3.7, and so does $H(\lambda)$. Hence, (6.8) admits a unique solution $w \in C^2(a,b)$. By the comparison principle (Theorem 5.3.1), we have $w = u_L$ in $[a,b]$ and thus $u_L \in C^2(\mathbf{R})$. ■

Theorem 6.3.2. *We assume* (6.2), (6.4), *and*(6.7). *Let $a < b$ be arbitrary. Then we have*

(i) $\|u_L\|_{C^1[a,b]} \leq C,$
(ii) $\|u_L\|_{C^2[a,b]} \leq C,$

where the constant $C > 0$ depends only on a, b.

Proof.

(1) By Lemma 6.2.1 (iii), we have

$$\left|\frac{u_L(x+h)-u_L(x)}{h}\right| \le \frac{C}{|h|^2}|h|^k + (1+|x+h|^k+|x|^k) \le C',$$

for $x \in [a,b]$, $|h| \le 1, h \neq 0$, where the constant $C' > 0$ depends only on a, b. Letting $h \to 0$, we have

$$|u_L'(x)| \le C', \quad x \in [a,b].$$

Therefore (i) is immediate from Lemma 6.2.1 (i).

(2) We rewrite (6.8) as

$$u_L'' = \frac{2}{\sigma^2}(H(u_L') + \alpha u_L - f).$$

It is easily seen that

$$0 \ge \min_{0\le p\le L}(u_L' p + |p|^k) \ge \min_{p\ge 0}(u_L' p + |p|^k) \ge (1-k)\left(\frac{|u_L'|}{k}\right)^{k/(k-1)}.$$

Hence

$$\|u_L\|_{C^2[a,b]} \le C(\||u_L'|^{k/(k-1)}\|_{C[a,b]} + \|f\|_{C[a,b]} + \|u_L\|_{C^1[a,b]}).$$

Thus, by (i), we obtain (ii). ■

Theorem 6.3.3. *We assume* (6.2), (6.4), *and* (6.7). *Then there exists a solution* $v \in C^2(\mathbf{R})$ *of* (6.3) *such that*

$$0 \le v(x) \le C(1+|x|^k), \quad x \in \mathbf{R}. \tag{6.9}$$

Proof. By Theorem 6.3.2, we apply the Ascoli–Arzelà theorem. Then, by extracting a subsequence, we find $v \in C^1(\mathbf{R})$ such that

$$u_L \to v, \quad u_L' \to v' \quad \text{locally uniformly in } \mathbf{R}.$$

By Theorem 6.3.2, we choose $L > 0$ such that

$$|u_L'(x)| \le C < kL^{k-1}, \quad x \in [a,b],$$

for any $a < b$. Hence, by an elementary calculation,

$$\begin{aligned}\min_{0\le p\le L}(u_L'(x)p + |p|^k) &= \Psi(u_L'(x)) \;\to\; \Psi(v'(x)) \\ &= \min_{p\ge 0}(v'(x)p + |p|^k), \quad x \in [a,b],\end{aligned}$$

and then

$$u_L'' = \frac{2}{\sigma^2}(H(u_L') + \alpha u_L - f) \to \frac{2}{\sigma^2}(\xi v' - \Psi(v') + \alpha v - f) \quad \text{on } [a,b].$$

Since

$$-\int v'\phi' dx = -\lim_{L\to\infty}\int u_L'\phi' dx = \lim_{L\to\infty}\int u_L''\phi dx$$
$$= \int \frac{2}{\sigma^2}\{\xi v' - \Psi(v') + \alpha v - f\}\phi dx \quad \text{for } \phi \in C_c^\infty(a,b),$$

we get

$$\frac{1}{2}\sigma^2 v'' = \xi v' - \Psi(v') + \alpha v - f \in L^2(a,b)$$

in the weak sense. Furthermore, choosing $\varphi_n \in C^2(\mathbf{R})$ such that $\varphi_n(x) \nearrow x^-$ and $\varphi_n'(x) \nearrow 1_{\{x\le 0\}}$, we note by the chain rule that

$$-\int \Psi_0(\varphi_n(u_L'))\phi' dx = \int \Psi_0'(\varphi_n(u_L'))\varphi_n'(u_L')u_L''\, \phi\, dx,$$

where $\Psi_0(x) = (1-k)(x/k)^{k/(k-1)}$. Letting $L \to \infty$ and $n \to \infty$, we have

$$-\int \Psi(v')\phi' dx = -\int \Psi_0((v')^-)\phi' dx = \int \Psi_0'((v')^-)1_{\{v'\le 0\}}v''\, \phi\, dx.$$

By differentiation and (6.7)

$$\frac{1}{2}\sigma^2 v''' = \xi v'' - \Psi_0'((v')^-)1_{\{v'\le 0\}}v'' + \alpha v' - f' \in L^2(a,b),$$

which implies $v \in W^{3,2}(a,b)$. By the Sobolev imbedding theorem, we deduce $v \in C^2[a,b]$ and thus $v \in C^2(\mathbf{R})$. Therefore, we see that v solves (6.3). Furthermore, by Lemma 6.2.1 (ii), v satisfies (6.9). ∎

6.4 Optimal Production Planning

Consider the equation,

$$dX^*(t) = [V(X^*(t)) - \xi]dt + \sigma dB(t), \quad X^*(0) = x, \tag{6.10}$$

where $V(x)$ is the minimizer of $\min_{p\ge 0}(v'(x)p + |p|^k)$, that is, $V(x) = \psi(v'(x))$. For the existence of $\{X^*(t)\}$, we need the following in Lemma 6.4.1.

Lemma 6.4.1. *We assume* (6.2), (6.4), *and* (6.7). *Then* u_L *of* (6.6) *is convex.*

Proof. For any $\varepsilon > 0$, there exist $p, \hat{p} \in \mathcal{A}_L$ such that

$$E[\int_0^\infty e^{-\alpha t}\{f(X(t)) + |p(t)|^k\}dt] < u_L(x) + \varepsilon,$$
$$E[\int_0^\infty e^{-\alpha t}\{f(\hat{X}(t)) + |\hat{p}(t)|^k\}dt] < u_L(\hat{x}) + \varepsilon,$$

on the same probability space, where $\{\hat{X}(t)\}$ is the solution of (6.1) corresponding to $\{\hat{p}(t)\}$ with $\hat{X}(0) = \hat{x}$. For $0 \le \gamma \le 1$, we define

$$\bar{X}(t) = \gamma X(t) + (1-\gamma)\hat{X}(t) \quad \text{and} \quad \bar{p} = \gamma p + (1-\gamma)\hat{p}.$$

Then

$$d\bar{X}(t) = (\bar{p}(t) - \xi)dt + \sigma dB(t), \quad \bar{X}(0) = \gamma x + (1-\gamma)\hat{x}.$$

Clearly, $\bar{p}$ belongs to $\mathcal{A}_L$. Hence, by the convexity of f and $|p|^k$,

$$\begin{aligned} u_L(\gamma x + (1-\gamma)\hat{x}) &\le E[\int_0^\infty e^{-\alpha t}\{f(\bar{X}(t)) + |\bar{p}(t)|^k\}dt] \\ &\le \gamma E[\int_0^\infty e^{-\alpha t}\{f(X(t)) + |p(t)|^k\}dt] \\ &\quad + (1-\gamma)E[\int_0^\infty e^{-\alpha t}\{f(\hat{X}(t)) + |\hat{p}(t)|^k\}dt] \\ &\le \gamma u_L(x) + (1-\gamma)u_L(\hat{x}) + \varepsilon. \end{aligned}$$

Letting $\varepsilon \to 0$, we obtain the convexity of u_L. ■

Lemma 6.4.2. *We assume* (6.2), (6.4), *and* (6.7). *Then there exists a unique solution* $\{X^*(t)\}$ *to* (6.10) *such that for any integer* $m \ge 2$

$$E[|X^*(t)|^m] \le |x|^m + Ct, \tag{6.11}$$

where the constant $C > 0$ *depends only on* m.

Proof.

(1) We claim that there exists $d \in \mathbf{R}$ such that $v'(d) = 0$. By Lemma 6.4.1 and Theorem 6.2.3, v is convex. Since v' is nondecreasing, the limit $v'(\infty)$ exists. Suppose $v'(\infty) < \infty$. Clearly,

$$v'(x)/x \to 0, \quad v(x)/x^2 \to 0 \quad \text{as } x \to \infty.$$

Hence $\int_0^x v''(x)dx/x \to 0$, so that

$$\liminf_{x\to\infty} v''(x) = 0.$$

Also, by (6.2), $f(x)/x^2$ does not coverge to 0 as $x \to \infty$. We rewrite (6.3) as

$$\frac{1}{2}\sigma^2\frac{v''(x)}{x^2} - \xi\frac{v'(x)}{x^2} + \frac{\Psi(v'(x))}{x^2} - \alpha\frac{v(x)}{x^2} = -\frac{f(x)}{x^2}.$$

Then, passing to the limit, we get a contradiction. Thus $v'(\infty) = \infty$. Similarly, $v'(-\infty) = -\infty$. Therefore, we obtain d as required.

(2) Let $\delta > 0$ and we approximate (6.10) by

$$dX_\delta(t) = [V_\delta(X_\delta(t)) - \xi]dt + \sigma dB(t), \quad X_\delta(0) = x, \tag{6.12}$$

where $V_\delta(x) = \psi_\delta(v'(x))$ and $\psi_\delta(x) = (\psi(x-\delta) - \psi(-\delta))^+$. It is clear that

$$\begin{cases} \psi_\delta(x): \ \text{locally Lipschitz, nonincreasing,} \\ \psi_\delta(x) \nearrow \psi(x) \quad \text{as } \delta \searrow 0, \end{cases}$$

and then

$$\begin{cases} V_\delta(x): \text{ locally Lipschitz, nonincreasing,} \\ V_\delta(x) \nearrow V(x) \quad \text{as } \delta \searrow 0. \end{cases} \tag{6.13}$$

By Theorem 1.9.1, there exists a solution $\{X_\delta(t)\}$ of (6.12) up to explosion time e. We set $Z_n(t) = X_\delta(t \wedge \rho_n)$ for $\rho_n = \inf\{t : |X_\delta(t)| \geq n\}, = \infty$ if $\{ \} = \phi$. Then

$$Z_n(t) = x + \int_0^t 1_{\{s \leq \rho_n\}}(V_\delta(Z_n(s)) - \xi)ds + \int_0^t 1_{\{s \leq \rho_n\}}\sigma dB(s), \quad \text{a.s.}$$

By (1), we have $(x - d)V(x) \leq 0$, so that $(x - d)V_\delta(x) \leq 0$. By Doob's maximal inequality, for $0 \leq r \leq T$,

$$\begin{aligned} &E[\sup_{0\leq t\leq r}(Z_n(t) - d)^2] - (x - d)^2 \\ &\leq E[\sup_{0\leq t\leq r}\int_0^t 1_{\{s\leq\rho_n\}}\{2(Z_n(s) - d)(V_\delta(Z_n(s)) - \xi) + \sigma^2\}ds] \\ &\quad + E[\sup_{0\leq t\leq r}|\int_0^t 1_{\{s\leq\rho_n\}}2(Z_n(s) - d)\sigma dB(s)|] \\ &\leq E[\int_0^r (Z_n(s) - d)^2 ds] + (\xi^2 + \sigma^2)r + 4|\sigma|(E[\int_0^r (Z_n(s) - d)^2 ds])^{1/2} \\ &\leq C_1 E[\int_0^r \sup_{0\leq t\leq s}(Z_n(t) - d)^2 ds] + C_2, \end{aligned}$$

where $C_1 = (1 + 4|\sigma|)$, $C_2 = (\xi^2 + \sigma^2)T + 4|\sigma|$. By Gronwall's inequality,

$$E[\sup_{0\leq t\leq T}(Z_n(t) - d)^2] \leq \{C_2 + (x - d)^2\}e^{C_1 T} =: C_3.$$

Hence

$$\begin{aligned} P(\rho_n \leq T) &= P(\sup_{0\leq t\leq T}|Z_n(t)| \geq n) \leq \frac{1}{n^2}E[\sup_{0\leq t\leq T}|Z_n(t)|^2] \\ &\leq 2(d^2 + C_3)/n^2 \to 0 \quad \text{as } n \to \infty \quad \text{and } T \to \infty. \end{aligned}$$

Therefore, $\{X_\delta(t)\}$ has no explosion, that is, $P(\mathrm{e} < \infty) = 0$. By the monotonicity of $V_\delta(x)$, $\{X_\delta(t)\}$ is a unique strong solution of (6.12). By the above calculation, we obtain

$$E[\sup_{0\leq t\leq T}(X_\delta(t) - d)^2] \leq C_3, \qquad \text{for each } T > 0. \tag{6.14}$$

(3) We show the existence of a unique solution $\{X^*(t)\}$ to (6.10). We set $Y_\delta(t) = X_\delta(t) - \sigma B(t)$. Then, by (6.12),

$$dY_\delta(t) = [V_\delta(Y_\delta(t) + \sigma B(t)) - \xi]dt, \quad Y_\delta(0) = x.$$

As in the proof of Theorem 1.10.1, using a localizing argument, we have

$$E[(Y_\delta(t) - Y_\varepsilon(t))^+] = E[\int_0^t 1_{\{Y_\delta(s)-Y_\varepsilon(s)\geq 0\}}\{V_\delta(Y_\delta(s) + \sigma B(s)) \\ - V_\varepsilon(Y_\varepsilon(s) + \sigma B(s))\}ds]$$

for $0 < \varepsilon < \delta$. Hence, by (6.13)

$$\begin{aligned} E[(Y_\delta(t) - Y_\varepsilon(t))^+] &\leq E[\int_0^t 1_{\{Y_\delta(s)-Y_\varepsilon(s)\geq 0\}}\{V_\delta(Y_\delta(s) + \sigma B(s)) \\ &\quad - V_\delta(Y_\varepsilon(s) + \sigma B(s))\}ds] \\ &\leq C_\delta E[\int_0^t 1_{\{Y_\delta(s)-Y_\varepsilon(s)\geq 0\}}|Y_\delta(s) - Y_\varepsilon(s)|ds] \\ &\leq C_\delta E[\int_0^t (Y_\delta(s) - Y_\varepsilon(s))^+ ds], \end{aligned}$$

where the constant $C_\delta > 0$ depends only on δ. Thus, by Gronwall's inequality

$$Y_\delta(t) - Y_\varepsilon(t) \leq 0, \quad t \geq 0, \quad \text{a.s.},$$

that is,

$$X_\delta(t) \leq X_\varepsilon(t), \quad t \geq 0, \quad \text{a.s.}$$

Define $X^\star(t) = \lim_{\delta\searrow 0} X_\delta(t)$ for each $t \geq 0$. By (6.14) and Fatou's lemma, we have

$$\sup_{0\leq t\leq T} E[X^\star(t)^2] < \infty, \qquad \text{for each } T > 0.$$

By (6.12) and (6.13)

$$x + \int_0^t [V_\delta(X_\varepsilon(s)) - \xi]ds + \sigma B(t) \leq X_\varepsilon(t) \leq x \\ + \int_0^t [V(X_\varepsilon(s)) - \xi]ds + \sigma B(t), \quad \text{a.s.}$$

Letting $\varepsilon \searrow 0$ and then $\delta \searrow 0$, we get

$$X^\star(t) = x + \int_0^t [V(X^\star(s)) - \xi]ds + \sigma B(t), \quad \text{a.s.},$$

for each $t \geq 0$. Therefore, we deduce that $\{X^\star(t)\}$ has a continuous modification $\{X^*(t)\}$, which satisfies (6.10).

(4) We set $g(x) = |x|^m$ and $G(x) = g'(x)(V(x) - \xi) + \frac{1}{2}\sigma^2 g''(x)$. We recall Young's inequality,

$$|ab| \leq \frac{|a|^p}{p} + \frac{|b|^q}{q} \quad \text{for} \quad 1 < p, q < \infty, \ \frac{1}{p} + \frac{1}{q} = 1.$$

Then we find a constant $C_\varepsilon > 0$, depending on $\varepsilon > 0$, such that

$$\frac{1}{2}\sigma^2 g''(x) = \frac{1}{2}\sigma^2 m(m-1)|x|^{m-2} \le m\varepsilon|x|^{m-1} + C_\varepsilon.$$

Taking $\varepsilon = \xi/2$, we have for $x \ge d \vee 0$,

$$\begin{aligned} G_0(x) &:= g'(x)(V(x)-\xi) + m\varepsilon|x|^{m-1} \\ &= m|x|^{m-1}\{sgn(x)(V(x)-\xi)+\varepsilon\} \\ &= m|x|^{m-1}(\varepsilon-\xi) \le 0. \end{aligned}$$

Also, choosing $c < 0$ such that $V(c) \ge \xi + \varepsilon$, we have for $x < c$

$$\begin{aligned} G_0(x) &= m|x|^{m-2}x(V(x)-\xi) + m\varepsilon|x|^{m-1} \\ &= m|x|^{m-2}x(V(x)-\xi-\varepsilon) \le 0. \end{aligned}$$

Hence $G_0(x)$ is bounded above and so is $G(x)$. Thus, by Itô's formula,

$$E[g(Z_n(t))] = g(x) + E[\int_0^t G(Z_n(s))ds] \le |x|^m + Ct$$

for some constant $C > 0$. Letting $n \to \infty$ and $\delta \searrow 0$, we obtain (6.11). ■

Theorem 6.4.3. *We assume* (6.2), (6.4), *and* (6.7). *Then the optimal production rate* $p^* = \{p^*(t)\}$ *is given by*

$$p^*(t) = V(X^*(t)).$$

Proof.

(1) Let $v \in C^2(\mathbf{R})$ be the solution of (6.3) satisfying (6.9). Then

$$\frac{1}{2}\sigma^2 v''(x) - \xi v'(x) + (v'(x)V(x) + |V(x)|^k) - \alpha v(x) = -f(x), \quad x \in \mathbf{R}.$$

By Itô's formula,

$$\begin{aligned} e^{-\alpha t}v(X^*(t)) - v(x) &= \int_0^t e^{-\alpha s}\{-\alpha v(x) + v'(x)(V(x)-\xi) \\ &\quad + \frac{1}{2}\sigma^2 v''(x)\}\Big|_{x=X^*(s)} ds + M_t, \\ &= -\int_0^t e^{-\alpha s}\{f(X^*(s)) + |V(X^*(s))|^k\}ds + M_t, \\ &\qquad t \ge 0, \quad \text{a.s.}, \end{aligned}$$

where $M_t = \int_0^t e^{-\alpha t} v'(X^*(t))\sigma dB_t$. Let $\{\tau_n\}$ be a sequence of localizing stopping times $\tau_n \uparrow \infty$ of the local martingale $\{M_t\}$. Since $E[M_{t\wedge\tau_n}] = E[M_0] = 0$, we have

$$E[e^{-\alpha(t\wedge\tau_n)} v(X^*(t\wedge\tau_n))] = v(x) - E[\int_0^{t\wedge\tau_n} e^{-\alpha s}\{f(X^*(s)) + |V(X^*(s))|^k\}ds].$$

By virtue of (6.11)

$$\begin{aligned} E[\sup_{0\le t\le T} |X^*(t)|^m] &\le |x|^m + E[\sup_{0\le t\le T} \int_0^t G(X^*(s))ds] \\ &+ E[\sup_{0\le t\le T} |\int_0^t m|X^*(s)|^{m-1} sgn(X^*(s))\sigma dB(s)|] \\ &\le (|x|^m + CT) + 2m|\sigma|(E[\int_0^T |X^*(t)|^{2(m-1)}dt])^{1/2} < \infty, \quad \forall m \ge 2. \end{aligned}$$

Hence

$$E[\sup_{0\le t\le T} e^{-\alpha t}|X^*(t)|^m] \le E[\sup_{0\le t\le T} |X^*(t)|^m] < \infty, \qquad \text{for each } T > 0.$$

Thus, by (6.9)

$$E[\sup_n e^{-\alpha(t\wedge\tau_n)} v(X^*(t\wedge\tau_n))] \le CE[\sup_{0\le s\le t} e^{-\alpha s}(1 + |X^*(s)|^k)] < \infty.$$

Letting $n \to \infty$, by the dominated convergence theorem, we get

$$E[e^{-\alpha t} v(X^*(t))] = v(x) - E[\int_0^t e^{-\alpha s}\{f(X^*(s)) + |V(X^*(s))|^k\}ds].$$

By (6.9) and (6.11), we have $\lim_{t\to\infty} E[e^{-\alpha t} v(X^*(t))] = 0$. Therefore,

$$v(x) = E[\int_0^\infty e^{-\alpha s}\{f(X^*(s)) + |p^*(s)|^k\}ds = J_x(p^*),$$

and hence p^* belongs to $\mathcal{A}$.

(2) By (6.3) we have

$$\frac{1}{2}\sigma^2 v''(x) - \xi v'(x) + (v'(x)p + |p|^k) - \alpha v(x) \ge -f(x), \ \ \forall x \in \mathbf{R}, p \ge 0.$$

Let $p = \{p(t)\} \in \mathcal{A}$ be arbitrary. Since

$$E[\int_0^T |p(t)|^k dt] \le e^{\alpha T} E[\int_0^T e^{-\alpha t}|p(t)|^k dt] < \infty,$$

we get

$$E[\sup_{0\le t\le T}|X(t)|^k] \le 2^{k-1}E[\sup_{0\le t\le T}|\int_0^t (p(s)-\xi)ds|^k + \sup_{0\le t\le T}|\sigma B(t)|^k]$$
$$\le 2^{k-1}\{E[\int_0^T |p(t)-\xi|^k dt]\cdot T^{k-1}$$
$$+(\frac{k}{k-1})^k|\sigma|^k T^{k/2}\} < \infty.$$

By the same line as (1), we have

$$E[e^{-\alpha t}v(X(t))] \ge v(x) - E[\int_0^t e^{-\alpha s}\{f(X(s)) + |p(s)|^k\}ds].$$

We may consider $J_x(p) < \infty$ and by (6.2),

$$E[\int_0^\infty e^{-\alpha t}|X(t)|^k dt] \le (E[\int_0^\infty e^{-\alpha t} f(X(t))dt] + c_2/\alpha)/c_1 < \infty.$$

Hence, by a contradiction argument, we see that

$$\liminf_{t\to\infty} E[e^{-\alpha t}|X(t)|^k] = 0.$$

Therefore,

$$0 \ge v(x) - \limsup_{t\to\infty} E[\int_0^t e^{-\alpha s}\{f(X(s)) + |p(s)|^k\}ds],$$

that is,

$$v(x) \le E[\int_0^\infty e^{-\alpha s}\{f(X(s)) + |p(s)|^k\}ds] = J_x(p).$$

Combining (1) with this relation, we conclude that p^* is optimal.

(3) By (1) and (2), we notice that v has a representation, that is,

$$v(x) = \inf_{p\in\mathcal{A}} J_x(p).$$

This implies the uniqueness of the solution of (6.3) in C^2 satisfying (6.9).

■

7

Optimal Consumption/Investment Models

This chapter is concerned with the consumption and portfolio selection problems formulated by R. C. Merton [112] in mathematical finance. We consider the optimization problems to maximize the expected utility function with respect to consumption rates $c(t)$ and portfolio processes $\pi(t)$ for the vector process $S(t)$ of prices of N risky assets $S_1(t), \ldots, S_N(t)$ and general utility functions $U(c)$. We try to find a connection between the HJB equations and the linear differential equations by using the Legendre transform. We study the existence of the smooth solutions of the HJB equations from the point of view of viscosity solutions.

7.1 The Model

Consider the consumption and portfolio selection problem. Define the following quantities:

$S_0(t) =$ price of the riskless asset at time t.
$S(t) =$ vector process of prices of the N risky assets $S_1(t), \ldots, S_N(t)$ at time t.
$B(t) =$ the N-dimensional standard Brownian motion.
$X(t) =$ total wealth at time t.
$\pi(t) =$ fraction of wealth in the risky assets at time t.
$r =$ return on the riskless asset.
$b =$ vector of expected return on the N risky assets.
$\sigma =$ volatility $N \times N$ matrix of risky assets, $\sigma \neq 0$.
$c(t) =$ consumption rate at time t.
$U(c) =$ the utility function.

We assume that

$$\begin{aligned} dS_0(t) &= rS_0(t)dt, \\ dS(t) &= \mathrm{diag}(S_1(t), \ldots, S_N(t))\{bdt + \sigma dB(t)\}, \quad t \geq 0, \end{aligned}$$

on a complete probability space $(\Omega, \mathcal{F}, P)$ endowed with the natural filtration $\mathcal{F}_t$ generated by $\sigma(B(s), s \le t)$. The change in wealth is given by

$$\frac{dX(t)}{X(t)} = \{1 - \sum_{i=1}^{N} \pi_i(t)\}\frac{dS_0(t)}{S_0(t)} + \cdots + \pi_N(t)\frac{dS_N(t)}{S_N(t)},$$

if $\pi_i(t), 1 - \sum_{i=1}^{N} \pi_i(t) \ge 0$ for $\pi(t) = (\pi_1(t), \ldots, \pi_N(t))^\top$. This can be rewritten as

$$dX(t) = rX(t)dt + \pi(t)^\top X(t)[(b - r\mathbf{1})dt + \sigma dB(t)], \quad X(0) = x > 0,$$

where $\pi(t)^\top$ is the transpose of $\pi(t)$ and $\mathbf{1}$ denotes the N-dimensional vector of ones.

Given a policy $\{c(t), \pi(t)\}$, the wealth process $\{X_t\}$ evolves according to the one-dimensional stochastic differential equation,

$$\begin{aligned} dX(t) &= [rX(t) - c(t)]dt + \pi(t)^\top (b - r\mathbf{1})dt \\ &\quad + \pi(t)^\top \sigma dB(t), \quad X(0) = x > 0. \end{aligned} \tag{7.1}$$

The policy $\{c(t), \pi(t)\}$ is admissible if

$\{c(t)\}$ is $\{\mathcal{F}_t\}$-progressively measurable, nonnegative, $\int_0^T c(s)ds < \infty$ a.s. for each $T > 0$,
$\{\pi(t)\}$ is $\{\mathcal{F}_t\}$-progressively measurable, $\mathbf{R}^N$-valued, $\int_0^T |\pi(s)|^2 ds < \infty$ a.s. for each $T > 0$,
$X(t) \ge 0$, for all $t \ge 0$ a.s.

We denote by $\mathcal{A}$ the set of all admissible policies $\{c(t), \pi(t)\}$.

The objective is to find an optimal policy $\{c^*(t), \pi^*(t)\}$ so as to maximize the expected utility function with discount factor $\alpha > 0$:

$$J(c, \pi) = E[\int_0^{\theta(X)} e^{-\alpha t} U(c(t))dt], \tag{7.2}$$

over $(c, \pi) = \{c(t), \pi(t)\} \in \mathcal{A}$ for the exit time $\theta(X) = \inf\{t \ge 0 : X(t) = 0\}$ of $\{X(t)\}$ from $(0, \infty)$. We assume that the utility function $U(c)$ possesses the following properties:

$$\begin{cases} U(c) \text{ is a } C[0, \infty) \text{ and } C^2(0, \infty)\text{-function}, \\ U''(c) < 0, \\ U'(0+) = U(\infty) = \infty, \quad U'(\infty) = U(0) = 0, \\ 0 \le U(c) \le K(c) := C_0(1 + c^\gamma) \quad \text{for some constants } C_0 > 0, \quad 0 < \gamma < 1. \end{cases} \tag{7.3}$$

Then $U'(x)$ has the inverse $I(x) := (U')^{-1}(x)$, which fulfills

$$\begin{cases} I(0+) = \infty, \quad I(\infty) = 0, \\ \tilde{U}'(x) = -I(x), \\ \tilde{U}''(x) = -I'(x) = -1/U''(I(x)) > 0, \end{cases} \tag{7.4}$$

where $\tilde{U}(x) = \max_{c>0}\{U(c) - cx\}, x > 0$.

7.2 HARA Utility

Before going into details, we give the example, based on work by Merton, concerning the consumption/investment models in the case when $N = 1$ and $U(c)$ is the HARA utility function:

$$U(c) = c^\gamma/\gamma, \quad 0 < \gamma < 1.$$

Consider the optimization problem (7.2) subject to (7.1). The HJB equation associated with this problem is given by

$$\begin{cases} -\alpha v + \max_{\pi \in \mathbf{R}} \left[\frac{1}{2}\pi^2\sigma^2 v'' + \pi(b-r)v' \right] + rxv' + \tilde{U}(v') = 0, & x > 0, \\ v(0) = 0. \end{cases}$$

Proposition 7.2.1. *We assume*

$$\alpha > r\gamma + \frac{(b-r)^2\gamma}{2\sigma^2(1-\gamma)}.$$

Then the HJB equation has a solution of the form:

$$v(x) = hx^\gamma$$

for a suitable choice of $h > 0$.

Proof. Substitute $v(x) = hx^\gamma$ into the HJB equation. Note that

$$\max_\pi[\ldots] = -\frac{(b-r)^2(v')^2}{2\sigma^2 v''} = \frac{(b-r)^2 h\gamma}{2\sigma^2(1-\gamma)} x^\gamma,$$

$$\text{Maximizer } \Pi_m(x) = -\frac{(b-r)v'}{\sigma^2 v''} = \frac{(b-r)}{\sigma^2(1-\gamma)} x,$$

and also

$$\max_{c \geq 0}[\ldots] = \left((v')^{1/(\gamma-1)} \right)^\gamma /\gamma - (v')^{1/(\gamma-1)} v' = (v')^{\gamma/(\gamma-1)}(1/\gamma - 1)$$
$$= x^\gamma (h\gamma)^{1/(\gamma-1)}(1-\gamma)h,$$

$$\text{Maximizer } C_m(x) = (v')^{1/(\gamma-1)} = (h\gamma)^{1/(\gamma-1)} x.$$

Then, dividing both sides by x^γ and h, we have

$$-\alpha + r\gamma + \frac{(b-r)^2\gamma}{2\sigma^2(1-\gamma)} + (h\gamma)^{1/(\gamma-1)}(1-\gamma) = 0,$$

that is,

$$h\gamma = ([\alpha - r\gamma - \frac{(b-r)^2\gamma}{2\sigma^2(1-\gamma)}]/(1-\gamma))^{\gamma-1}.$$

■

Proposition 7.2.2. *We make the assumption of Proposition 7.2.1. Then the optimal policy* $\{c^*(t), \pi^*(t)\}$ *is given by*

$$c^*(t) = C_m(X^*(t)) = (h\gamma)^{1/(\gamma-1)}X^*(t), \quad \pi^*(t) = \Pi_m(X^*(t)) = \frac{(b-r)}{\sigma^2(1-\gamma)}X^*(t),$$

where

$$\begin{aligned} dX^*(t) &= [rX^*(t) - C_m(X^*(t))]dt + \Pi_m(X^*(t))[(b-r)dt + \sigma dB_t], \\ &= \left(r - (h\gamma)^{1/(\gamma-1)} + \frac{(b-r)^2}{\sigma^2(1-\gamma)}\right)X^*(t)dt + \frac{(b-r)}{\sigma(1-\gamma)}X^*(t)dB(t), \\ &\qquad X^*(0) = x > 0. \end{aligned}$$

Proof. By the HJB equation, we have

$$-\alpha v + \left[\frac{1}{2}\pi^2\sigma^2 v'' + \pi(b-r)v'\right] + rxv' + [-cv'] \le -U(c), \quad \forall \pi, c,$$

$$-\alpha v + \left[\frac{1}{2}\Pi_m^2\sigma^2 v'' + \Pi_m(b-r)v'\right] + rxv' + [U(C_m) - C_m v'] = 0.$$

By Itô's formula,

$$\begin{aligned} e^{-\alpha T}v(X(T)) = v(x) + \int_0^T & e^{-\alpha t}(-\alpha v + [(rx - c(t)) + \pi(t)(b-r)]v' \\ &+ \frac{1}{2}\pi(t)^2\sigma^2 v'')|_{x=X(t)}dt + \int_0^T e^{-\alpha t}v'(X_t)\pi(t)\sigma dB(t), \quad a.s. \end{aligned}$$

Since the last term is a local martingale, we see by the localizing argument that

$$E[e^{-\alpha T\wedge\theta}v(X(T\wedge\theta)) + \int_0^{T\wedge\theta} e^{-\alpha t}U(c(t))dt] \le v(x).$$

Passing to the limit, we get

$$E[\int_0^\theta e^{-\alpha t}U(c(t))dt] \le v(x).$$

Recall by the HJB equation that

$$-\alpha v + \left[\frac{1}{2}\Pi_m^2\sigma^2 v'' + \Pi_m(b-r)v'\right] + rxv' + [U(C_m) - C_m v']\Big|_{x=X^*(t)} = 0.$$

It is easy to see that $\{\int_0^t e^{-\alpha s}v'(X^*(s))\pi^*(s)\sigma dB(s)\}$ is a martingale. By Itô's formula,

$$\begin{aligned} E[e^{-\alpha T}v(X^*(T))] &= v(x) + E[\int_0^T e^{-\alpha t}(-\alpha v + [(rx - C_m) + \Pi_m(b-r)]v' \\ &\quad + \frac{1}{2}\Pi_m^2\sigma^2 v'')|_{x=X^*(t)}dt] \\ &= v(x) - E[\int_0^T e^{-\alpha t}U(C_m(X^*(t)))dt]. \end{aligned}$$

Passing to the limit, we have

$$v(x) = E[\int_0^\infty e^{-\alpha t} U(c^*(t))dt].$$

It is clear that $X^*(t) > 0$ a.s., and $\{c^*(t), \pi^*(t)\}$ belongs to $\mathcal{A}_{ad}$.

To complete the proof, it suffices to show that

$$\lim_{T\to\infty} E[e^{-\alpha T}(X^*(T))^\gamma] = 0.$$

Indeed, we apply Itô's formula to

$$dX^*(t) = \left(r - (h\gamma)^{1/(\gamma-1)} + \frac{(b-r)^2}{\sigma^2(1-\gamma)}\right) X^*(t)dt + \frac{(b-r)}{\sigma(1-\gamma)} X^*(t)dB(t).$$

Then

$$E[e^{-\alpha T}(X^*(T))^\gamma] = x^\gamma + E[\int_0^T e^{-\alpha t} G(X^*(t))dt],$$

where

$$\begin{aligned}
G(x) &= -\alpha x^\gamma + \gamma x^{\gamma-1}\left(r - (h\gamma)^{1/(\gamma-1)} + \frac{(b-r)^2}{\sigma^2(1-\gamma)}\right)x \\
&\quad + \frac{1}{2}\gamma(\gamma-1)x^{\gamma-2}\frac{(b-r)^2}{\sigma^2(1-\gamma)^2}x^2 \\
&= [-\alpha + r\gamma - \gamma(h\gamma)^{1/(\gamma-1)} + \frac{1}{2}\gamma\frac{(b-r)^2}{\sigma^2(1-\gamma)}]x^\gamma \\
&= [-(1-\gamma)(h\gamma)^{1/(\gamma-1)} - \gamma(h\gamma)^{1/(\gamma-1)}]x^\gamma \quad \text{(by Proposition 7.2.1)} \\
&= [-(h\gamma)^{1/(\gamma-1)}]x^\gamma.
\end{aligned}$$

This yields that

$$E[e^{-\alpha T}(X^*(T))^\gamma] = x^\gamma exp\{-(h\gamma)^{1/(\gamma-1)}T\} \to 0.$$ ■

7.3 HJB Equations

In this section, returning to the general case, we study the HJB equation associated with the optimization problem:

$$\begin{cases} -\alpha v + \max_{\pi\in\mathbf{R}^N}[\frac{1}{2}|\sigma^\top\pi|^2 v'' + \pi^\top(b - r\mathbf{1})v'] + rxv' + \tilde{U}(v') = 0, \quad x > 0, \\ v(0) = 0. \end{cases} \tag{7.5}$$

Let $\tilde{v}(x)$ be the Legendre transform of $-v(-x)$, that is, $\tilde{v}(x) = \max_{y>0}\{v(y) - yx\}$. Suppose that (7.5) has a solution $v \in C^2(0,\infty)$ with $v'' < 0$, and there is y_x

attaining the maximum of $\tilde{v}(x)$. Formally, we have

$$\begin{aligned}\tilde{v}(x) &= v(y_x) - y_x x, \quad v'(y_x) = x,\\ \tilde{v}'(x) &= v'(y_x)y_x' - (y_x' x + y_x) = -y_x,\\ \tilde{v}''(x) &= -y_x' = -\frac{1}{v''(y_x)} > 0.\end{aligned}$$

We assume that

$$\begin{cases}\sigma \text{ is nonsingular,}\\ \nu := |\sigma^{-1}(b - r\mathbf{1})| \neq 0.\end{cases} \tag{7.6}$$

Then

$$\begin{aligned}\max_{\pi\in\mathbf{R}^N}[\frac{1}{2}|\sigma^\top\pi|^2 v'' + \pi^\top(b - r\mathbf{1})v'] &= \frac{v''}{2}\min_{\pi\in\mathbf{R}^N}[|\sigma^\top\pi + \frac{v'}{v''}\sigma^{-1}(b - r\mathbf{1})|^2\\ &\quad - |\sigma^{-1}(b - r\mathbf{1})|^2(\frac{v'}{v''})^2]\\ &= -\frac{\nu^2}{2}\frac{(v')^2}{v''}.\end{aligned}$$

Thus (7.5) can be reduced to

$$-\alpha\tilde{v}(x) + \frac{\nu^2}{2}x^2\tilde{v}''(x) + (\alpha - r)x\tilde{v}'(x) + \tilde{U}(x) = 0, \quad x > 0.$$

Now, we consider to find a solution u of

$$-\alpha u(x) + \frac{\nu^2}{2}x^2 u''(x) + (\alpha - r)xu'(x) + \tilde{U}(x) = 0, \quad x > 0. \tag{7.7}$$

Theorem 7.3.1. *Assume* (7.3) *and* (7.6). *Furthermore, we assume*

$$\alpha > r\gamma + \frac{\nu^2}{2}(\frac{\gamma}{1-\gamma}). \tag{7.8}$$

Then there exists a solution $u \in C^2(0,\infty)$ *of* (7.7) *such that*

(i) $0 \le u(x) \le C_1(1 + x^{-\gamma/(1-\gamma)})$, *for some constant* $C_1 > 0$,
(ii) $u(\infty) = 0, \quad u(0+) = \infty$,
(iii) $u''(x) > 0, x > 0$,
(iv) $u'(x) < 0$.

Proof.

(1) Let $\{Z(t)\}$ be the solution of

$$dZ(t) = (\alpha - r)Z(t)dt + \nu Z(t)dB_1(t), \quad Z(0) = x > 0,$$

where $\{B_1(t)\}$ is a one-dimensional standard Brownian motion. Recall that $Z(t) = x\exp\{(\alpha - r)t + \nu B_1(t) - \nu^2 t/2\}$. Then

$$E[e^{-\alpha t}Z(t)^{-\gamma/(1-\gamma)}] = x^{-\gamma/(1-\gamma)}\exp\{(-\alpha + r\gamma + \frac{1}{2}\nu^2\frac{\gamma}{1-\gamma})\frac{1}{1-\gamma}t\}. \tag{7.9}$$

Hence, by (7.8)

$$E[\int_0^\infty e^{-\alpha t} Z(t)^{-\gamma/(1-\gamma)} dt] < \infty.$$

Define

$$u(x) = E[\int_0^\infty e^{-\alpha t} \tilde{U}(Z(t)) dt]. \tag{7.10}$$

By a simple calculation, we get

$$0 \leq \tilde{U}(x) \leq \tilde{K}(x) = C_0 + C_0(1-\gamma)(C_0\gamma)^{\gamma/(1-\gamma)} x^{-\gamma/(1-\gamma)},$$

which implies (i). Also, for any $0 < a < b$

$$E[\int_0^\infty e^{-\alpha t} \sup_{x \in [a,b]} \tilde{U}(Z(t)) dt] < \infty.$$

By the dominated convergence theorem, u is continuous on $(0, \infty)$. By the strong Markov property,

$$\begin{aligned} u(x) &= E[\int_0^\tau e^{-\alpha t} \tilde{U}(Z(t)) dt + E[\int_\tau^\infty e^{-\alpha t} \tilde{U}(Z(t)) dt | \mathcal{F}_\tau]] \\ &= E[\int_0^\tau e^{-\alpha t} \tilde{U}(Z(t)) dt + e^{-\alpha \tau} u(Z(\tau))] \end{aligned}$$

for any stopping time $\tau \in \mathcal{S}_b$. This implies that the dynamic programming principle holds for u, where the class of admissible policies is interpreted as $\{0\}$. In view of Chapter 4, u is a viscosity solution of (7.7). Consider the boundary value problem:

$$\begin{cases} -\alpha w(x) + \dfrac{\nu^2}{2} x^2 w''(x) + (\alpha - r) x w'(x) + \tilde{U}(x) = 0, & \text{in } (a, b), \\ w(a) = u(a), \quad w(b) = u(b). \end{cases} \tag{7.11}$$

Since $\tilde{U}$ is convex, we see by Theorem 5.3.7 that (7.11) has a unique solution $w \in C[a, b] \cap C^2(a, b)$. By Theorem 5.3.1, we obtain the uniqueness result on (7.11) in the viscosity sense. Therefore, we deduce $u = w \in C^2(a, b)$ and hence $u \in C^2(0, \infty)$.

(2) By (7.3) and (7.4), we recall that

$$\tilde{U}(0+) \geq \sup_{c>0} U(c) = \infty,$$

$$\lim_{x \to \infty} \tilde{U}(x) = \lim_{x \to \infty} \{U(I(x)) - x I(x)\} = 0.$$

By (7.10) and the monotonicity of $\tilde{U}(x)$, we obtain (ii).

(3) By (7.10), u is convex. Suppose $u''(x_0) = 0$ for some $x_0 > 0$. Then, $\zeta(x) := \frac{\nu^2}{2} x^2 u''(x)$ attains its minimum at x_0. By (7.4), (7.7) and (1), $u'' \in C^2(0, \infty)$. Hence

$$0 = \zeta'(x_0) = \frac{\nu^2}{2} \{2 x_0 u''(x_0) + x_0^2 u'''(x_0)\},$$

which yields $u'''(x_0) = 0$. From (7.7) it follows that

$$0 \le \zeta''(x_0) = -\tilde{U}''(x_0) + \alpha u''(x_0) - (\alpha - r)\{x_0 u'''(x_0) + 2u''(x_0)\},$$

and thus

$$\tilde{U}''(x_0) \le 0.$$

This is contrary to (7.4). Therefore we get (iii).

(4) Suppose $u'(\bar{x}) \ge 0$ for some $\bar{x} > 0$. Then, by (iii), $u'(x) > 0$ for all $x > \bar{x}$. This contradicts (ii). Therefore, we obtain (iv). ∎

Theorem 7.3.2. *We make the assumptions of Theorem 7.3.1. Let $u_i \in C^2(0, \infty)$, $i = 1, 2$, be two solutions of (7.7) such that*

$$0 \le u_i(x) \le C_2(1 + x^{-\gamma/(1-\gamma)}) \tag{7.12}$$

for some constant $C_2 > 0$. Then we have

$$u_1 = u_2.$$

Proof. By (7.7) and Itô's formula,

$$\begin{aligned} e^{-\alpha t} u_i(Z(t)) &= u_i(x) + \int_0^t e^{-\alpha s}\{-\alpha u_i(x) + \frac{\nu^2}{2} x^2 u_i''(x) \\ &\quad + (\alpha - r) x u_i'(x)\}|_{x=Z(t)} ds + M(t) \\ &= u_i(x) - \int_0^t e^{-\alpha s} \tilde{U}(Z(s)) ds + M(t), \quad \text{a.s.,} \end{aligned}$$

and hence

$$u_i(x) = E[e^{-\alpha(t\wedge\tau_n)} u_i(Z(t \wedge \tau_n)) + \int_0^{t\wedge\tau_n} e^{-\alpha s} \tilde{U}(Z(s)) ds],$$

where $\{\tau_n\}$ is a sequence of localizing stopping times $\tau_n \nearrow \infty$ of the local martingale $M(t) := \nu \int_0^t e^{-\alpha s} Z(t) u_i'(Z(s)) dB_1(s)$. By (7.8), we get

$$\begin{aligned} &x^{2\eta} E[(e^{-\alpha(t\wedge\tau_n)} Z(t \wedge \tau_n)^{-\eta})^2] \\ &= E[\exp\{(-\alpha + r\gamma + \frac{\nu^2}{2}\gamma)\frac{2\eta}{\gamma}(t \wedge \tau_n) - 2\nu\eta B(t \wedge \tau_n)\}] \\ &\le E[\exp\{-2\nu\eta B(t \wedge \tau_n)\}] \\ &\le E[\exp\{-2\nu\eta B(t \wedge \tau_n) - 2\nu^2\eta^2(t \wedge \tau_n)\} \exp\{2\nu^2\eta^2 t\}] \\ &= \exp\{2\nu^2\eta^2 t\}, \end{aligned}$$

where $\eta := \gamma/(1-\gamma)$. Hence, by (7.12), $\{e^{-\alpha(t\wedge\tau_n)} u_i(Z(t \wedge \tau_n))\}$ is uniformly integrable. Letting $n \to \infty$, we have

$$u_i(x) = E[e^{-\alpha t} u_i(Z(t)) + \int_0^t e^{-\alpha s} \tilde{U}(Z(s)) ds].$$

Passing to the limit, by (7.9) and (7.10), we obtain $u_i = u$. ∎

Theorem 7.3.3. *We make the assumptions of Theorem 7.3.1. Then there exists a solution $v \in C[0,\infty) \cap C^2(0,\infty)$ of (7.5) such that*

(i) $0 \le v(x) \le C_3(1+x^\gamma)$, *for some constant* $C_3 > 0$,
(ii) $v''(x) < 0, \; x > 0$,
(iii) $v'(x) > 0$.

Proof. Let u be the solution of (7.7) given in Theorem 7.3.1 and we set $v(x) = \inf_{y>0}\{u(y) + yx\}$ for $x \ge 0$. We notice that v is nondecreasing and

$$v(0) = v(0+) \le \inf_{y>0} u(y) = 0.$$

By Theorem 7.3.1, we have

$$0 \le v(x) \le C_1 + \inf_{y>0}\{C_1 y^{-\gamma/(1-\gamma)} + yx\},$$

which implies (i). Since

$$\lim_{y\to\infty} u(y) + yx \ge \lim_{y\to\infty} yx = \infty, \quad x > 0,$$

the map $y \to u(y) + yx$ attains its minimum at some $y_0 > 0$. By Theorem 7.3.1 (iii), y_0 can be uniquely determined. Let $m(x) = y_0$ for $x > 0$. It is obvious that

$$\begin{aligned} m(x) &= (-u')^{-1}(x) \in C^1(0,\infty), \\ v(x) &= u(m(x)) + m(x)x. \end{aligned} \tag{7.13}$$

Hence

$$\begin{aligned} v'(x) &= u'(m(x))m'(x) + \{m'(x)x + m(x)\} = m(x), \\ v''(x) &= m'(x) = -\frac{1}{u''(m(x))}, \end{aligned} \tag{7.14}$$

which implies (ii) and (iii). Substituting $m(x)$ into x in (7.7), by (7.13) and (7.14), we deduce that v solves (7.5). ∎

7.4 Optimal Policies

We present a synthesis of optimal policies for the optimization problem. We consider the stochastic differential equation,

$$\begin{aligned} dX^*(t) &= [rX^*(t) - I(v'(X^*(t)))]dt \\ &\quad + \{-\frac{v'(X^*(t))}{v''(X^*(t))}(\sigma\sigma^\top)^{-1}(b - r\mathbf{1})\}^\top[(b - r\mathbf{1})dt + \sigma dB(t)], \\ X^*(0) &= x > 0. \end{aligned} \tag{7.15}$$

Our objective is to prove the following theorem.

Theorem 7.4.1. *We assume* (7.3), (7.6), *and* (7.8). *Then the optimal policy* $\{c^*(t), \pi^*(t)\}$ *is given by*

$$c^*(t) = I(v'(X^*(t))),$$

$$\pi^*(t) = -\frac{v'(X^*(t))}{v''(X^*(t))}(\sigma\sigma^\top)^{-1}(b - r\mathbf{1}).$$

Proof.

(1) We claim that there exists a unique positive solution $\{X^*(t)\}$ to (7.15). By (7.4) and (7.7), we observe that $f := -u'$ solves

$$-rf(x) + \frac{\nu^2}{2}x^2 f''(x) + (\alpha - r + \nu^2)xf'(x) + I(x) = 0.$$

By (7.13), we recall that

$$m'(x) = \frac{1}{f'(m(x))}, \quad m''(x) = -\frac{f''(m(x))}{(f'(m(x)))^3} = -f''(m(x))(m'(x))^3.$$

Substituting $m(x)$ into x, we get

$$rx + \frac{\nu^2}{2}m(x)^2\frac{m''(x)}{(m'(x))^3} - (\alpha - r + \nu^2)\frac{m(x)}{m'(x)} - I(m(x)) = 0. \quad (7.16)$$

By (7.14), setting $\lambda := \sigma^{-1}(b - r\mathbf{1})$, we rewrite (7.15) as

$$\begin{aligned} dX^*(t) &= [rX^*(t) - I(m(X^*(t)))]dt \\ &\quad + \{-\frac{m(X^*(t))}{m'(X^*(t))}(\sigma^\top)^{-1}\lambda\}^\top[(b - r\mathbf{1})dt + \sigma dB(t)], \\ X^*(0) &= x > 0. \end{aligned}$$

Then, by (7.16) and Itô's formula,

$$\begin{aligned} dm(X^*(t)) &= \Big(m'(x)[rx - I(m(x)) + \{-\frac{m(x)}{m'(x)}(\sigma^\top)^{-1}\lambda\}^\top(b - r\mathbf{1})] \\ &\quad + \frac{1}{2}m''(x)|\{\frac{m(x)}{m'(x)}(\sigma^\top)^{-1}\lambda\}^\top\sigma|^2\Big)|_{x=X^*(t)}\, dt \\ &\quad + \{-m(X^*(t))(\sigma^\top)^{-1}\lambda\}^\top\sigma dB(t) \\ &= (\alpha - r)m(X^*(t))dt - m(X^*(t))\lambda^\top dB(t). \end{aligned}$$

Hence we have

$$m(X^*(t)) = \mu(t), \quad \text{a.s.},$$

where $\{\mu(t)\}$ is the unique solution of

$$d\mu(t) = (\alpha - r)\mu(t)dt - \mu(t)\lambda^\top dB(t), \quad \mu(0) = m(x).$$

Therefore we can verify that $\{f(\mu(t))\}$ is a unique positive solution to (7.15), and $\{c^*(t), \pi^*(t)\}$ belongs to $\mathcal{A}$.

(2) We note that the pair $(\bar{c}(x), \bar{\pi}(x))$ of maximizers in (7.5) is given by

$$\begin{cases} \bar{c}(x) = I(v'(x)), \\ \bar{\pi}(x) = -\dfrac{v'(x)}{v''(x)}(\sigma\sigma^\top)^{-1}(b - r\mathbf{1}). \end{cases}$$

By (7.5) and Itô's formula, we have

$$\begin{aligned} E[e^{-\alpha(t\wedge\tau_n)}v(X^*(t\wedge\tau_n))] &= v(x) + E[\int_0^{t\wedge\tau_n} e^{-\alpha s}\{-\alpha v(x) + rxv'(x) \\ &\quad - c^*(s)v'(x) + \pi^*(s)^\top(b - r\mathbf{1})v'(x) \\ &\quad + \frac{1}{2}|\sigma^\top\pi^*(s)|^2 v''(x)\}|_{x=X^*(s)}ds + \bar{M}(t\wedge\tau_n)] \\ &= v(x) - E[\int_0^{t\wedge\tau_n} e^{-\alpha s}U(c^*(s))ds] \end{aligned} \qquad (7.17)$$

for a sequence of localizing stopping times $\tau_n \nearrow \infty$ of the local martingale $\bar{M}(t) := \int_0^t e^{-\alpha s}v'(X^*(s))\pi^*(t)^\top\sigma dB(s)$. By (7.8), we take $p \in (1, 1/\gamma)$ such that

$$\alpha > rp\gamma + \frac{\nu^2}{2}(\frac{p\gamma}{1 - p\gamma}).$$

Then, by Theorem 7.3.3,

$$v(x)^p \leq \bar{v}(x) := 2^p C_3^p(1 + x^{p\gamma}).$$

Furthermore,

$$-\alpha\bar{v} + rx\bar{v}' - \frac{\nu^2}{2}\frac{(\bar{v}')^2}{\bar{v}''} - c\bar{v}' \leq 0, \qquad \text{for } c \geq 0.$$

By the localizing argument as in (7.17), we get

$$E[e^{-\alpha(t\wedge\tau_n)}\bar{v}(X^*(t\wedge\tau_n))] \leq \bar{v}(x).$$

Therefore, by Hölder's inequality

$$\begin{aligned} E[e^{-\alpha(t\wedge\tau_n)}v(X^*(t\wedge\tau_n))] &= E[e^{-\alpha(t\wedge\tau_n)/q}\cdot e^{-\alpha(t\wedge\tau_n)/p}v(X^*(t\wedge\tau_n))] \\ &\leq E[e^{-\alpha(t\wedge\tau_n)}]^{1/q}E[e^{-\alpha(t\wedge\tau_n)}v(X^*(t\wedge\tau_n))^p]^{1/p} \\ &\leq E[e^{-\alpha(t\wedge\tau_n)}]^{1/q}\bar{v}(x)^{1/p} \\ &\to 0 \quad \text{as } n\to\infty \quad \text{and } t\to\infty, \end{aligned}$$

where $1/p + 1/q = 1$. Passing to the limit in (7.17) and taking into account $\theta(X^*) = \infty$ a.s., we obtain

$$E[\int_0^{\theta(X^*)} e^{-\alpha t} U(c^*(t))dt] = v(x).$$

By the same calculation as above, we get

$$E[\int_0^{\theta(X)} e^{-\alpha t} U(c(t))dt] \leq v(x), \qquad \text{for all } (c, \pi) \in \mathcal{A}.$$

■

8

Optimal Exploitation of Renewable Resources

We study the stochastic optimization problem of renewable resources. We consider planning authorities who want to determine the amount of harvest from publicly owned resources.

The authorities expoit at the rate $c(t)$ from the stock of resource until the resource is totally exhaused. The stock $X(t)$ is renewable if some positive level of stock $X(t)$ can be maintaines by the rate of increase $f(X(t))$ of this stock indefinitely. The growth rate of stock $X(t)$ coincides with the totality of $f(X(t))$ and the white noise process $dB(t)/dt$, which is interpreted as the uncertain fluctuation of the resource, when there is no exploitation of the stock. The natural resource allows us to exploit if its position remains to be nonnegative. The authorities' aim is to maximize the expected discounted utility of exploitation with a utility function $U(c)$.

We analyze the associated HJB equation (8.7) and further show the C^2-regularity of the viscosity solution. We present the existence of an optimal policy in a feedback form or a stochastic version of Hotelling's rule.

8.1 The Model

Consider a mathematical problem of the renewable natural resource with uncertainty. Define the following quantities:

$X(t)$ = remaining stock of the renewable resource at time t.
$f(X(t))$ = rate of renewable change of the stock at time t.
$c(t)$ = flow of exploitation (or consumption) at time t.
$B(t)$ = the standard Brownian motion.
σ = the nonzero diffusion constant.
$U(c)$ = the utility function.

We assume that the stock $X(t)$ evolves according to the stochastic differential equation

$$dX(t) = [f(X(t)) - c(t)]dt + \sigma X(t)dB(t), \qquad X(0) = x \geq 0, \qquad (8.1)$$

on a complete probability space $(\Omega, \mathcal{F}, P)$ endowed with the natural filtration $\mathcal{F}_t$ generated by $\sigma(B(s), s \leq t)$, and $f(x)$ satisfies the following:

$$\begin{cases} \text{There is } C_f > 0 \text{ such that } |f(x) - f(y)| \leq C_f|x-y|, & x, y \geq 0, \\ f(x) : \text{concave on } [0,\infty), \quad f(x) \geq f(0) = 0. \end{cases} \tag{8.2}$$

The policy $c = \{c(t)\}$ is admissible if $\{c(t)\}$ is a nonnegative $\{\mathcal{F}_t\}$-progressively measurable process such that

$$\int_0^T c(t)dt < \infty, \quad T \geq 0, \quad \text{a.s.} \tag{8.3}$$

and

$$\text{there exists a unique nonnegative solution } \{X(t)\} \text{ of (8.1), for } x \geq 0. \tag{8.4}$$

We denote by $\mathcal{A}$ the set of admissible policies for $x \geq 0$. By the comparison theorem, it is easily seen that 0 belongs to $\mathcal{A}$.

The objective is to find an optimal policy $c^* = \{c^*(t)\}$ so as to maximize the discounted utility of exploitation

$$J_x(c) = E[\int_0^\theta e^{-\alpha t} U(c(t))\, dt], \tag{8.5}$$

over $c \in \mathcal{A}$ for the first exit time $\theta = \theta(X) = \inf\{t \geq 0 : X(t) = 0\}$ of $\{X(t)\}$ from $(0, \infty)$. Here $\alpha > 0$ is a discount rate and $U(c)$ is a utility function in $C^2(0,\infty) \cap C[0,\infty)$, which is assumed to have the following properties:

$$\begin{cases} U''(c) < 0, \\ U'(\infty) = U(0) = 0, \quad U'(0+) = U(\infty) = \infty. \end{cases} \tag{8.6}$$

The HJB equation associated with this optimization problem is given by

$$\begin{cases} \alpha v(x) - \frac{1}{2}\sigma^2 x^2 v''(x) - f(x)v'(x) - \tilde{U}(v'(x)) = 0, & x > 0, \\ v(0) = 0, \end{cases} \tag{8.7}$$

where $\tilde{U}(x)$ is the convex dual of $U(c)$, that is, $\tilde{U}(x) = \max_{c>0}\{U(c) - cx\}$ for $0 < x < \infty$.

To see that $\mathcal{A}$ is sufficiently rich, we consider the equation,

$$dS(t) = [f(S(t)) - c(t)1_{\{t \leq \theta(S)\}}]dt + \sigma S(t)dB(t), \quad S(0) = x \geq 0, \tag{8.8}$$

for any $c = \{c(t)\}$ satisfying (8.3).

Proposition 8.1.1. *We assume* (8.2). *Then there exists a unique nonnegative solution* $\{S(t)\}$ *of* (8.8).

Proof.

(1) Let $E[\int_0^T c(t)^2 dt] < \infty$ for each $T > 0$. By the same line as Theorem 1.7.2, using the contraction mapping theorem, we find a unique solution $\{\hat{S}(t)\}$ of

$$d\hat{S}(t) = [f(\hat{S}(t)^+) - c(t)]dt + \sigma \hat{S}(t) dB(t), \quad t \geq 0, \quad \hat{S}(0) = x \geq 0,$$

where $x^+ = \max(x, 0)$. Also,

$$E[\sup_{0 \leq t \leq T} |\hat{S}(t)|^2] < \infty, \qquad \text{for each} \quad T > 0.$$

Define

$$S(t) = \hat{S}(t \wedge \theta(\hat{S})) = \hat{S}(t) 1_{\{t \leq \theta(\hat{S})\}} \geq 0.$$

Then $\theta(S) = \theta(\hat{S})$ a.s. Hence

$$\begin{aligned} S(t) &= x + \int_0^{t \wedge \theta(\hat{S})} [f(\hat{S}(s)^+) - c(s)] ds + \int_0^{t \wedge \theta(\hat{S})} \sigma \hat{S}(s) dB(s) \\ &= x + \int_0^t [f(\hat{S}(s)) - c(s)] 1_{\{s \leq \theta(\hat{S})\}} ds + \int_0^t \sigma \hat{S}(s) 1_{\{s \leq \theta(\hat{S})\}} dB(s) \\ &= x + \int_0^t [f(S(s)) - c(s) 1_{\{s \leq \theta(S)\}}] ds + \int_0^t \sigma S(s) dB(s), \quad t \geq 0, \ \text{a.s.} \end{aligned}$$

Thus $\{S(t)\}$ solves (8.8).

(2) Let $c = \{c(t)\}$ satisfy (8.3). We set $c_n(t) = c(t) 1_{\{c(t) \leq n\}}$ for $n \in \mathbf{N}$. Clearly, $c_n(t) \nearrow c(t)$. By (1), there is a nonnegative solution $\{S_n(t)\}$ of

$$dS_n(t) = [f(S_n(t)) - c_n(t) 1_{\{t \leq \theta(S_n)\}}] dt + \sigma S_n(t) dB(t), \quad S_n(0) = x \geq 0.$$

By the same argument as the proof of the comparison theorem in Section 1.1.10, we have

$$\begin{aligned} E[(S_{n+1}(t) - S_n(t))^+] = E[\int_0^t & 1_{\{S_{n+1}(s) - S_n(s) > 0\}} [f(S_{n+1}(s)) \\ & - c_{n+1}(s) 1_{\{s \leq \theta(S_{n+1})\}} \\ & - f(S_n(s)) + c_n(s) 1_{\{s \leq \theta(S_n)\}}] ds] \\ \leq C_f E[\int_0^t & (S_{n+1}(s) - S_n(s))^+ ds]. \end{aligned}$$

This yields that $0 \leq S_{n+1}(t) \leq S_n(t)$ for all $t \geq 0$ and $\theta(S_{n+1}) \leq \theta(S_n)$ a.s. Define

$$\tilde{S}(t) = \lim_{n \to \infty} S_n(t) \quad \text{and} \quad \tilde{\theta} = \lim_{n \to \infty} \theta(S_n).$$

Then, for each $t \geq 0$,

$$\tilde{S}(t) = x + \int_0^t [f(\tilde{S}(s)) - c(s) 1_{\{s \leq \tilde{\theta}\}}] ds + \int_0^t \sigma \tilde{S}(s) dB(s), \quad \text{a.s.}$$

Hence $\{\tilde{S}(t)\}$ has a continuous modification, also denoted by $\{\tilde{S}(t)\}$. Thus we deduce that $S(t) := \tilde{S}(t \wedge \theta(\tilde{S}))$ solves (8.8).

(3) Let $\{S^{(i)}(t)\}$, $i = 1, 2$, be two solutions of (8.8). Then, by (8.2) and Gronwall's inequality, we have

$$S^{(1)}(t \wedge \theta(S^{(1)}) \wedge \theta(S^{(2)})) = S^{(2)}(t \wedge \theta(S^{(1)}) \wedge \theta(S^{(2)})), \quad \text{a.s.}$$

Therefore we have $\theta(S^{(1)}) = \theta(S^{(2)})$ and hence $S^{(1)}(t) = S^{(2)}(t)$ for all $t \geq 0$ a.s. ■

Proposition 8.1.2. *We assume* (8.2). *Then we have*

(i) $E[\sup_{0 \leq t \leq T} X(t)^2] < \infty$ *for each* $T > 0$,
(ii) $X(t) = 0$ *for* $t \geq \theta$ *on* $\{\theta < \infty\}$, *a.s.*,
(iii) $X(t) = S(t)$, $t \geq 0$, *a.s.* *for* $c \in \mathcal{A}$.

Proof.

(1) Let $\{N(t)\}$ be the solution of

$$dN(t) = C_f N(t)dt + \sigma N(t)dB(t), \quad N(0) = x \geq 0.$$

By (8.2), (8.4), and the comparison theorem, we have

$$0 \leq X(t) \leq N(t), \quad t \geq 0, \quad \text{a.s.}$$

Thus, applying Theorem 1.7.2 to $\{N(t)\}$, we obtain (i).

(2) Let $r \geq 0$ and we set $X_r(t) = X(t + r)$, $B_r(t) = B(t + r) - B(r)$. By (8.1), we see that

$$\begin{aligned} X_r(t) &= X_r(0) + \int_r^{t+r} [f(X(s)) - c(s)]ds + \int_r^{t+r} \sigma X(s)dB(s) \\ &= X_r(0) + \int_0^t [f(X_r(s)) - c(s + r)]ds + \int_0^t \sigma X_r(s)dB_r(s), \quad \text{a.s.} \end{aligned}$$

Let $\{\hat{X}_r(t)\}$ be the solution of

$$d\hat{X}_r(t) = C_f \hat{X}_r(t)dt + \sigma \hat{X}_r(t)dB_r(t), \quad \hat{X}_r(0) = X(r).$$

Recall that $\{(B_r(t), \mathcal{F}_{r+t})\}_{t \geq 0}$ is a Brownian motion. By the comparison theorem,

$$X_r(t) \leq \hat{X}_r(t) = X(r)\exp\{C_f t + \sigma B_r(t) - \frac{1}{2}\sigma^2 t\}, \quad t \geq 0, \quad \text{a.s.}$$

By the continuity of $r \to X_r(t)$, $\hat{X}_r(t)$, we have

$$X(t + \theta) = X_\theta(t) \leq \hat{X}_\theta(t) = 0, \quad \text{a.s. on } \{\theta < \infty\}.$$

Since $(s - \theta)^+$ is an $\{\mathcal{F}_{t+\theta}\}_{t \geq 0}$-stopping time for each $s \geq 0$, we deduce by the continuity of $X(t + \theta)$ that

$$X((s - \theta)^+ + \theta) = 0 \quad \text{on } \{\theta < \infty\} \text{ a.s.},$$

which implies (ii).

(3) By the same calculation as Proposition 8.1.1, we see that $X(t \wedge \theta)$ satisfies (8.8). Also, by (ii), $X(t) = X(t \wedge \theta)$. Thus, by uniqueness, we obtain (iii). ∎

Proposition 8.1.3. *We assume* (8.2). *Then we have*

$$J_x(c) = E[\int_0^\infty e^{-\alpha t} U(c(t))dt], \quad \text{for} \quad c \in \mathcal{A}.$$

Proof. Let $\hat{c}(t) = c(t)1_{\{t \le \theta\}}$. Then it is obvious that

$$J_x(c) = E[\int_0^\infty e^{-\alpha t} U(\hat{c}(t))dt].$$

By Proposition 8.1.2 (ii), we have

$$\int_0^t \hat{c}(s)ds = \int_0^t c(s)ds, \quad t \ge 0, \quad \text{a.s.}$$

Hence

$$E[\int_\theta^\infty e^{-\alpha t} c(t)dt] \le E[\int_\theta^\infty c(t)dt] = 0.$$

Thus, using $U(0) = 0$, we get

$$E[\int_\theta^\infty e^{-\alpha t} U(c(t))dt] = 0,$$

which implies

$$E[\int_0^\infty e^{-\alpha t} U(\hat{c}(t))dt] = E[\int_0^\infty e^{-\alpha t} U(c(t))dt].$$ ∎

8.2 Viscosity Solutions of the HJB Equations

We consider the viscosity solution v of the HJB equation (8.7) and the integrability of (8.5). We approximate (8.7) by

$$\begin{cases} \alpha u(x) - \frac{1}{2}\sigma^2 x^2 u''(x) - f(x)u'(x) - \tilde{U}_L(u'(x)) = 0, & x > 0, \\ u(0) = 0 \end{cases} \tag{8.9}$$

for each $L > 0$, where $\tilde{U}_L(x) = \max_{0 < c \le L} \{U(c) - cx\}, \ 0 \le x < \infty$.

Definition 8.2.1. *Let $v \in C[0, \infty)$. Then v is a viscosity solution of* (8.7) *if the following relations hold, together with $v(0) = 0$:*

$$\alpha v(x) \le \frac{1}{2}\sigma^2 x^2 q + f(x)p + \tilde{U}(p), \ \forall (p, q) \in J^{2,+}v(x), \quad \forall x > 0,$$

$$\alpha v(x) \ge \frac{1}{2}\sigma^2 x^2 q + f(x)p + \tilde{U}(p), \ \forall (p, q) \in J^{2,-}v(x), \quad \forall x > 0,$$

where $J^{2,+}v(x)$ and $J^{2,-}v(x)$ are the second-order superjet and subjet in Definition 4.4.1.

Define

$$u_L(x) = \sup_{c \in \mathcal{C}_L} E[\int_0^{\theta} e^{-\alpha t} U(c(t))\, dt], \tag{8.10}$$

where $\mathcal{C}_L = \{c = \{c(t)\} :$ progressively measurable, $0 \le c(t) \le L, \forall t \ge 0\}$ and the supremum is taken over all systems $(\Omega, \mathcal{F}, P, \{\mathcal{F}_t\}; \{B(t)\}, \{c(t)\})$. Let $\{c(t)\} \in \mathcal{C}_L$ and $\{X(t)\}$ be the solution of (8.1) with $f(x^+)$ replacing $f(x)$ to $\{c(t)\}$. Then $\{c(t)1_{\{t \le \theta(X)\}}\}$ belongs to $\mathcal{A}_L := \mathcal{C}_L \cap \mathcal{A}$. By Proposition 8.1.3, we see that

$$u_L(x) = \sup_{c \in \mathcal{A}_L} E[\int_0^{\infty} e^{-\alpha t} U(c(t))dt].$$

We make the following assumptions:

There exists a concave function $\varphi \in C^2(0,\infty) \cap C[0,\infty)$ such that

$$\begin{cases} \alpha\varphi(x) \ge \frac{1}{2}\sigma^2 x^2 \varphi''(x) + f(x)\varphi'(x) + \tilde{U}(\varphi'(x)), \quad x > 0, \\ \varphi(x) \text{ is Hölder continuous with exponent } \gamma \in (0,1), \\ \varphi'(x) > 0, \quad \varphi(0) = 0, \end{cases} \tag{8.11}$$

$$\alpha \ge \frac{1}{2}\sigma^2\gamma + C_f\gamma. \tag{8.12}$$

Lemma 8.2.2. *We assume* (8.11) *and* (8.12). *Then we have*

(i) $0 \le u_L(x) \le \varphi(x)$, *and*
(ii) $|u_L(x) - u_L(y)| \le K|x-y|^{\gamma}, \quad x, y \ge 0,$

for some constant $K > 0$, independent of L.

Proof.

(1) By Itô's formula and (8.11), we have

$$\begin{aligned} 0 &\le e^{-\alpha(t\wedge\theta)}\varphi(X(t\wedge\theta)) \\ &= \varphi(x) + \int_0^{t\wedge\theta} e^{-\alpha s}\{-\alpha\varphi(X(s)) + [f(X(s)) - c(s)]\varphi'(X(s)) \\ &\quad + \frac{1}{2}\sigma^2 X(s)^2\varphi''(X(s))\}ds + \int_0^{t\wedge\theta} e^{-\alpha s}\sigma X(s)\varphi'(X(s))dB(s) \\ &\le \varphi(x) - \int_0^{t\wedge\theta} e^{-\alpha s} U(c(s))ds + \int_0^{t\wedge\theta} e^{-\alpha s}\sigma X(s)\varphi'(X(s))dB(s), \quad \text{a.s.} \end{aligned}$$

Note that $\varphi(0) \le \varphi'(x)(0-x)+\varphi(x)$, and $E[e^{2\gamma\sigma B(t)}] = e^{2\gamma^2\sigma^2 t}$. Hence, by (8.12)

$$\begin{aligned}E[\int_0^\infty e^{-2\alpha s}|X(s)\varphi'(X(s))|^2 ds] &\le E[\int_0^\infty e^{-2\alpha s}\varphi(X(s))^2 ds]\\ &\le CE[\int_0^\infty e^{-2\alpha s}(1+|X(s)|^{2\gamma})ds]\\ &\le CE[\int_0^\infty e^{-2\alpha s}(1+N(s)^{2\gamma})ds]\\ &= C\int_0^\infty e^{-2\alpha s}(1+x^{2\gamma}e^{2\{\gamma C_f+(\gamma^2-\gamma/2)\sigma^2\}s})\\ &\quad \times ds < \infty,\end{aligned}$$

for some constant $C>0$. This yields that $\{\int_0^t e^{-\alpha s}X(s)\varphi'(X(s))dB(s)\}$ is a martingale, and hence

$$E[\int_0^{t\wedge\theta} e^{-\alpha s}X(s)\varphi'(X(s))dB(s)] = 0.$$

Therefore, we deduce (i). We remark by (8.11) that $J_x(c) \le \varphi(x) < \infty$.

(2) Let $\{Y(t)\}$ be the solution of (8.1) with $Y(0)=y$ and $\bar{\theta}=\theta(Y)$. By (8.10), we have

$$\begin{aligned}u_L(x)-u_L(y) &= \sup_{c\in\mathcal{C}_L} E[\int_0^\theta e^{-\alpha t}U(c(t))dt] - \sup_{\bar{c}\in\mathcal{C}_L} E[\int_0^{\bar{\theta}} e^{-\alpha t}U(\bar{c}(t))dt]\\ &\le \sup_{c\in\mathcal{C}_L} E[\int_0^\theta e^{-\alpha t}U(c(t))dt - \int_0^{\bar{\theta}} e^{-\alpha t}U(c(t))dt]\\ &\le \sup_{c\in\mathcal{C}_L} E[\int_{\theta\wedge\bar{\theta}}^\theta e^{-\alpha t}U(c(t))dt].\end{aligned}$$

By (8.11), there is $K>0$ such that

$$|\varphi(x)-\varphi(y)| \le K|x-y|^\gamma, \quad x,y\ge 0.$$

By the same line as (1), we see that

$$\begin{aligned}E[\int_{\theta\wedge\bar{\theta}}^\theta e^{-\alpha t}U(c(t))dt] &\le E[e^{-\alpha(\theta\wedge\bar{\theta})}\varphi(X(\theta\wedge\bar{\theta})) - e^{-\alpha\theta}\varphi(X(\theta))]\\ &= E[\{e^{-\alpha\bar{\theta}}\varphi(X(\bar{\theta})) - e^{-\alpha\theta}\varphi(X(\theta))\}1_{\{\bar{\theta}<\theta\}}]\\ &\le E[\{e^{-\alpha\bar{\theta}}\varphi(X(\bar{\theta})) - e^{-\alpha\bar{\theta}}\varphi(Y(\bar{\theta}))\}1_{\{\bar{\theta}<\theta\}}]\\ &\le K\sup_{c\in\mathcal{C}_L} E[e^{-\alpha\bar{\theta}}\left|X(\bar{\theta})-Y(\bar{\theta})\right|^\gamma 1_{\{\bar{\theta}<\theta\}}].\end{aligned}$$

We set $Z=X-Y$. It is clear that

$$dZ(t) = (f(X(t))-f(Y(t)))dt + \sigma Z(t)dB(t), \quad Z(0)=x-y.$$

Now we define

$$h_\beta(x) = (x^2 + \beta)^{\frac{\gamma}{2}}, \quad \beta > 0.$$

From (8.12) it follows that

$$\begin{aligned}
&-\alpha h_\beta(x) + C_f \left| x h'_\beta(x) \right| + \frac{1}{2}\sigma^2 x^2 h''_\beta(x) \\
&\le h_\beta(x)(-\alpha + C_f\gamma + \frac{1}{2}\sigma^2\gamma) \le 0, \quad x \in \mathbf{R}.
\end{aligned}$$

Using Itô's formula, we have

$$\begin{aligned}
&E[e^{-\alpha(t\wedge\bar{\theta}\wedge\theta)} h_\beta(Z(t \wedge \bar{\theta} \wedge \theta))] \\
&= h_\beta(x-y) + E[\int_0^{t\wedge\bar{\theta}\wedge\theta} e^{-\alpha s}\{-\alpha h_\beta(Z(s)) \\
&+ h'_\beta(Z(s))(f(X(s)) - f(Y(s))) + \frac{1}{2}\sigma^2 Z(s)^2 h''_\beta(Z(s))\} ds] \\
&\le h_\beta(x-y).
\end{aligned}$$

Letting $t \to \infty$ and $\beta \to 0$, we get

$$E[e^{-\alpha(\bar{\theta}\wedge\theta)} |Z(\bar{\theta} \wedge \theta)|^\gamma] \le |x-y|^\gamma,$$

which implies (ii). ■

Theorem 8.2.3. *We assume* (8.2), (8.6), (8.11), *and* (8.12). *Then the value function $u_L(x)$ is a viscosity solution of* (8.9).

Proof. By (8.10), it is easy to see that $u_L(0) = 0$. In view of Theorem 4.3.1, the viscosity property of u_L follows from the dynamic programming principle for u_L, that is,

$$u_L(x) = \sup_{c\in\mathcal{A}_L} E[\int_0^\tau e^{-\alpha s} U(c(s)) ds + e^{-\alpha\tau} u_L(X(\tau))], \quad \forall x \ge 0, \tag{8.13}$$

for any stopping time $\tau \in \mathcal{S}_b$, where the supremum is taken over all systems $(\Omega, \mathcal{F}, P, \{\mathcal{F}_t\}; \{B(t)\}, \{c(t)\})$. By the optional sampling theorem, we may consider that τ is nonrandom. By the same line as Theorem 3.2.5, we outline the proof of (8.13) on the Polish space Ω.

(1) Let $\bar{u}$ be the right-hand side of (8.13). By Theorem 3.2.5 (1), for $c = \{c(t)\} \in \mathcal{A}_L$, we have

$$J_{X(\tau)}(\tilde{c}) = E[\int_0^\infty e^{-\alpha t} U(\tilde{c}(t) dt | \mathcal{F}_\tau] = e^{\alpha\tau} E[\int_\tau^\infty e^{-\alpha t} U(c(t)) dt | \mathcal{F}_\tau], \quad \text{a.s.},$$

and hence

$$\begin{aligned}
J_x(c) &= E[\int_0^\tau e^{-\alpha t} U(c(t)) dt + \int_\tau^\infty e^{-\alpha t} U(c(t)) dt] \\
&= E[\int_0^\tau e^{-\alpha t} U(c(t)) dt + e^{-\alpha\tau} J_{X(\tau)}(\tilde{c})],
\end{aligned}$$

where $\tilde{c}$ is the shifted process of c by τ, that is, $\tilde{c}(t) = c(t + \tau)$. Therefore,

$$J_x(c) \le E[\int_0^\tau e^{-\alpha t} U(c(t))dt + e^{-\alpha\tau} u_L(X(\tau))].$$

Taking the supremum, we deduce $u_L \le \bar{u}$.

(2) Let $\varepsilon > 0$ be arbitrary. We consider a sequence $\{S_j : j = 1, \ldots, n+1\}$ of disjoint subsets of $[0, \infty)$ such that

$$\operatorname{diam}(S_j) < \delta, \quad \bigcup_{j=1}^{n} S_j = [0, R) \quad \text{and} \quad S_{n+1} = [R, \infty),$$

for $\delta, R > 0$ chosen later. We take $x_j \in S_j$ and $c^{(j)} \in \mathcal{A}_L$ with respect to x_j such that

$$u_L(x_j) - \varepsilon \le J_{x_j}(c^{(j)}), \quad j = 1, \ldots, n+1. \tag{8.14}$$

By the same argument as Lemma 8.2.2 (2), we note that

$$|J_x(c^{(j)}) - J_y(c^{(j)})| \le K|x - y|^\gamma, \quad x, y \ge 0,$$

for some constant $K > 0$. We choose $\delta > 0$ such that $K\delta^\gamma < \varepsilon$. Then

$$\left|J_{x_j}(c^{(j)}) - J_{X(\tau)}(c^{(j)})\right| \le K|x_j - X(\tau)|^\gamma < K\delta^\gamma < \varepsilon$$
$$\text{if} \quad X(\tau) \in S_j, \ j = 1, \ldots, n.$$

Hence, by Lemma 8.2.2 (ii) and (8.14), we have

$$\begin{aligned} J_{X(\tau)}(c^{(j)}) &= J_{X(\tau)}(c^{(j)}) - J_{x_j}(c^{(j)}) + J_{x_j}(c^{(j)}) \\ &\ge -\varepsilon + u_L(x_j) - \varepsilon \\ &\ge -2\varepsilon + u_L(X(\tau)) - \varepsilon \quad \text{if} \quad X(\tau) \in S_j, \ j = 1, \ldots, n. \end{aligned} \tag{8.15}$$

By definition, we find $c \in \mathcal{A}_L$ such that

$$\bar{u}(x) - \varepsilon \le E[\int_0^\tau e^{-\alpha t} U(c(t))dt + e^{-\alpha\tau} u_L(X(\tau))].$$

Define

$$c^\tau(t) = c(t)1_{\{t<\tau\}} + c^{(j)}(t - \tau)1_{\{0 \le t-\tau \le \theta(X^j)\}}$$
$$\text{if} \quad X(\tau) \in S_j, \ j = 1, \ldots, n+1,$$

on the same probability space, where $\{X^j(t)\}$ is the nonnegative solution of

$$dX^j(t) = [f(X^j(t)) - c^{(j)}(t)1_{\{t \le \theta(X^j)\}}]dt + \sigma X^j(t)dB_\tau(t),$$
$$X^j(0) = X(\tau)$$

for the Brownian motion $\{B_\tau(t)\}$ in Proposition 8.1.2 (2). It is easy to see that $\{c^\tau(t)\}$ belongs to $\mathcal{C}_L$. Let $\{X^\tau(t)\}$ be the solution of

$$dX^\tau(t) = [f(X^\tau(t)^+) - c^\tau(t)]dt + \sigma X^\tau(t)dB(t), \quad X^\tau(0) = x \ge 0.$$

Clearly, $X^\tau(t) = X(t) \geq 0$ a.s. if $t < \tau$. Furthermore, for each $j = 1, \ldots, n+1$, we have on $\{X(\tau) \in S_j\}$

$$\begin{aligned} X^\tau(t+\tau) &= X(\tau) + \int_\tau^{t+\tau} f(X^\tau(s)^+)ds - \int_0^t c^\tau(s+\tau)ds \\ &\quad + \int_\tau^{t+\tau} \sigma X^\tau(s)dB(s) \\ &= X(\tau) + \int_0^t [f(X^\tau(s+\tau)^+) - c^{(j)}(s)1_{\{s \leq \theta(X^j)\}}]ds \\ &\quad + \int_0^t \sigma X^\tau(s+\tau)dB_\tau(s), \quad \text{a.s.} \end{aligned}$$

Hence $X^\tau(t+\tau) = X^j(t) \geq 0$ for all $t \geq 0$ a.s. on $\{X(\tau) \in S_j\}$. Thus $\{c^\tau(t)\}$ belongs to $\mathcal{A}_L$.

Now, we choose $R > 0$ such that

$$\begin{aligned} \sup_{c \in \mathcal{A}_L} E[u_L(X(\tau))1_{\{X(\tau) \geq R\}}] &\leq \frac{U(L)}{\alpha} \sup_{c \in \mathcal{A}_L} P(X(\tau) \geq R) \\ &\leq \frac{U(L)}{\alpha} P(N(\tau) \geq R) \leq \frac{U(L)}{\alpha R^2} E[N(\tau)^2] \\ &< \varepsilon, \end{aligned}$$

where $N(\tau)$ is as in Proposition 8.1.2 (1). By Proposition 8.1.3 and Theorem 3.2.5 (1), we get

$$\begin{aligned} J_{X(\tau)}(\tilde{c}^\tau) &= E[\int_0^\infty e^{-\alpha t} U(c^\tau(t+\tau))dt | \mathcal{F}_\tau] \\ &= E[\int_0^\infty e^{-\alpha t} U(c^{(j)}(t)1_{\{t \leq \theta(X^j)\}})dt | \mathcal{F}_\tau] \\ &= J_{X(\tau)}(c^{(j)}) \quad a.s. \text{ on } \{X(\tau) \in S_j\}. \end{aligned}$$

Hence, by (8.15), we have

$$\begin{aligned} E[\int_\tau^\infty e^{-\alpha t} U(c^\tau(t))dt] &= E[E[\int_\tau^\infty e^{-\alpha t} U(c^\tau(t))dt | \mathcal{F}_\tau]] \\ &= E[e^{-\alpha\tau} J_{X(\tau)}(\tilde{c}^\tau)] \\ &= E[e^{-\alpha\tau} \sum_{j=1}^{n+1} J_{X(\tau)}(c^{(j)}) 1_{\{X(\tau) \in S_j\}}] \\ &\geq E[e^{-\alpha\tau} \sum_{j=1}^{n} \{u_L(X(\tau)) - 3\varepsilon\} 1_{\{X(\tau) \in S_j\}}] \\ &\geq E[e^{-\alpha\tau} \{u_L(X(\tau)) - u_L(X(\tau)) 1_{\{X(\tau) \in S_{n+1}\}}\}] - 3\varepsilon \\ &\geq E[e^{-\alpha\tau} u_L(X(\tau))] - 4\varepsilon. \end{aligned}$$

Thus

$$\begin{aligned} u_L(x) &\geq E[\int_0^\tau e^{-\alpha t} U(c^\tau(t))dt + \int_\tau^\infty e^{-\alpha t} U(c^\tau(t))dt] \\ &\geq E[\int_0^\tau e^{-\alpha t} U(c(t))dt + e^{-\alpha\tau} u_L(X(\tau))] - 4\varepsilon \\ &\geq \bar{u}(x) - 5\varepsilon. \end{aligned}$$

Therefore, letting $\varepsilon \to 0$, we obtain $\bar{u} \leq u_L$. ∎

Theorem 8.2.4. *We make the assumptions of Theorem 8.2.3. Then there exists a viscosity solution v of* (8.7) *such that* $0 \leq v \leq \varphi$.

Proof. Since u_L is nondecreasing and $u_L \leq \varphi$, we can define v by $v(x) = \lim_{L\to\infty} u_L(x)$. By Lemma 8.2.2, we have $v \in C[0, \infty)$ and $0 \leq v \leq \varphi$. Thus, by Dini's theorem, we see the locally uniform convergence of u_L. It is obvious that

$$\tilde{U}_L \to \tilde{U} \quad \text{locally uniformly} \quad \text{in } (0, \infty) \quad \text{as } L \to \infty.$$

Therefore, passing to the limit in (8.9), we deduce by Theorem 4.5.1 that v is a viscosity solution of (8.7). ∎

8.3 Concavity and Regularity

In this section, we study the concavity and C^2-regularity of the viscosity solution v of (8.7).

Proposition 8.3.1. *We assume* (8.2), (8.6), (8.11), *and* (8.12). *Then v is concave.*

Proof. It suffices to show that u_L is concave. Let $x_i \geq 0$, $i = 1, 2$, and $0 \leq \xi \leq 1$. For any $\varepsilon > 0$, there exists $c_i \in \mathcal{A}_L$ such that

$$u_L(x_i) - \varepsilon < E[\int_0^{\theta(X_i)} e^{-\alpha t} U(c_i(t))dt],$$

where $\{X_i(t)\}$ is the solution of (8.1) corresponding to c_i with $X_i(0) = x_i$ on the same probability space. Define $\{\bar{X}(t)\}$ and $\{\tilde{X}(t)\}$ by

$$\begin{aligned} d\bar{X}(t) &= [f(\bar{X}(t)) - (\xi c_1(t) + (1-\xi)c_2(t))]dt + \sigma \bar{X}(t)dB(t), \\ &\qquad\qquad\qquad\qquad \bar{X}(0) = \xi x_1 + (1-\xi)x_2, \\ \tilde{X}(t) &= \xi X_1(t) + (1-\xi)X_2(t). \end{aligned}$$

Clearly,

$$\begin{aligned} d(\tilde{X}(t) - \bar{X}(t)) = [\xi f(X_1(t)) + (1-\xi)f(X_2(t)) - f(\bar{X}(t))]dt \\ + \sigma(\tilde{X}(t) - \bar{X}(t))dB(t). \end{aligned}$$

By the same line as the proof of the comparison theorem, we have

$$E[(\tilde{X}(t) - \bar{X}(t))^+] = E[\int_0^t 1_{\{\tilde{X}(s) > \bar{X}(s)\}}[\xi f(X_1(s)) + (1-\xi) f(X_2(s)) - f(\bar{X}(s))]ds + \int_0^t 1_{\{\tilde{X}(s) > \bar{X}(s)\}} \sigma(\tilde{X}(s) - \bar{X}(s)) dB(s)].$$

By (8.2) and Proposition 8.1.2 (i)

$$\begin{aligned} E[(\tilde{X}(t) - \bar{X}(t))^+] &\le E[\int_0^t [f(\tilde{X}(s)) - f(\bar{X}(s))] 1_{\{\tilde{X}(s) \ge \bar{X}(s)\}} ds] \\ &\le C_f E[\int_0^t |\tilde{X}(s) - \bar{X}(s)| 1_{\{\tilde{X}(s) \ge \bar{X}(s)\}} ds] \\ &\le C_f E[\int_0^t (\tilde{X}(s) - \bar{X}(s))^+ ds]. \end{aligned}$$

By Gronwall's inequality and continuity

$$0 \le \tilde{X}(t) \le \bar{X}(t), \quad t > 0, \quad \text{a.s.}$$

Hence $\{\xi c_1(t) + (1-\xi) c_2(t)\}$ belongs to $\mathcal{A}_L$ and $\theta(\bar{X}) \ge \theta(X_1), \theta(X_2)$. Thus

$$\begin{aligned} & u_L(\xi x_1 + (1-\xi) x_2) \\ &\ge E[\int_0^{\theta(\bar{X})} e^{-\alpha s} U(\xi c_1(s) + (1-\xi) c_2(s)) ds] \\ &\ge \xi E[\int_0^{\theta(\bar{X})} e^{-\alpha s} U(c_1(s)) ds] + (1-\xi) E[\int_0^{\theta(\bar{X})} e^{-\alpha s} U(c_2(s)) ds] \\ &\ge \xi E[\int_0^{\theta(X_1)} e^{-\alpha s} U(c_1(s)) ds] + (1-\xi) E[\int_0^{\theta(X_2)} e^{-\alpha s} U(c_2(s)) ds] \\ &> \xi u_L(x_1) + (1-\xi) u_L(x_2) - \varepsilon. \end{aligned}$$

Therefore, letting $\varepsilon \to 0$, we obtain the concavity of u_L. ■

To show the smoothness of v, we need the following.

Lemma 8.3.2. *Let $h \in C[0,\infty)$ be nonnegative and concave. Then the right-derivative d^+h is nonnegative on $(0,\infty)$, and h is nondecreasing.*

Proof. Let $0 < a < b$ be arbitrary. We will show that $d^+h \ge 0$ on (a,b) and $h(x_2) \ge h(x_1)$ for $a < x_1 < x_2 < b$.

(1) We claim that there exists a right-continuous nonincreasing function ψ on (a,b) such that

$$h(y) - h(x) = \int_x^y \psi(t) dt, \quad x, y \in (a,b).$$

By Theorem 4.4.4 (1), we observe that h is Lipschitz on $[a,b]$. Hence h is absolutely continuous on $[a,b]$, so that h can be rewritten as

$$h(x) = h(a) + \int_a^x \psi_0(t)dt, \quad x \in [a,b],$$

for some $\psi_0 \in L^1(a,b)$ (cf. H. L. Royden [139]). Also, by concavity,

$$\frac{h(x+r)-h(x)}{r} \geq \frac{h(x+s)-h(x)}{s}, \quad 0 < r < s, \quad x \in (a,b).$$

Hence the right-derivative $d^+h(x) := \lim_{r\to 0+} \frac{h(x+r)-h(x)}{r}$ exists. By Lebesgue's Theorem,

$$d^+h(x) = \psi_0(x), \quad \text{a.e.}$$

By concavity again,

$$\frac{h(x)-h(x-s)}{s} \geq \frac{h(x)-h(x-r)}{r}$$
$$\geq \frac{h(x+r)-h(x)}{r}, \quad 0 < r < s, \quad x \in (a,b).$$

Taking $s = x_2 - x_1 > 0$ in these two inequalities, we have

$$d^+h(x_1) \geq \frac{h(x_2)-h(x_1)}{x_2 - x_1} \geq d^+h(x_2),$$

that is, d^+h is nonincreasing. Define $\psi(x) = \lim_{r\to 0+} d^+h(x+r)$ for $x \in (a,b)$. Since a monotone function can have only a countable number of discontinuities, we get $\psi(x) = d^+h(x)$ a.s. on (a,b). Furthermore, by the right continuity of ψ, we have

$$d^+h(x) = \lim_{r\to 0+} \frac{1}{r} \int_x^{x+r} \psi(t)dt = \psi(x), \quad \text{for all } x \in (a,b).$$

Therefore, ψ has the desired property.

(2) Let $c \in (a,b)$ be fixed. For $x \geq c$, by (1), we have

$$h(x) = h(c) + \int_c^x d^+h(t)dt \leq h(c) + d^+h(c)(x-c).$$

For $x < c$,

$$h(x) = h(c) - \int_x^c d^+h(t)dt \leq h(c)$$
$$- \int_x^c d^+h(c)dt \leq h(c) + d^+h(c)(x-c).$$

(3) We claim that $d^+h(x) \geq 0$ for all $x \in (a,b)$. If not, there is $c \in (a,b)$ such that $d^+h(c) < 0$. For any $k > b$, by (2), we have

$$h(x) \leq h(c) + d^+h(c)(x-c), \quad a < x < k.$$

Thus $\lim_{k\to\infty} h(k) = -\infty$, which is a contradiction. Clearly, we obtain

$$h(x_2) = h(x_1) + \int_{x_1}^{x_2} d^+h(t)dt \geq h(x_1), \quad x_2 > x_1. \qquad \blacksquare$$

Theorem 8.3.3. *We assume* (8.2), (8.6), (8.11), *and* (8.12). *Then there exists a solution* $v \in C^2(0,\infty) \cap C[0,\infty)$ *of* (8.7) *such that*

$$v'(x) > 0, \quad \text{for} \;\; x > 0. \tag{8.16}$$

Proof.

(1) We claim that

$$d^+v(x) > 0, \quad \text{for} \;\; x > 0. \tag{8.17}$$

Since $v \geq 0$, Proposition 8.3.1 and Lemma 8.3.2 yield that $d^+v(x) \geq 0$. Suppose that $d^+v(x_0) = 0$ for some $x_0 > 0$. By Alexandrov's theorem, there is $x_1 > x_0$ such that $v(x)$ is twice differentiable at x_1. For $(p,q) \in J^{2,+}v(x_1) \cap J^{2,-}v(x_1)$, by Lemma 8.3.2, we have

$$0 \leq p = v'(x_1) \leq d^+v(x_0) = 0.$$

Hence

$$\alpha v(x_1) \geq \frac{1}{2}\sigma^2 x_1^2 q + f(x_1)p + \tilde{U}(p) = \infty,$$

which is a contradiction. Thus we obtain (8.17).

(2) Let $I(x)$ be the maximizer of $\tilde{U}(x)$ given by $I(x) = (U')^{-1}(x)$. Clearly, $\tilde{U}(x) = U(I(x)) - xI(x)$. By (8.6), we have

$$I'(x) = \frac{1}{U''(I(x))} < 0.$$

Thus

$$I \in C^1(0,\infty), \tag{8.18}$$

and

$$\tilde{U}'(x) = -I(x).$$

(3) Consider the boundary value problem:

$$\begin{cases} \alpha w - \frac{1}{2}\sigma^2 x^2 w'' - f(x)w' - \tilde{U}(w' \vee d^+v(b)) = 0 & \text{in} \;\; (a,b), \\ w(a) = v(a), \quad w(b) = v(b). \end{cases} \tag{8.19}$$

By (1) and (2), $\tilde{U}'(x)$ is bounded on $[d^+v(b),\infty)$ and

$$\text{the map } x \to \tilde{U}(x \vee d^+v(b)) \text{ is Lipschitz.} \tag{8.20}$$

By uniform ellipticity, Theorem 5.3.7 yields that (8.19) has a unique solution $w \in C^2(a,b)$. On the other hand, for $(p,q) \in J^{2,-}v(x)$, $x \in (a,b)$

$$\begin{aligned}&\alpha v - \frac{1}{2}\sigma^2 x^2 q - f(x)p - \tilde{U}(p \vee d^+ v(b))\\ &\quad \geq \alpha v - \frac{1}{2}\sigma^2 x^2 q - f(x)p - \tilde{U}(p) \geq 0.\end{aligned}$$

For $(p,q) \in J^{2,+}v(x)$, $x \in (a,b)$, by Proposition 8.3.1 and Definition 4.4.1, we have

$$\begin{aligned}d^+ v(y)(y-x) &\leq v(y) - v(x)\\ &\leq p(y-x) + \tfrac{1}{2}q(y-x)^2 + o(|y-x|^2)\\ &\text{as } y \to x,\ x < y.\end{aligned}$$

Dividing both sides by $y - x$ and letting $y \to x$, by concavity, we get

$$d^+ v(b) \leq d^+ v(x) \leq p.$$

Hence

$$\begin{aligned}&\alpha v - \frac{1}{2}\sigma^2 x^2 q - f(x)p - \tilde{U}(p \vee d^+ v(b))\\ &\quad = \alpha v - \frac{1}{2}\sigma^2 x^2 q - f(x)p - \tilde{U}(p) \leq 0.\end{aligned}$$

Thus v is a viscosity solution of (8.19). By Theorem 5.3.1, taking into account (8.2) and (8.20), we see that the viscosity solution of (8.19) is unique. Therefore we deduce $v = w$. Since $[a,b]$ is arbitrary, we obtain $v \in C^2(0,\infty)$ and (8.16). ■

8.4 Optimal Exploitation

We give a synthesis of the optimal policy $c^* = \{c^*(t)\}$ for the optimization problem (8.5) subject to (8.1). We consider the stochastic differential equation,

$$dX^*(t) = [f(X^*(t)) - I \circ v'(X^*(t))]dt + \sigma X^*(t)dB(t), \qquad X^*(0) = x \geq 0, \tag{8.21}$$

where $I(x)$ is as in (8.18) and $I \circ v'(0) = 0$.

Our objective is to prove the following theorem.

Theorem 8.4.1. *We assume* (8.2), (8.6), (8.11), *and* (8.12). *Then the optimal exploitation* $\{c^*(t)\}$ *is given by*

$$c^*(t) = I \circ v'(X^*(t)).$$

Before going into the proof, we need the following lemmas.

Lemma 8.4.2. *We make the assumptions of Theorem 8.4.1. Then there exists a unique nonnegative solution* $\{X^*(t)\}$ *of* (8.21).

Proof.

(1) We claim that

$$v'(0+) = \infty. \tag{8.22}$$

Suppose $0 \le v'(0+) < \infty$. By concavity and (8.16), $0 < v'(1) \le v'(0+)$. Also, by (8.7), $\lim_{x\to 0+} x^2 v''(x)$ exists. Hence L'Hospital's rule gives

$$\lim_{x\to 0+} x^2 v''(x) = \lim_{x\to 0+} \frac{x^2 v'(x)}{x} = 0.$$

Letting $x \to 0+$ in (8.7), we have

$$\tilde{U}(v'(0+)) = \max_{c>0}\{U(c) - cv'(0+)\} = -f(0+)v'(0+) = 0,$$

and then

$$U(c) = cv'(0+), \quad \forall c \in [0, I \circ v'(0+)].$$

This is contrary to (8.6). Thus we get (8.22) and, by monotonicity,

$$0 = I \circ v'(0+) \le I \circ v'(x), \quad x > 0.$$

(2) Define

$$g(x) = \begin{cases} I \circ v'(x) & \text{if } x > 0, \\ 0 & \text{if } x \le 0, \end{cases}$$

and

$$g_m(x) = g(x \vee \frac{1}{m}) - g(\frac{1}{m})$$

for each $m \in \mathbf{N}$. By (8.18) and Theorem 8.3.3, $g_m(x)$ is locally Lipschitz on $\mathbf{R}$. By Theorem 1.9.1 (3), there exists a solution $\{Y_m(t)\}$ of

$$dY_m(t) = [f(Y_m(t)^+) - g_m(Y_m(t))]dt + \sigma Y_m(t)dB(t), \\ \text{for} \quad t < \mathsf{e},\ Y_m(0) = x > 0, \tag{8.23}$$

where e is the explosion time of $Y_m(t)$, that is, $\mathsf{e} = \inf\{t \ge 0 : |Y_m(t)| = \infty\}$. Let $\{M(t)\}$ be the unique solution of

$$dM(t) = f(M(t)^+)dt + \sigma M(t)dB(t), \quad M(0) = x > 0.$$

We set $Z_n(t) = Y_m(t \wedge \tau_n)$, $M_n(t) = M(t \wedge \tau_n)$ for $\tau_n = \inf\{t \geq 0 : |Y_m(t)| \geq n\}$. Applying the proof of the comparison theorem, we have

$$\begin{aligned}
E[(Z_n(t) - M_n(t))^+] &= E[\int_0^t 1_{\{Z_n(s) > M_n(s)\}}[f(Z_n(s)^+) - g_m(Z_n(s)) \\
&\quad - f(M_n(s)^+)]1_{\{s \leq \tau_n\}} ds \\
&\quad + \int_0^t 1_{\{Z_n(s) > M_n(s)\}} \sigma(Z_n(s) - M_n(s)) 1_{\{s \leq \tau_n\}} dB(s)] \\
&\leq C_f E[\int_0^t (Z_n(s) - M_n(s))^+ ds].
\end{aligned}$$

Similarly,

$$\begin{aligned}
E[(Z_n(t))^-] &= x^- + E[\int_0^t 1_{\{Z_n(s) < 0\}}[-f(Z_n(s)^+) + g_m(Z_n(s))]1_{\{s \leq \tau_n\}} ds] \\
&\leq E[\int_0^t 1_{\{Z_n(s) < 0\}} g_m(Z_n(s)) 1_{\{Z_n(s) \geq 1/m\}} ds] = 0.
\end{aligned}$$

Hence

$$0 \leq Z_n(t) \leq M_n(t), \quad t \geq 0, \quad \text{a.s.}$$

Letting $n \to \infty$, we get

$$0 \leq Y_m(t) \leq M(t) \quad \text{on} \quad \{t < \mathsf{e}\}, \quad \text{a.s.}$$

Thus we see that $\mathsf{e} = \infty$ a.s. We note that $g_{m+1} \geq g_m$ and

$$0 \leq g(x) - g_m(x) \leq g(\frac{1}{m}) \to 0 \quad \text{as} \quad m \to \infty.$$

Then we have $Y_{m+1}(t) \leq Y_m(t)$ a.s. We define

$$X^*(t) = \lim_{m \to \infty} Y_m(t).$$

Passing to the limit in (8.23), we deduce that $\{X^*(t)\}$ satisfies (8.21). Clearly, zero satisfies (8.21) if $x = 0$. Thus $\{c^*(t)\}$ belongs to $\mathcal{A}$.

(3) Let $\{X^{(i)}(t)\}, i = 1, 2$, be two nonnegative solutions of (8.21). Then $\chi(t) := X^{(1)}(t) - X^{(2)}(t)$ satisfies

$$\begin{aligned}
d\chi(t)^2 &= 2\chi(t)[f(X^{(1)}(t)) - f(X^{(2)}(t)) - g(X^{(1)}(t)) + g(X^{(2)}(t))]dt \\
&\quad + 2\sigma\chi(t)^2 dB(t) + \sigma^2\chi(t)^2 dt, \quad \chi(0) = 0.
\end{aligned}$$

By (2), we note that

$$(g(x) - g(y))(x - y) \geq 0.$$

Hence

$$\begin{aligned} E[\chi(t)^2] &\leq E[\int_0^t \{2\chi(s)(f(X^{(1)}(s)) - f(X^{(2)}(s))) + \sigma^2 \chi(s)^2\} ds] \\ &\leq (2C_f + \sigma^2) E[\int_0^t \chi(s)^2 ds]. \end{aligned}$$

This implies that $X^{(1)}(t) = X^{(2)}(t)$ for all $t \geq 0$ a.s. ■

Lemma 8.4.3. *For any $c \in \mathcal{A}$, we have*

$$E[e^{-\alpha t} \varphi(X(t))] \to 0 \quad as \quad t \to \infty.$$

Proof. By Proposition 8.1.2 (1), we have

$$X(t) \leq N(t) = x \exp\{(C_f - \frac{1}{2}\sigma^2)t + \sigma B(t)\}.$$

By Lemma 2.4.3, we recall that $E[\exp\{\sigma\gamma B(t) - \frac{1}{2}\sigma^2\gamma^2 t\}] = 1$. By (8.11) and (8.12), we deduce

$$\begin{aligned} E[e^{-\alpha t}\varphi(X(t))] &\leq E[e^{-\alpha t}\varphi(N(t))] \leq KE[e^{-\alpha t}|N(t)|^\gamma] + e^{-\alpha t}\varphi(0) \\ &= Kx^\gamma \exp\{(-\alpha + \frac{1}{2}\sigma^2\gamma^2 + (C_f - \frac{1}{2}\sigma^2)\gamma)t\} + e^{-\alpha t}\varphi(0) \\ &\to 0 \quad \text{as} \quad t \to \infty. \end{aligned}$$

■

Proof of Theorem 8.4.1. Let $\tau_n^* = \inf\{t \geq 0 : X^*(t) \leq 1/n\}$ and $\theta^* = \theta(X^*) = \inf\{t \geq 0 : X^*(t) = 0\}$. It is clear that $\tau_n^* \nearrow \theta^*$. By concavity,

$$0 < v'(x)x \leq v(x), \quad x > 0.$$

By Theorem 8.2.4, we note that

$$\begin{aligned} E[\int_0^t (e^{-\alpha s} v'(X^*(s))X^*(s))^2 ds] &\leq E[\int_0^t (e^{-\alpha s} v(X^*(s)))^2 ds] \\ &\leq E[\int_0^t (e^{-\alpha s} \varphi(X^*(s)))^2 ds] \\ &\leq E[\int_0^t (e^{-\alpha s} \varphi(N(s)))^2 ds] < \infty. \end{aligned}$$

This yields that $\{\int_0^{t\wedge\tau_n^*} e^{-\alpha s} v'(X_s^*)X_s^* dB(s)\}$ is a martingale. Hence, by (8.7) and Itô's formula,

$$\begin{aligned}E[e^{-\alpha(t\wedge\tau_n^*)}v(X^*(t\wedge\tau_n^*))] &= v(x) + E[\int_0^{t\wedge\tau_n^*} e^{-\alpha s}\{-\alpha v(X^*(s)) \\ &\quad + f(X^*(s))v'(X^*(s)) - c^*(s)v'(X^*(s)) \\ &\quad + \frac{1}{2}\sigma^2(X^*(s))^2 v''(X^*(s))\}ds] \\ &= v(x) - E[\int_0^{t\wedge\tau_n^*} e^{-\alpha s}U(c^*(s))ds].\end{aligned}$$

Letting $n \to \infty$ and then $t \to \infty$, by Lemma 8.4.3, we have

$$E[\int_0^{\theta^*} e^{-\alpha t}U(c^*(t))dt] = v(x).$$

By the same calculation as above, we obtain

$$E[\int_0^{\theta} e^{-\alpha t}U(c(t))dt] \le v(x)$$

for any $c \in \mathcal{A}$. ∎

8.5 Examples

In this section, we present some examples of $\varphi(x)$ satisfying (8.11), which is needed for the existence of an optimal policy.

Example 8.5.1. We consider the simple growth function $f(x) = kx, k > 0$, and the utility function $U(x) = x^\gamma/\gamma, 0 < \gamma < 1$. It is clear that (8.2) and (8.6) are fulfilled. We set $\varphi(x) = Qx^\gamma/\gamma$ for a positive constant Q. Then $\tilde{U}(x) = (1/\gamma - 1)x^{\gamma/(\gamma-1)}$ and (8.11) is satisfied if

$$\alpha \ge \sigma^2\gamma(\gamma-1)/2 + k\gamma + (1-\gamma)Q^{1/(\gamma-1)}.$$

In particular, we obtain the explicit solution of (8.7) given by $v(x) = Q_0x^\gamma/\gamma$, taking $Q_0 > 0$ such that

$$\alpha = \sigma^2\gamma(\gamma-1)/2 + k\gamma + (1-\gamma)Q_0^{1/(\gamma-1)}.$$

Example 8.5.2. Consider the modified logistic function,

$$f(x) = \begin{cases} kx & \text{if } x < 0, \\ kx(1-x) & \text{if } 0 \le x \le 1, \\ -kx + k & \text{if } x > 1, \end{cases}$$

for $k > 0$. Clearly, $f(x)$ satisfies (8.2). Let $U(x)$ be the same utility function as Example 8.1, and we take $\varphi(x) = Qx^\gamma/\gamma$ for Q in Example 8.1. Then (8.11) is fulfilled since $kx \ge f(x)$ for $x \ge 0$.

Example 8.5.3. Consider

$$f(x) = \begin{cases} 1 - e^{-x} & \textit{if } x \geq 0, \\ 0 & \textit{if } x < 0, \end{cases}$$

which satisfies (8.2). For the utility function $U(x)$ of Example 8.1, we take $\varphi(x) = x^{\gamma}/\gamma$. Since $x \geq f(x)$ for $x \geq 0$, we get (8.11) if $\alpha \geq \sigma^2\gamma(\gamma - 1)/2 + 1$.

In this case, we consider the deterministic equation of (8.1) with $c(t) = 0$ and $\sigma = 0$:

$$\frac{dX}{dt} = 1 - e^{-X(t)}, \qquad X(0) = x > 0.$$

By the transformation of $L(t) = 1 - e^{-X(t)}$, we remark that this equation is equivalent to the logistic equation,

$$\frac{dL}{dt} = L(1 - L).$$

9

Optimal Consumption Models in Economic Growth

We study the optimal consumption problem in the neoclassical theory of capital accumulation and economic growth under uncertainty. We consider two factor markets – one for labor and one for capital services.

There are many identical households, each with a utility function given by $U(c)$ for consumption c. Households supply the uncertain amount of labor $Y(t)$. There are many identical firms, each with the same technology for production. Firms rent the services of capital and labor to produce output. The increase of the capital stock $X(t)$ coincides with the totality of the production $F(X(t), Y(t))$ net of the consumption $c(t)$ and the depreciation $\lambda X(t)$ with rate λ until the capital stock vanishes. The decision that the household has to make is how much to consume or how to maximize the expected discounted utility of consumption with a utility function $U(c)$.

We show the existence of a classical solution of the HJB equation (9.6) associated with the stochastic optimization problem, and then give an optimal consumption policy in terms of its solution.

9.1 The Model

Consider the neoclassical growth model of the Solow-type under uncertainty, in Merton [113]. Define the following quantities:

$Y(t) =$ labor supply at time t.
$X(t) =$ capital stock at time t.
$F(x, y) =$ constant-returns-to-scale production function producing the commodity for the capital stock $x \geq 0$ and the labor force $y > 0$.
$B(t) =$ the standard Brownian motion.
$\kappa =$ the constant rate of growing labor force, $\kappa \neq 0$.
$\sigma =$ the nonzero diffusion constant.
$\lambda =$ the nonnegtaive constant rate of depreciation.
$c(t) =$ consumption rate per person at time t.

$c(t)Y(t) =$ the totality of consumption rate.
$U(c) =$ the utility function for the consumption rate $c \geq 0$.

We assume that the labor supply $Y(t)$ and the capital stock $X(t)$ evolve according to the stochastic differential equation

$$dY(t) = \kappa Y(t)dt + \sigma Y(t)dB(t), \qquad Y(0) = y > 0, \tag{9.1}$$

and

$$\frac{d}{dt}X(t) = F(X(t), Y(t)) - \lambda X(t) - c(t)Y(t), \qquad X(0) = x \geq 0, \tag{9.2}$$

on a complete probability space $(\Omega, \mathcal{F}, P)$, endowed with the natural filtration $\{\mathcal{F}_t\}$ generated by $\sigma(B(s), s \leq t)$. The production function $F(x, y)$ is usually accompanied by the assumptions:

$$\begin{cases} F(x, y) \text{ is a nonnegative, } C([0, \infty) \times (0, \infty)) \text{ and } C^2((0, \infty)^2) \text{ function,} \\ F_x > 0,\ F_y > 0,\ F_{xx} < 0,\ F_{yy} < 0 \quad \text{in} \quad (0, \infty)^2, \\ F(0+, y) = F(x, 0+) = 0, \qquad x, y > 0, \\ F(rx, ry) = rF(x, y), \qquad r > 0. \end{cases} \tag{9.3}$$

The objective is to find an optimal consumption policy $c^* = \{c^*(t)\}$ so as to maximize the expected utility function with discount factor $\alpha > 0$:

$$J(c) = E[\int_0^{\theta(X)} e^{-\alpha t} U(c(t))\, dt] \tag{9.4}$$

over the class $\mathcal{A}$ of all consumption policies $c = \{c(t)\}$ such that

$$\begin{cases} c(t) \text{ is nonnegative } \{\mathcal{F}_t\}\text{-progressively measurable,} \\ \int_0^T c(t)dt < \infty \qquad \text{for each } T \geq 0 \text{ a.s.,} \\ \text{there exists a unique nonnegative solution } \{X(t)\} \text{ of (9.2) for } x \geq 0, \end{cases}$$

where $\theta(X)$ is the first exit time of $\{X(t)\}$ from $(0, \infty)$, that is, $\theta(X) = \inf\{t \geq 0 : X(t) = 0\}$. The utility function $U(c)$ is assumed to have the following properties:

$$\begin{cases} U \in C[0, \infty) \cap C^2(0, \infty), \\ U''(c) < 0, \\ U'(\infty) = U(0+) = 0, \qquad U'(0+) = U(\infty) = \infty. \end{cases} \tag{9.5}$$

The HJB equation associated with this optimization problem is given by

$$\begin{aligned} &\alpha v(x, y) - \{\frac{1}{2}\sigma^2 y^2 v_{yy}(x, y) + \kappa y v_y(x, y) + (F(x, y) - \lambda x)v_x(x, y) \\ &\quad + \tilde{U}(v_x(x, y)y)\} = 0, \quad x > 0,\ y > 0, \\ &v(0, y) = 0, \quad y > 0, \end{aligned} \tag{9.6}$$

where $\tilde{U}(x)$ is the convex dual of $U(x)$, that is, $\tilde{U}(x) = \max_{c>0} \{U(c) - cx\}$.

Example 9.1.1. *The standard production functions between capital and labor are given by*

$$F^1(x,y) = Ax^a y^b, \quad A, a, b > 0, \quad a+b=1,$$
$$F^2(x,y) = A(ax^{-\xi} + by^{-\xi})^{-1/\xi}, \quad A, a, b > 0, \quad a+b=1, \quad \xi > 0,$$

called, respectively, the Cobb–Douglas and the CES (Constant Elasticity of Substitution) production function (cf. M. Allen [4]). Clearly, F^1 and F^2 satisfy (9.3) *and*

$$\begin{cases} F_x^1(0+,y) = \infty, \quad F_x^2(0+,y) < \infty, \\ F_x^1(\infty,y) = F_x^2(\infty,y) = 0, \quad y > 0. \end{cases}$$

Since

$$\lim_{\xi \to 0} \frac{\log(a(x/y)^{-\xi} + b)}{\xi} = -a\log(x/y),$$

we obtain

$$\lim_{\xi \to 0} F^2(x,y) = F^1(x,y).$$

9.2 HJB Equations

We consider the solution $v(x,y)$ of (9.6) of the form:

$$v(x,y) = u(z), \quad z = \frac{x}{y}.$$

By substitution, $u(z)$ solves the one-dimensional HJB equation,

$$\begin{aligned} \alpha u(z) &= \frac{1}{2}\sigma^2 z^2 u''(z) + (f(z) - \mu z)u'(z) + \tilde{U}(u'(z)), \quad z > 0, \\ u(0) &= 0, \end{aligned} \tag{9.7}$$

where $\mu = \kappa + \lambda - \sigma^2$ and $f(z) = F(z,1)$. For approximation, we study the HJB equation,

$$\begin{aligned} \alpha w_L(z) &= \frac{1}{2}\sigma^2 z^2 w_L''(z) + (f(z) - \mu z)w_L'(z) + \tilde{U}_L(w_L'(z)), \quad z > 0, \\ w_L(0) &= 0, \end{aligned} \tag{9.8}$$

or, equivalently,

$$\begin{aligned} (\alpha + \frac{1}{\varepsilon})w_L(z) &= \frac{1}{2}\sigma^2 z^2 w_L''(z) + (f(z) - \mu z)w_L'(z) \\ &\quad + \tilde{U}_L(w_L'(z)) + \tfrac{1}{\varepsilon}w_L(z), \quad z > 0, \\ w_L(0) &= 0, \end{aligned} \tag{9.9}$$

where $\tilde{U}_L(x) = \max_{0<c\le L}\{U(c) - cx\}$, $L \ge 1$.

From now on, we deal with (9.7) under the milder conditions on $\kappa, \lambda, \sigma^2$ than those in Chapter 8. Furthermore, we assume

$$F_x(0+, 1) < \infty. \tag{9.10}$$

Hence, it is clear that

$$f(z) : \text{Lipschitz, concave, } f(0) := f(0+) = 0, \quad C_f := f'(0+) < \infty. \tag{9.11}$$

We note that (9.8) is the HJB equation associated with the stochastic Ramsey problem so as to maximize

$$\bar{J}(c) = E[\int_0^{\theta(Z)} e^{-\alpha t} U(c(t))dt] \tag{9.12}$$

subject to

$$dZ(t) = [f(Z(t)) - \mu Z(t) - c(t)]dt - \sigma Z(t)dB(t), \quad Z(0) = z \geq 0, \tag{9.13}$$

over the class $\bar{\mathcal{A}}$ of all $c \in \mathcal{A}$ such that (9.13) has a nonnegative solution $\{Z(t)\}$ for $z \geq 0$. We notice that the optimization problem (9.12) is the same as in Chapter 8. Thus (9.12) can be rewritten as

$$\bar{J}(c) = E[\int_0^{\infty} e^{-\alpha t} U(c(t))dt]. \tag{9.14}$$

We will show the existence of a unique solution w_L of (9.9) and

$$w_L \to u \text{ as } L \to \infty.$$

9.3 Viscosity Solutions

In this section, we show the existence and uniqueness results on the viscosity solution u of (9.7).

Definition 9.3.1. *Let $u \in C[0, \infty)$ and $u(0) = 0$. Then u is a viscosity solution of (9.7) if the following relations hold:*

$$\alpha u(z) \leq \frac{1}{2}\sigma^2 z^2 q + (f(z) - \mu z)p + \tilde{U}(p), \quad \forall (p, q) \in J^{2,+}u(z), \quad \forall z > 0,$$

$$\alpha u(z) \geq \frac{1}{2}\sigma^2 z^2 q + (f(z) - \mu z)p + \tilde{U}(p), \quad \forall (p, q) \in J^{2,-}u(z), \quad \forall z > 0,$$

where $J^{2,+}$ and $J^{2,-}$ are the second-order superjet and subjet in Definition 4.4.1.

9.3.1 Existence

We begin with the HJB equation (9.9) associated with the optimization problem

$$w_L(z) = \sup_{c \in \mathcal{C}_L} E[\int_0^{\theta(Z)} e^{-(\alpha + \frac{1}{\varepsilon})t} \{U(c(t)) + \frac{1}{\varepsilon} w_L(Z(t))\}dt], \tag{9.15}$$

for $\varepsilon > 0$, where $\mathcal{C}_L$ denotes the class of all $\{\mathcal{F}_t\}$-progressively measurable processes $c = \{c(t)\}$ such that $0 \le c(t) \le L$, for all $t \ge 0$, $\{Z(t)\}$ is the solution of (9.13) with $f(x^+)$ replacing $f(x)$, for $c \in \mathcal{C}_L$, and the supremum is taken over all of the admissible control systems. Since $\{c(t)1_{\{t \le \theta(Z)\}}\}$ belongs to $\mathcal{A}_L := \mathcal{C}_L \cap \bar{\mathcal{A}}$, we may consider that $\{Z(t)\}$ is nonnegative. We assume that

$$\alpha + \mu > 0. \tag{9.16}$$

Taking $\beta, \bar{\beta}$ such that

$$0 < \beta < \alpha, \quad 0 < \bar{\beta} < \beta + \mu, \tag{9.17}$$

by concavity, we can choose $A > 0$ such that

$$f(z) < \bar{\beta} z + A. \tag{9.18}$$

We observe by (9.11), (9.17), and (9.18) that $\varphi(z) := z + \bar{A}$ fulfills

$$\begin{aligned} &-\beta\varphi(z) + \frac{1}{2}\sigma^2 z^2 \varphi''(z) + (f(z) - \mu z)\varphi'(z) + \tilde{U}(\varphi'(z)) \\ &\le -\beta\bar{A} + A + \tilde{U}(1) < 0, \quad z \ge 0, \end{aligned} \tag{9.19}$$

for some constant $\bar{A} > 0$. Furthermore, we assume that there exists a concave function $\psi \in \mathcal{B} \cap C^2(0, \infty)$ such that

$$\begin{aligned} &-\alpha\psi(z) + \frac{1}{2}\sigma^2 z^2 \psi''(z) + (f(z) - \mu z)\psi'(z) + \tilde{U}(\psi'(z)) \le 0, \quad z > 0, \\ &\psi'(z) > 0, \quad \psi(0) = 0, \end{aligned} \tag{9.20}$$

where $\mathcal{B}$ denotes the normed space of all $h \in C[0, \infty)$ satisfying the relation:

$$\begin{aligned} &\text{There exists } C_\rho > 0, \text{ for any } \rho > 0, \text{ such that} \\ &|h(z) - h(y)| \le C_\rho |z - y| + \rho(\varphi(z) + \varphi(y)), \quad z, y \in [0, \infty), \end{aligned} \tag{9.21}$$

with norm $\|h\| := \sup_{z \ge 0} |h(z)|/\varphi(z) < \infty$. For (9.20), see (8.11) of Chapter 8 and the related examples.

Proposition 9.3.2. *The class $\mathcal{B}$ is a Banach space.*

Proof. We will show the completeness. Let $\{h_n\}$ be a Cauchy sequence in $\mathcal{B}$. Then $\{h_n/\varphi\}$ is a Cauchy sequence in the Banach space $C_b[0, \infty)$ of all bounded continuous functions on $[0, \infty)$ with the uniform norm. Hence there exists $g \in C_b[0, \infty)$ such that

$$\sup_{z \ge 0} |h_n(z)/\varphi(z) - g(z)| \to 0 \quad \text{as } n \to \infty.$$

Define

$$h = g \cdot \varphi.$$

Clearly, $\|h\| < \infty$ and $\|h_n - h\| \to 0$ as $n \to \infty$. For any $\rho > 0$, we choose $k \in \mathbf{N}$ such that $\|h_k - h\| \le \rho/2$. Furthermore, by (9.21), we take $\bar{C}_\rho > 0$ such that

$$|h_k(z) - h_k(y)| \le \bar{C}_\rho |z - y| + \rho(\varphi(z) + \varphi(y))/2.$$

Then we have

$$\begin{aligned} |h(z) - h(y)| &\le |h(z) - h_k(z)| + |h_k(z) - h_k(y)| + |h_k(y) - h(y)| \\ &\le \|h - h_k\|(\varphi(z) + \varphi(y)) + |h_k(z) - h_k(y)| \\ &\le \bar{C}_\rho |z - y| + \rho(\varphi(z) + \varphi(y)), \end{aligned}$$

which implies (9.21). Therefore, h belongs to $\mathcal{B}$. ■

Remark 9.3.3. *By Young's inequality, we can show that x^γ belongs to $\mathcal{B}$ for any $0 < \gamma < 1$.*

Theorem 9.3.4. *We assume* (9.3), (9.5), (9.10), (9.16), *and* (9.20). *Then there exists a unique solution $w_L \in \mathcal{B}$ of* (9.15) *for sufficiently small $\varepsilon > 0$. Furthermore, we have $w_L(0) = 0$ and $0 \le w_L \le \varphi$.*

Proof.

(1) We take the supremum below in the sense of (9.15). By virtue of Proposition 3.2.3, we may consider that every process is defined on the same probability space, if necessary. We claim that

$$\sup_{c \in \mathcal{C}_L} E\left[\int_0^{\theta(Z)} e^{-(\beta+\frac{1}{\varepsilon})t}\{U(c(t)) + \frac{1}{\varepsilon}\varphi(Z(t))\}dt\right] \le \varphi(z). \tag{9.22}$$

By (9.13) and the comparison theorem, for $c \in \mathcal{C}_L$, we have

$$E\left[\int_0^t (Z(s)\varphi'(Z(s)))^2 ds\right] = E\left[\int_0^t Z(s)^2 ds\right] < \infty.$$

This yields that $M_\varepsilon(t) := \int_0^t e^{-(\beta+\frac{1}{\varepsilon})s}\varphi'(Z(s))\sigma Z(s) dB(s)$ is a martingale. Hence, by (9.19), Itô's formula gives

$$\begin{aligned} 0 &\le E[e^{-(\beta+\frac{1}{\varepsilon})\tau}\varphi(Z(\tau))] \\ &= \varphi(z) + E\Big[\int_0^\tau e^{-(\beta+\frac{1}{\varepsilon})s}\{-(\beta + \frac{1}{\varepsilon})\varphi(Z(s)) \\ &\quad + (f(Z(s)) - \mu Z(s) - c(s))\varphi'(Z(s)) \\ &\quad + \frac{1}{2}\sigma^2 Z(s)^2\varphi''(Z(s))\}ds + M_\varepsilon(\tau)\Big] \\ &\le \varphi(z) - E\Big[\int_0^\tau e^{-(\beta+\frac{1}{\varepsilon})s}\{U(c(s)) + \frac{1}{\varepsilon}\varphi(Z(s))\}ds\Big], \end{aligned} \tag{9.23}$$

for any $\tau \in \mathcal{S}_b$. Taking $\tau = t \wedge \theta(Z)$ and letting $t \to \infty$, we obtain (9.22).

(2) We claim that

$$\sup_{c\in\mathcal{C}_L} E[e^{-(\beta+\frac{1}{\varepsilon})\tau}|Z(\tau\wedge\theta(Z)) - S(\tau\wedge\theta(S))|1_{\{\theta(S)\le\theta(Z)\}}] \le 2|z-s|, \tag{9.24}$$

for any stopping time τ and sufficiently small $\varepsilon > 0$, where $\{S(t)\}$ is the response to $c \in \mathcal{C}_L$ with $S(0) = s$. Without loss of generality, we may assume that τ is bounded. Setting $V(t) = Z(t) - S(t)$, we have

$$\begin{aligned} dV(t) &= [f(Z(t)^+) - f(S(t)^+) - \mu V(t)]dt - \sigma V(t)dB(t) \\ &\le (C_f + |\mu|)|V(t)|dt - \sigma V(t)dB(t), \quad V(0) = z - s. \end{aligned}$$

Let $\{K(t)\}$ be the nonnegative solution of

$$dK(t) = (C_f + |\mu|)K(t)dt - \sigma K(t)dB(t), \quad K(0) = |z-s|.$$

By the comparison theorem,

$$\max\{V(t), -V(t)\} \le K(t), \quad \forall t \ge 0, \quad \text{a.s.}$$

Thus, choosing $1/\varepsilon > C_f + |\mu| - \beta$, we get

$$\begin{aligned} &E[e^{-(\beta+\frac{1}{\varepsilon})\tau}|Z(\tau\wedge\theta(Z)) - S(\tau\wedge\theta(S))|1_{\{\tau\le\theta(S)\le\theta(Z)\}}] \\ &= E[e^{-(\beta+\frac{1}{\varepsilon})\tau}|V(\tau)|1_{\{\tau\le\theta(S)\le\theta(Z)\}}] \\ &\le E[e^{-(\beta+\frac{1}{\varepsilon})\tau}K(\tau)] \\ &= |z-s|E[\exp\{-(\beta+\frac{1}{\varepsilon})\tau + (C_f+|\mu|)\tau - \sigma B(\tau) - \frac{1}{2}\sigma^2\tau\}] \\ &\le |z-s|E[\exp\{-\sigma B(\tau) - \frac{1}{2}\sigma^2\tau\}] \\ &= |z-s|. \end{aligned}$$

Next, setting $\hat{Z}(t) = Z(t \vee \theta(S))$, we have on $\{\theta(S) < \infty\}$

$$\begin{aligned} \hat{Z}(t) &= Z(\theta(S)) + \int_{\theta(S)}^{t\vee\theta(S)} [f(Z(r)^+) - \mu Z(r)]dr - \int_{\theta(S)}^{t\vee\theta(S)} c(r)dr \\ &\quad - \int_{\theta(S)}^{t\vee\theta(S)} \sigma Z(r)dB(r) \\ &\le Z(\theta(S)) + \int_{\theta(S)}^{t} [f(\hat{Z}(r)^+) - \mu\hat{Z}(r)]dr - \int_{\theta(S)}^{t} \sigma\hat{Z}(r)dB(r), \quad \text{a.s.} \end{aligned}$$

Hence

$$\begin{aligned}
&\hat{Z}(t \wedge \theta(Z)) \\
&\quad \le Z(\theta(S)) + \int_{\theta(S)}^{t\wedge\theta(Z)} (C_f + |\mu|)|\hat{Z}(r)|dr - \int_{\theta(S)}^{t\wedge\theta(Z)} \sigma \hat{Z}(r) dB(r) \\
&\quad = V(\theta(S)) + \int_{\theta(S)}^{t} (C_f + |\mu|)\hat{Z}(r)1_{\{r\le\theta(Z)\}}dr \\
&\qquad - \int_{\theta(S)}^{t} \sigma \hat{Z}(r)1_{\{r\le\theta(Z)\}}dB(r) \\
&\quad \le K(\theta(S)) + \int_{\theta(S)}^{t} (C_f + |\mu|)\hat{Z}(r\wedge\theta(Z))dr \\
&\qquad - \int_{\theta(S)}^{t} \sigma \hat{Z}(r\wedge\theta(Z))dB(r) \\
&\qquad \text{on} \quad \{\theta(S) < \infty, \theta(S) \le \theta(Z)\}, \quad \text{a.s.}
\end{aligned}$$

Since

$$\begin{aligned}
K(t \vee \theta(S)) = K(\theta(S)) &+ \int_{\theta(S)}^{t} (C_f + |\mu|)K(r\vee\theta(S))dr \\
&- \int_{\theta(S)}^{t} \sigma K(r\vee\theta(S))dB(r) \\
&\text{on} \quad \{\theta(S) < \infty, \theta(S) \le \theta(Z)\}, \quad \text{a.s.,}
\end{aligned}$$

we see, by the comparison theorem, that

$$\hat{Z}(t\wedge\theta(Z)) \le K(t\vee\theta(S)) \ \text{ on } \ \{\theta(S) < \infty, \theta(S) \le \theta(Z)\} \in \mathcal{F}_{\theta(S)}, \ \text{ a.s.,}$$

and then

$$0 \le \hat{Z}(\tau) \le K(\tau) \quad \text{on} \quad \{\theta(S) \le \tau \le \theta(Z)\}, \quad \text{a.s.}$$

Therefore, we have

$$\begin{aligned}
&E[e^{-(\beta+\frac{1}{\varepsilon})\tau}|Z(\tau\wedge\theta(Z)) - S(\tau\wedge\theta(S))|1_{\{\theta(S)\le\tau\le\theta(Z)\}}] \\
&\quad = E[e^{-(\beta+\frac{1}{\varepsilon})\tau}|Z(\tau) - S(\theta(S))|1_{\{\theta(S)\le\tau\le\theta(Z)\}}] \\
&\quad = E[e^{-(\beta+\frac{1}{\varepsilon})\tau}\hat{Z}(\tau)1_{\{\theta(S)\le\tau\le\theta(Z)\}}] \\
&\quad \le E[e^{-(\beta+\frac{1}{\varepsilon})\tau}K(\tau)] \\
&\quad \le |z - s|,
\end{aligned}$$

which implies (9.24).

(3) We define

$$\begin{cases} \mathcal{T}_L h(z) = \sup_{c\in\mathcal{C}_L} T_c h(z), \\ T_c h(z) = E[\int_0^{\theta(Z)} e^{-(\alpha+\frac{1}{\varepsilon})t}\{U(c(t)) + \frac{1}{\varepsilon}h(Z(t))\}dt], \\ \qquad\qquad \text{for} \quad c \in \mathcal{C}_L, \quad h \in \mathcal{B},\ z \geq 0, \end{cases} \tag{9.25}$$

and show that

$$\mathcal{T}_L : \mathcal{B}_\varphi \to \mathcal{B}_\varphi, \tag{9.26}$$

where $\mathcal{B}_\varphi$ is the closed subset of $\mathcal{B}$ defined by $\mathcal{B}_\varphi = \{h \in \mathcal{B} :\ 0 \leq h \leq \varphi, h(0) = 0\}$. By (9.22), it is easy to see that $\mathcal{T}_L h(0) = 0 \leq \mathcal{T}_L h \leq \varphi$ for $h \in \mathcal{B}_\varphi$ and $\|\mathcal{T}_L h\| < \infty$. Taking into account (9.21), we have

$$\begin{aligned} &\max(T_c h(z) - T_c h(s), \mathcal{T}_L h(z) - \mathcal{T}_L h(s)) \\ &\leq \sup_{c\in\mathcal{C}_L} E[\int_0^{\theta(Z)} e^{-(\alpha+\frac{1}{\varepsilon})t}\{U(c(t)) + \frac{1}{\varepsilon}h(Z(t))\}dt \\ &\quad - \int_0^{\theta(S)} e^{-(\alpha+\frac{1}{\varepsilon})t}\{U(c(t)) + \frac{1}{\varepsilon}h(S(t))\}dt] \\ &\leq \sup_{c\in\mathcal{C}_L} E[\int_{\theta(Z)\wedge\theta(S)}^{\theta(Z)} e^{-(\alpha+\frac{1}{\varepsilon})t}\ U(c(t))dt \\ &\quad + \int_0^\infty e^{-(\alpha+\frac{1}{\varepsilon})t}\frac{1}{\varepsilon}|h(Z(t\wedge\theta(Z)))) - h(S(t\wedge\theta(S)))|dt] \\ &\leq \sup_{c\in\mathcal{C}_L} E[\int_{\theta(Z)\wedge\theta(S)}^{\theta(Z)} e^{-(\alpha+\frac{1}{\varepsilon})t}\ U(c(t))dt] \\ &\quad + \frac{C_\rho}{\varepsilon}\sup_{c\in\mathcal{C}_L} E[\int_0^\infty e^{-(\alpha+\frac{1}{\varepsilon})t}|Z(t\wedge\theta(Z)) - S(t\wedge\theta(S))|dt] \\ &\quad + \frac{\rho}{\varepsilon}\sup_{c\in\mathcal{C}_L} E[\int_0^\infty e^{-(\alpha+\frac{1}{\varepsilon})t}\{\varphi(Z(t\wedge\theta(Z))) + \varphi(S(t\wedge\theta(S)))\}dt] \\ &\equiv J_1 + J_2 + J_3. \end{aligned}$$

Let $\{N(t)\}$ be the solution of

$$dN(t) = (C_f + |\mu|)N(t)dt - \sigma N(t)dB(t), \quad N(0) = z \geq 0.$$

By (9.13) and the comparison theorem,

$$E[Z(t)^2] \leq E[N(t)^2] = z^2\exp\{(2(C_f + |\mu|) + \sigma^2)t\}.$$

By Doob's maximal inequality, we can take sufficiently small $\varepsilon \in (0, 1)$ such that

$$E[\sup_t |M_\varepsilon(t)|] \le 2|\sigma| E[\int_0^\infty e^{-2(\beta+\frac{1}{\varepsilon})s} Z(s)^2 ds]^{1/2}$$
$$\le 2|\sigma| z (\int_0^\infty e^{(-2(\beta+1/\varepsilon)+C)t} dt)^{1/2} < \infty.$$

Thus, by the optional sampling theorem,

$$E[M_\varepsilon(\theta(Z))] = E[M_\varepsilon(\theta(Z) \wedge \theta(S))] = 0.$$

By the same line as (9.23), using (9.20), we get

$$E[\int_{\theta(Z)\wedge\theta(S)}^{\theta(Z)} e^{-(\alpha+\frac{1}{\varepsilon})t}\, U(c(t))dt]$$
$$\le E[e^{-(\alpha+\frac{1}{\varepsilon})(\theta(Z)\wedge\theta(S))}\psi(Z(\theta(Z)\wedge\theta(S))) - e^{-(\alpha+\frac{1}{\varepsilon})\theta(Z)}\psi(Z(\theta(Z)))]$$
$$= E[\{e^{-(\alpha+\frac{1}{\varepsilon})(\theta(Z)\wedge\theta(S))}\psi(Z(\theta(Z)\wedge\theta(S)))$$
$$- e^{-(\alpha+\frac{1}{\varepsilon})\theta(Z)}\psi(Z(\theta(Z)))\}1_{\{\theta(S)<\theta(Z)\}}]$$
$$\le E[\{e^{-(\alpha+\frac{1}{\varepsilon})\theta(S)}\psi(Z(\theta(S)) - e^{-(\alpha+\frac{1}{\varepsilon})\theta(S)}\psi(S(\theta(S)))\}1_{\{\theta(S)<\theta(Z)\}}].$$

Hence, by (9.21), (9.23), and (9.24),

$$J_1 \le \sup_{c\in\mathcal{C}_L} E[e^{-(\alpha+\frac{1}{\varepsilon})\theta(S)}\{C_\rho|Z(\theta(S)) - S(\theta(S))|$$
$$+ \rho\Big(\varphi(Z(\theta(S))) + \varphi(S(\theta(S)))\Big)\}1_{\{\theta(S)<\theta(Z)\}}]$$
$$\le 2C_\rho|z-s| + \rho(\varphi(z)+\varphi(s)).$$

Moreover, by (9.24)

$$J_2 \le (4C_\rho|z-s|/\varepsilon)\int_0^\infty e^{-(\alpha-\beta)t}dt = 4C_\rho|z-s|/(\alpha-\beta)\varepsilon.$$

Also, by (9.22),

$$J_3 \le \rho(\varphi(z)+\varphi(0)) + \rho(\varphi(s)+\varphi(0)) \le 2\rho(\varphi(z)+\varphi(s)).$$

Therefore we have $\mathcal{T}_L h \in \mathcal{B}$, which implies (9.26).

(4) By (9.19) and (9.23), we see that

$$E[e^{-\beta(t\wedge\theta(Z))}\varphi(Z(t\wedge\theta(Z)))] \le \varphi(z).$$

Hence, for $h_1, h_2 \in \mathcal{B}_\varphi$,

$$\begin{aligned}
&|\mathcal{T}_L h_1(z) - \mathcal{T}_L h_2(z)| \\
&\le \sup_{c\in\mathcal{C}_L} E[\int_0^{\theta(Z)} e^{-(\alpha+\frac{1}{\varepsilon})t} \frac{1}{\varepsilon}|h_1(Z(t)) - h_2(Z(t))|dt] \\
&\le \sup_{c\in\mathcal{C}_L} E[\int_0^{\infty} e^{-(\alpha-\beta+\frac{1}{\varepsilon})t} \frac{1}{\varepsilon}\|h_1 - h_2\| e^{-\beta(t\wedge\theta(Z))}\varphi(Z(t\wedge\theta(Z)))dt] \\
&\le \frac{1}{(\alpha-\beta)\varepsilon + 1}\|h_1 - h_2\|\varphi(z).
\end{aligned}$$

By the contraction mapping theorem, $\mathcal{T}_L$ has a fixed point $w_L \in \mathcal{B}_\varphi$. Therefore we obtain (9.15). ■

Theorem 9.3.5. *We make the assumptions of Theorem 9.3.4. Then* $w_L \in \mathcal{B}$ *of* (9.15) *is a concave viscosity solution of* (9.8).

Proof.

(1) By virtue of (9.14), we rewrite (9.15) as

$$\begin{cases} w_L(z) = \sup_{c\in\mathcal{A}_L} T_c w_L(z), \\ T_c w_L(z) = E[\int_0^\infty e^{-\alpha t}\{U(c(t)) + \frac{1}{\varepsilon} w_L(Z(t))\}dt], \quad c \in \mathcal{A}_L. \end{cases}$$

We note by (9.23) that

$$E[e^{-(\alpha+\frac{1}{\varepsilon})\tau} w_L(Z(\tau))] \le E[e^{-(\alpha+\frac{1}{\varepsilon})\tau}\varphi(Z(\tau))] \le \varphi(z)$$

for any $\tau \in \mathcal{S}_b$. By Theorem 9.3.4 (3), there is $C_\rho > 0$, for any $\rho > 0$, such that

$$|T_c w_L(z) - T_c w_L(s)| + |w_L(z) - w_L(s)| \le C_\rho|z-s| + \rho(\varphi(z) + \varphi(s)), \quad z, s \ge 0, \quad c \in \mathcal{A}_L.$$

Hence, we can take $\delta > 0$, for any $\varepsilon' > 0$, such that whenever $|z-s| < \delta$,

$$\begin{aligned}
|T_c w_L(z) - T_c w_L(s)| + |w_L(z) - w_L(s)| &\le C_\rho\delta + \rho(\delta + 2\varphi(z)) \\
&< \frac{\varepsilon'}{2}\varphi(0) + \frac{\varepsilon'}{2}\varphi(z) \le \varepsilon'\varphi(z).
\end{aligned}$$

Thus, by analogy with Theorems 8.2.3 and 3.3.2, we can show that the dynamic programming principle holds for w_L, that is,

$$\begin{aligned}
w_L(z) = \sup_{c\in\mathcal{A}_L} E[&\int_0^\tau e^{-(\alpha+\frac{1}{\varepsilon})t}\{U(c(t)) + \frac{1}{\varepsilon} w_L(Z(t))\}dt \\
&+ e^{-(\alpha+\frac{1}{\varepsilon})\tau} w_L(Z(\tau))].
\end{aligned}$$

By (9.13), we see that

$$E[\sup_{0\le t\le h} |Z(t) - Z(0)|^2] \to 0 \quad \text{as } h \to 0, \quad \forall c \in \mathcal{A}_L.$$

By (9.21), letting $h \to 0$ and then $\rho \to 0$, we have

$$\begin{aligned}\sup_{0\le t\le h} E[|w_L(Z(t)) - w_L(Z(0))|\} &\le C_\rho \sup_{0\le t\le h} E[|Z(t) - Z(0)|] \\ &\quad + \rho \sup_{0\le t\le h} E[\varphi(Z(t)) + \varphi(Z(0))] \to 0.\end{aligned}$$

Therefore, by the same line as Theorem 4.3.1, we deduce that w_L is a viscosity solution of (9.9) and then (9.8).

(2) We claim that if $h \in \mathcal{B}_\varphi$ is concave, then so is $\mathcal{T}_L h$. Let $z_i \ge 0, i = 1, 2$, and $0 \le \xi \le 1$. For any $\rho > 0$, there exists $c_i \in \mathcal{A}_L$ such that

$$\mathcal{T}_L h(z_i) - \rho < E[\int_0^\infty e^{-(\alpha+\frac{1}{\varepsilon})t}\{U(c_i(t)) + \frac{1}{\varepsilon}h(Z_i(t))\}dt],$$

where $\{Z_i(t)\}$ denotes the solution $\{Z(t)\}$ corresponding to c_i with $Z_i(0) = z_i$. Define $\{\bar{Z}(t)\}$ and $\{\tilde{Z}(t)\}$ by

$$\begin{aligned}&d\bar{Z}(t) = [f(\bar{Z}(t)) - \mu\bar{Z}(t) - (\xi c_1(t) + (1-\xi)c_2(t))]dt - \sigma\bar{Z}(t)dB(t), \\ &\quad \bar{Z}(0) = \xi z_1 + (1-\xi)z_2, \\ &\quad \tilde{Z}(t) = \xi Z_1(t) + (1-\xi)Z_2(t).\end{aligned}$$

By the same line as Proposition 8.3.1, we see that

$$0 \le \tilde{Z}(t) \le \bar{Z}(t), \quad \text{a.s.},$$

and hence $\{\xi c_1(t) + (1-\xi)c_2(t)\}$ belongs to $\mathcal{A}_L$. Thus, by monotonicity and concavity

$$\begin{aligned}&\mathcal{T}_L h(\xi z_1 + (1-\xi)z_2) \\ &\quad \ge E[\int_0^\infty e^{-(\alpha+\frac{1}{\varepsilon})t}\{U(\xi c_1(t) + (1-\xi)c_2(t)) + \frac{1}{\varepsilon}h(\bar{Z}(t))\}dt] \\ &\quad \ge E[\int_0^\infty e^{-(\alpha+\frac{1}{\varepsilon})t}\{U(\xi c_1(t) + (1-\xi)c_2(t)) + \frac{1}{\varepsilon}h(\tilde{Z}(t))\}dt] \\ &\quad \ge \xi E[\int_0^\infty e^{-(\alpha+\frac{1}{\varepsilon})t}\{U(c_1(t)) + \frac{1}{\varepsilon}h(Z_1(t))\}dt] \\ &\qquad + (1-\xi)E[\int_0^\infty e^{-(\alpha+\frac{1}{\varepsilon})t}\{U(c_2(t)) + \frac{1}{\varepsilon}h(Z_2(t))\}dt] \\ &\quad > \xi\mathcal{T}_L h(z_1) + (1-\xi)\mathcal{T}_L h(z_2) - \rho,\end{aligned}$$

which implies that $\mathcal{T}_L h$ is concave.

(3) It follows from (2) that $\mathcal{T}_L 0$ is concave. Moreover, by induction, $\mathcal{T}_L^n 0$ is concave for any $n \ge 1$. By Theorem 9.3.4, we have

$$\mathcal{T}_L^n 0 \to w_L \quad \text{in } \mathcal{B} \quad \text{as} \quad n \to \infty.$$

Therefore we deduce that w_L is concave. ■

Theorem 9.3.6. *We make the assumptions of Theorem 9.3.4. Then there exists a concave viscosity solution u of* (9.7) *for sufficiently small $\varepsilon > 0$ such that $0 \le u \le \varphi$.*

Proof. Let $L < \bar{L}$ and $h, \bar{h} \in \mathcal{B}$, $h \le \bar{h}$. By (9.25), we have $\mathcal{T}_L h \le \mathcal{T}_L \bar{h} \le \mathcal{T}_{\bar{L}} \bar{h}$. Hence

$$w_L = \mathcal{T}_L^n w_L \le \mathcal{T}_{\bar{L}}^n w_L, \quad n \ge 1.$$

Letting $n \to \infty$, we get $w_L \le w_{\bar{L}} \le \varphi$. Thus $w_L(z)$ converges to some $u(z)$ for each $z \in [0, \infty)$ as $L \nearrow \infty$. It is clear that $u(0) = 0$, u is concave and then continuous on $(0, \infty)$.

Now, we show that $u(0+) = 0$. It follows from (9.15) and (9.23) that

$$u(z) \le \sup_{c \in \bar{\mathcal{A}}} E[\int_0^\infty e^{-(\alpha+\frac{1}{\varepsilon})t}\{U(c(t)) + \frac{1}{\varepsilon}u(Z(t))\}dt] \le \varphi(z), \quad z > 0.$$

For any $\rho > 0$, there exists $c \in \bar{\mathcal{A}}$ such that

$$\begin{aligned} u(z) - \rho &< E[\int_0^\infty e^{-(\alpha+\frac{1}{\varepsilon})t}\{U(c(t)) + \frac{1}{\varepsilon}u(Z(t))\}dt] \\ &= E[\int_0^{\theta(Z)} e^{-(\alpha+\frac{1}{\varepsilon})t}U(c(t))dt] + E[\int_0^\infty e^{-(\alpha+\frac{1}{\varepsilon})t}\frac{1}{\varepsilon}u(Z(t))\}dt]. \end{aligned}$$

Since

$$Z(t) \le N(t) = z\exp\{(C_f + |\mu|)t - \sigma B(t) - \frac{\sigma^2}{2}t\}, \quad \text{a.s.},$$

and $E[\sup_{0\le t\le s} N(t)^2] < \infty$ for each $s > 0$, we see by (9.13) that

$$\begin{aligned} E[e^{-(\alpha+\frac{1}{\varepsilon})\theta(Z)\wedge s}Z(\theta(Z)\wedge s)] = z + E[\int_0^{\theta(Z)\wedge s} & e^{-(\alpha+\frac{1}{\varepsilon})t}\{-(\alpha + \frac{1}{\varepsilon})Z(t) \\ &+ f(Z(t)) - \mu Z(t) - c(t)\}dt]. \end{aligned}$$

We choose $\varepsilon > 0$ sufficiently small such that $C_f + |\mu| < 1/\varepsilon$. Letting $s \to \infty$ and $z \to 0$, we get

$$\begin{aligned} E[\int_0^{\theta(Z)} e^{-(\alpha+\frac{1}{\varepsilon})t}c(t)dt] &\le z + E[\int_0^{\theta(Z)} e^{-(\alpha+\frac{1}{\varepsilon})t}f(Z(t))dt] \\ &\le z + E[\int_0^\infty e^{-(\alpha+\frac{1}{\varepsilon})t}C_f N(t)dt] \quad \to \quad 0. \end{aligned}$$

By (9.5), for any $\rho > 0$, there exists $C_\rho > 0$ such that

$$U(x) \le C_\rho x + \rho, \quad \forall x \ge 0.$$

Letting $z \to 0$ and then $\rho \to 0$, we obtain

$$E[\int_0^{\theta(Z)} e^{-(\alpha+\frac{1}{\varepsilon})t}U(c(t))dt] \le C_\rho E[\int_0^{\theta(Z)} e^{-(\alpha+\frac{1}{\varepsilon})t}c(t)dt] + \frac{\rho}{\alpha + \frac{1}{\varepsilon}} \quad \to \quad 0.$$

Thus

$$u(0+) \le E[\int_0^\infty e^{-(\alpha+\frac{1}{\varepsilon})t}\frac{1}{\varepsilon}u(0+)dt] = \frac{1}{\alpha\varepsilon+1}u(0+),$$

which implies $u(0+) = 0$. Therefore, $u \in C[0,\infty)$.

By the Dini Theorem, w_L converges to u locally uniformly on $[0,\infty)$. Therefore, by Theorem 4.5.1, we deduce that u is a concave viscosity solution of (9.7). ■

9.3.2 Comparison

We present a comparison theorem for the viscosity solution u of (9.7).

Theorem 9.3.7. *Let $f_i, i = 1, 2$, satisfy* (9.11) *and let $u_i \in C[0,\infty)$ be the concave viscosity solution of* (9.7) *for f_i replacing f such that $0 \le u_i \le \varphi$. Suppose*

$$f_1 \le f_2. \tag{9.27}$$

Then, under (9.5), (9.16), *and* (9.20), *we have*

$$u_1 \le u_2.$$

Proof.

(1) By (9.17), there exists $1 < \nu < 2$ such that

$$-\alpha + \frac{1}{2}\sigma^2\nu(\nu-1) + \bar{\beta}\nu - \mu\nu < 0.$$

Hence, by (9.11) and (9.18), $\varphi_\nu(z) := z^\nu + \bar{C}$ satisfies

$$\begin{aligned} -\alpha\varphi_\nu(z) &+ \frac{1}{2}\sigma^2 z^2\varphi_\nu''(z) + (f_2(z) - \mu z)\varphi_\nu'(z) \\ &\le (-\alpha + \frac{1}{2}\sigma^2\nu(\nu-1) + \bar{\beta}\nu - \mu\nu)z^\nu + A\nu z^{\nu-1} \\ &- \alpha\bar{C} < 0, \quad z > 0, \end{aligned} \tag{9.28}$$

for a suitable choice of $\bar{C} > 0$.

(2) Suppose that $u_1(z_0) - u_2(z_0) > 0$ for some $z_0 \in (0,\infty)$. Then there exists $\eta > 0$ such that

$$\sup_{z\ge 0}[u_1(z) - u_2(z) - 2\eta\varphi_\nu(z)] > 0.$$

Since

$$u_1(z) - u_2(z) - 2\eta\varphi_\nu(z) \le \varphi(z) - 2\eta\varphi_\nu(z) \to -\infty \text{ as } z \to \infty,$$

we find $\bar{z} \in (0,\infty)$ such that

$$\sup_{z\ge 0}[u_1(z) - u_2(z) - 2\eta\varphi_\nu(z)] = u_1(\bar{z}) - u_2(\bar{z}) - 2\eta\varphi_\nu(\bar{z}) > 0.$$

Define

$$\Psi_n(z, y) = u_1(z) - u_2(y) - \frac{n}{2}|z - y|^2 - \eta(\varphi_\nu(z) + \varphi_\nu(y)),$$

for any $n \in \mathbf{N}$. It is clear that

$$\Psi_n(z, y) \le \varphi(z) + \varphi(y) - \eta(\varphi_\nu(z) + \varphi_\nu(y)) \rightarrow \ -\infty \ as \ z + y \to \infty.$$

Hence we find $(z_n, y_n) \in [0, \infty)^2$ such that

$$\begin{aligned}\Psi_n(z_n, y_n) &= \sup\{\Psi_n(z, y) : (z, y) \in [0, \infty)^2\} \\ &= u_1(z_n) - u_2(y_n) - \frac{n}{2}|z_n - y_n|^2 - \eta(\varphi_\nu(z_n) + \varphi_\nu(y_n)) \\ &\ge u_1(\bar{z}) - u_2(\bar{z}) - 2\eta\varphi_\nu(\bar{z}) > 0, \qquad (9.29)\end{aligned}$$

from which

$$\begin{aligned}\frac{n}{2}|z_n - y_n|^2 &\le u_1(z_n) - u_2(y_n) - \eta(\varphi_\nu(z_n) + \varphi_\nu(y_n)) \\ &\le \varphi(z_n) + \varphi(y_n) - \eta(\varphi_\nu(z_n) + \varphi_\nu(y_n)).\end{aligned}$$

Thus we deduce that the sequences $\{z_n + y_n\}$ and $\{n|z_n - y_n|^2\}$ are bounded, and

$$|z_n - y_n| \le C/\sqrt{n} \to 0 \ as \ n \to \infty$$

for some constant $C > 0$. Moreover, extracting a subsequence, we have

$$z_n, \ y_n \ \ \to \hat{z} \in [0, \infty) \ as \ n \to \infty. \qquad (9.30)$$

By (9.29), we recall that

$$\begin{aligned}\Psi_n(z_n, y_n) &= u_1(z_n) - u_2(y_n) - \frac{n}{2}|z_n - y_n|^2 - \eta(\varphi_\nu(z_n) + \varphi_\nu(y_n)) \\ &\ge u_1(z_n) - u_2(z_n) - 2\eta\varphi_\nu(z_n),\end{aligned}$$

and hence

$$\begin{aligned}\frac{n}{2}|z_n - y_n|^2 &\le u_2(z_n) - u_2(y_n) + \eta(\varphi_\nu(z_n) \\ &\quad - \varphi_\nu(y_n)) \ \ \to \ 0 \text{ as } n \ \to \ \infty. \qquad (9.31)\end{aligned}$$

Passing to the limit in (9.29), we get

$$u_1(\hat{z}) - u_2(\hat{z}) - 2\eta\varphi_\nu(\hat{z}) > 0, \quad \hat{z} \ne 0. \qquad (9.32)$$

(3) Let $V_1(z) = u_1(z) - \eta\varphi_\nu(z)$ and $V_2(y) = u_2(y) + \eta\varphi_\nu(y)$. By (9.30), we may consider $(z_n, y_n) \in (0, \infty)^2$. Applying Ishii's lemma to

$$\Psi_n(z, y) = V_1(z) - V_2(y) - \frac{n}{2}|z - y|^2,$$

we obtain $q_1, q_2 \in \mathbf{R}$ such that

$$\begin{aligned} &(n(z_n - y_n), q_1) \in \bar{J}^{2,+} V_1(z_n), \\ &(n(z_n - y_n), q_2) \in \bar{J}^{2,-} V_2(y_n), \\ &\begin{pmatrix} q_1 & 0 \\ 0 & -q_2 \end{pmatrix} \le 3n \begin{pmatrix} 1 & -1 \\ -1 & 1 \end{pmatrix}. \end{aligned} \tag{9.33}$$

Recall that

$$J^{2,+}u_1(z) = \{(p + \eta\varphi_\nu'(z), q + \eta\varphi_\nu''(z)) : (p,q) \in J^{2,+}V_1(z)\},$$
$$J^{2,-}u_2(y) = \{(p - \eta\varphi_\nu'(y), q - \eta\varphi_\nu''(y)) : (p,q) \in J^{2,-}V_2(y)\}.$$

Hence

$$(p_1, \bar{q}_1) := (n(z_n - y_n) + \eta\varphi_\nu'(z_n), q_1 + \eta\varphi_\nu''(z_n)) \in \bar{J}^{2,+}u_1(z_n),$$
$$(p_2, \bar{q}_2) := (n(z_n - y_n) - \eta\varphi_\nu'(y_n), q_2 - \eta\varphi_\nu''(y_n)) \in \bar{J}^{2,-}u_2(y_n).$$

By Definition 9.3.1, we have

$$-\alpha u_1(z_n) + \frac{1}{2}\sigma^2 z_n^2 \bar{q}_1 + (f_1(z_n) - \mu z_n)p_1 + \tilde{U}(p_1) \ge 0,$$
$$-\alpha u_2(y_n) + \frac{1}{2}\sigma^2 y_n^2 \bar{q}_2 + (f_2(y_n) - \mu y_n)p_2 + \tilde{U}(p_2) \le 0.$$

Putting these inequalities together, we get

$$\begin{aligned} &\alpha[u_1(z_n) - u_2(y_n)] \\ &\quad \le \frac{1}{2}\sigma^2(z_n^2\bar{q}_1 - y_n^2\bar{q}_2) \\ &\qquad + \{(f_1(z_n) - \mu z_n)p_1 - (f_2(y_n) - \mu y_n)p_2\} + \{\tilde{U}(p_1) - \tilde{U}(p_2)\} \\ &\quad \equiv I_1 + I_2 + I_3. \end{aligned}$$

(4) We consider the case when there are infinitely many $z_n \ge y_n$. By (9.31) and (9.33), it is easy to see that

$$\begin{aligned} I_1 &\le \frac{1}{2}\sigma^2\{3n|z_n - y_n|^2 + \eta(z_n^2\varphi_\nu''(z_n) + y_n^2\varphi_\nu''(y_n))\} \\ &\to \sigma^2\eta\hat{z}^2\varphi_\nu''(\hat{z}) \quad \text{as } n \to \infty. \end{aligned}$$

By monotonicity and $p_1 \ge p_2$,

$$I_3 \le 0.$$

By (9.27), we have $f_1(z_n)(z_n - y_n) \le f_2(z_n)(z_n - y_n)$. Hence

$$\begin{aligned} I_2 &\le (f_2(z_n) - f_2(y_n))n(z_n - y_n) - \mu n(z_n - y_n)^2 \\ &\quad + \eta\{(f_1(z_n) - \mu z_n)\varphi_\nu'(z_n) + (f_2(y_n) - \mu y_n)\varphi_\nu'(y_n)\} \\ &\to \eta(f_1(\hat{z}) + f_2(\hat{z}) - 2\mu\hat{z})\varphi_\nu'(\hat{z}) \quad \text{as } n \to \infty. \end{aligned}$$

Thus, by (9.27) and (9.28)

$$\alpha[u_1(\hat{z}) - u_2(\hat{z})] \leq 2\eta\{\frac{1}{2}\sigma^2\hat{z}^2\varphi_\nu''(\hat{z}) + (f_2(\hat{z}) - \mu\hat{z})\varphi_\nu'(\hat{z})\}$$
$$\leq 2\eta\alpha\varphi_\nu(\hat{z}).$$

This is contrary to (9.32).

(5) Suppose that there are infinitely many $y_n \geq z_n$. By concavity and Lemma 8.3.2, we have $u_i(y_n) \geq u_i(z_n)$, $i = 1, 2$, so that

$$u_1(y_n) - u_2(z_n) \geq u_1(z_n) - u_2(y_n).$$

Hence the maximum of $\Psi_n(z, y)$ is attained at (y_n, z_n). Interchanging y_n and z_n in (3), by the same argument as (4), we get a contradiction. Therefore, we obtain the assertion of the theorem. ∎

9.4 Classical Solutions

In this section, we will show the smoothness of the viscosity solution u of (9.7).

Theorem 9.4.1. *Under (9.3), (9.5), (9.10), (9.16) and (9.20), we have $u \in C^2(0, \infty)$, $u'(0+) = \infty$ and $u'(z) > 0$ for $z > 0$.*

Proof. By virtue of Theorem 8.3.3 (1), we see that $d^+u(z) > 0$ for $z > 0$, where $d^+u(x)$ is the right derivative of u. Let $[a, b] \subset (0, \infty)$ be arbitrary and we consider the boundary value problem:

$$\alpha w(z) = \tfrac{1}{2}\sigma^2 z^2 w''(z) + (f(z) - \mu z)w'(z) + \tilde{U}(w'(z) \vee d^+u(b)) \text{ in } (a, b),$$
$$w(a) = u(a), \quad w(b) = u(b), \tag{9.34}$$

We notice that (9.34) is reduced to (8.19) of Theorem 8.3.3. Thus we have $u \in C^2(0, \infty)$. Furthermore, by Lemma 8.4.2 (1), we obtain the desired result. ∎

Corollary 9.4.2. *Under the assumptions of Theorem 9.4.1, we have the solution $v \in C([0, \infty) \times (0, \infty)) \cap C^2((0, \infty)^2)$ of (9.6).*

Proof. We define v by $v(x, y) = u(x/y)$. Then the proof is immediate from Theorem 9.4.1. ∎

9.5 Optimal Policies

In this section we give a synthesis of the optimal consumption policy $c^* = \{c^*(t)\}$ for the optimization problem (9.4) subject to (9.1) and (9.2). We consider the equation

$$\frac{d}{dt}X^*(t) = F(X^*(t), Y(t)) - \lambda X^*(t) - c^*(t)Y(t), \quad X^*(0) = x \geq 0, \tag{9.35}$$

where

$$c^*(t) = (U')^{-1}(v_x(X^*(t), Y(t))Y(t))1_{\{t \leq \theta(X^*)\}}. \tag{9.36}$$

Our objective is to prove the following theorem.

Theorem 9.5.1. *We assume* (9.3), (9.5), (9.10), (9.16), *and* (9.20). *Then the optimal consumption* $c^* = \{c^*(t)\}$ *is given by* (9.36).

For the proof of Theorem 9.5.1, we need the following two lemmas.

Lemma 9.5.2. *Under the assumptions of Theorem 9.5.1, we have*

$$\liminf_{t\to\infty} E[e^{-\alpha t} v(X(t), Y(t))] = 0, \tag{9.37}$$

for each $c \in \mathcal{A}$.

Proof. We set $\Phi(x, y) = \varphi(x/y)$. By (9.19), we have

$$-\beta\Phi + \frac{1}{2}\sigma^2 y^2 \Phi_{yy} + \kappa y \Phi_y + (F(x, y) - \lambda x)\Phi_x + \tilde{U}(\Phi_x(x, y)y) \le 0,$$
$$x \ge 0, y > 0. \tag{9.38}$$

Hence, Itô's formula gives

$$E[e^{-\alpha t}\Phi(X(t), Y(t))] \le \Phi(x, y) + E[\int_0^t e^{-\alpha s}(-\alpha + \beta)\Phi(X(s), Y(s))ds],$$

from which

$$(\alpha - \beta)E[\int_0^\infty e^{-\beta s}\Phi(X(s), Y(s))ds] \le \Phi(x, y).$$

Thus we get

$$\liminf_{t\to\infty} E[e^{-\alpha t}\Phi(X(t), Y(t))] \le \liminf_{t\to\infty} E[e^{-\beta t}\Phi(X(t), Y(t))] = 0.$$

By Theorem 9.3.6 and Corollary 9.4.2

$$v(x, y) \le \Phi(x, y),$$

which implies (9.37). ∎

Lemma 9.5.3. *Under the assumptions of Theorem 9.5.1, there exists a unique nonnegative solution* $\{X^*(t)\}$ *of* (9.35).

Proof.

(1) Let $g(z) = f(z^+) - \lambda z^+ - (U')^{-1}(u'(z^+))$, $g(0) = 0$ and $G(x, y) = F(x^+, y) - \lambda x^+ - (U')^{-1}(v_x(x^+, y)y)y$ for $z, x \in \mathbf{R}$ and $y > 0$. Clearly, the map $z \to g(z)$ is continuous and $G(x, y) = g(x/y)y$. Also, for every $n \in \mathbf{N}$

$$\begin{cases} |g(z)| \le K_n|z|, \\ |g(z) - g(z')| \le K_n|z - z'| \quad \text{if } 1/n \le z, z' \le n, \end{cases}$$

where the constant $K_n > 0$ depends only on n. We set

$$\xi_n(x) = \begin{cases} n & \text{if } x > n, \\ x & \text{if } 1/n \le x \le n, \\ 1/n & \text{if } x < 1/n. \end{cases}$$

Then ξ_n is bounded and Lipschitz on $\mathbf{R}$. By the same line as Theorem 1.7.2, there exists a unique $\{\mathcal{F}_t\}$-adapted solution $\{X_n(t)\}$ of

$$X_n(t) = x + \int_0^t g \circ \xi_n(X_n(s)/Y(s))Y(s)ds, \quad \text{a.s.}, \quad x \geq 0.$$

By analogy with the proof of Theorem 1.9.1, we can show the following:

① If $n < m, \ X_n(t \wedge \tau_n) = X_m(t \wedge \tau_n)$ for all $t \geq 0$ a.s.,
where $\tau_n = \inf\{t \geq 0 : X_n(t)/Y(t) < 1/n \text{ or } X_n(t)/Y(t) > n\}$.

② τ_n is nondecreasing since $1/m < X_m(\tau_n)/Y(\tau_n) < m$.

③ $X_n(t) = X_m(t)$ for $t \leq \tau_n$, a.s.

④ Define $\mathsf{e} = \lim_{n\to\infty} \tau_n$ and $\chi(t) = \lim_{n\to\infty} X_n(t)$, a.s.
Then $g \circ \xi_n(X_n(t)/Y(t)) = g(\chi(t)/Y(t))$, for $t \leq \tau_n$.

⑤ The process $\{\chi(t)\}$ is $\{\mathcal{F}_t\}$-adapted and satisfies

$$\begin{aligned}\chi(t) &= x + \int_0^t g(\chi(s)/Y(s))Y(s)ds \\ &= x + \int_0^t G(\chi(s), Y(s))ds, \quad \text{on } \{t < \mathsf{e}\} \text{ a.s.}\end{aligned}$$

Now, we note by the concavity of f that $G(x, y) \leq \hat{C}(1 + x + y)$, for some $\hat{C} > 0$. Applying the comparison theorem, we obtain $\chi(t \wedge \tau_n) \leq \hat{X}(t \wedge \tau_n)$, for all $t \geq 0$ a.s., where

$$d\hat{X}(t) = \hat{C}(1 + \hat{X}(t) + Y(t))dt, \quad \hat{X}(0) = x \geq 0.$$

Hence $\chi(t) \leq \hat{X}(t) < \infty$ on $\{t < \mathsf{e}\}$ a.s. Hence $P(\chi(\mathsf{e}) = \infty) = 0$, that is, $\{\chi(t)\}$ has no explosion above. Define

$$X^*(t) = \chi(t \wedge \theta(\chi)) \geq 0.$$

It is clear that $\theta(X^*) = \theta(\chi)$ and

$$\begin{aligned}\frac{d}{dt}X^*(t) &= 1_{\{t\leq\theta(\chi)\}}G(\chi(t), Y(t)) \\ &= 1_{\{t\leq\theta(X^*)\}}[F(X^*(t), Y(t)) - \lambda X^*(t) - c^*(t)Y(t)].\end{aligned}$$

Therefore $\{X^*(t)\}$ solves (9.35). Clearly, $\{c^*(t)\}$ belongs to $\mathcal{A}$.

(2) To prove uniqueness, let $\{\tilde{X}(t)\}$ be another solution of (9.35). Since g is locally Lipschitz, we can show by Gronwall's inequality that

$$X^*(t \wedge \theta(X^*) \wedge \theta(\tilde{X})) = \tilde{X}(t \wedge \theta(X^*) \wedge \theta(\tilde{X})), \quad \text{a.s.}$$

Hence

$$\theta(X^*) = \theta(\tilde{X}), \quad X^*(t) = \tilde{X}(t) \quad \text{for} \quad t \leq \theta(X^*), \quad \text{a.s.}$$

This implies $X^*(t) = \tilde{X}(t)$ for all $t \geq 0$ a.s. ■

Proof of Theorem 9.5.1. We set $\zeta = s \wedge \theta(X^*)$ for any $s > 0$. By (9.6), (9.35), and Itô's formula, we have

$$\begin{aligned} E[e^{-\alpha\zeta_n} v(X^*(\zeta_n), Y(\zeta_n))] &= v(x,y) + E[\int_0^{\zeta_n} e^{-\alpha t}\{-\alpha v(x,y) \\ &\quad + v_x(x,y)[F(x,y) - \lambda x - c^*(t)y] + \kappa y v_y(x,y) \\ &\quad + \frac{1}{2}\sigma^2 y^2 v_{yy}(x,y)\Big|_{(x,y)=(X^*(t),Y(t))} \}dt + M(\zeta_n)] \\ &= v(x,y) - E[\int_0^{\zeta_n} e^{-\alpha t} U(c^*(t))dt], \end{aligned}$$

where $\zeta_n := \zeta \wedge \tau_n$ for a localizing sequence of stopping times $\tau_n \nearrow \infty$ of the local martingale $M(t) := \int_0^t e^{-\alpha r} v_y(X^*(r), Y(r))\sigma Y(r) dB(r)$. From (9.38) and Doob's maximal inequality, it follows that

$$\begin{aligned} E[\sup_n e^{-\alpha\zeta_n} v(X^*(\zeta_n), Y(\zeta_n))] &\le E[\sup_{0\le r\le s} e^{-\alpha r}\Phi(X^*(r), Y(r))] \\ &\le \Phi(x,y) + E[\sup_{0\le r\le s} |\int_0^r e^{-\alpha t}\Phi_y(X^*(t), Y(t))\sigma Y(t) dB(t)|] \\ &= \Phi(x,y) + E[\sup_{0\le r\le s} |\int_0^r e^{-\alpha t}\sigma Z^*(t) dB(t)|] \\ &\le \Phi(x,y) + 2|\sigma| E[\int_0^s (Z^*(t))^2 dt]^{1/2}, \end{aligned}$$

where $Z^*(t) = X^*(t)/Y(t)$. Since $\{Z^*(t)\}$ fulfills (9.13) with c^* replacing c, we have

$$E[\int_0^s (Z^*(t))^2 dt] < \infty.$$

Letting $n \to \infty$, by the dominated convergence theorem, we get

$$E[e^{-\alpha\zeta} v(X^*(\zeta), Y(\zeta))] = v(x,y) - E[\int_0^{\zeta} e^{-\alpha t} U(c^*(t))dt].$$

By (9.37) and $v(0, y) = 0$, we have

$$\liminf_{s\to\infty} E[e^{-\alpha\zeta} v(X^*(\zeta), Y(\zeta))] = \liminf_{s\to\infty} E[e^{-\alpha s} v(X^*(s), Y(s)) 1_{\{\theta(X^*)\ge s\}}] = 0.$$

Passing to the limit, we deduce

$$J(c^*) = E[\int_0^{\theta(X^*)} e^{-\alpha t} U(c^*(t))dt] = v(x,y).$$

By the same calculation as above, we obtain

$$J(c) = E[\int_0^{\theta(X)} e^{-\alpha t} U(c(t))dt] \le v(x,y), \quad \text{for} \quad c \in \mathcal{A}. \qquad \blacksquare$$

Remark 9.5.4. *For a consumption problem with the Cobb–Douglas production function, see H. Morimoto and X. Y. Zhou [126]*

10

Optimal Pollution Control with Long-Run Average Criteria

In this chapter, we study the theory of *optimal pollution management* in environmental economics. We consider a society consuming some good, which generates pollution as a byproduct of this consumption.

The pollution stock $X(t)$ is only gradually degraded, and its growth rate incorporates a random shock with mean of zero and constant standard deviation r. The social welfare is defined by the utility $U(c)$ of the consumption c net of the disutility $D(x)$ of pollution x. The objective of the social planner is to choose time paths for consumption to maximize the social welfare with long-run average criteria.

By using the vanishing discount technique, we solve the HJB equation (10.6) associated with the long-run average problem as the limit equation when the discount rate β converges to zero. The optimal consumption policy is shown to exist in a feedback form, and the maximum value is independent of the initial condition $X(0) > 0$.

10.1 The Model

Consider a society consuming a homogeneous good and accumlating pollution. Define the following quantities:

$X(t) =$ stock of pollution at time t.
$r =$ the constant rate of pollution decay, $r > 0$.
$L =$ the upper bound of the maximum flow of pollusion, $L > 0$.
$c(t) =$ flow of pollution (or consumption) at time t.
$B(t) =$ the standard Brownian motion.
$\sigma =$ the nonzero diffusion constant.
$U(c) =$ the social utility function.
$D(x) =$ the social disutility of the pollution stock x.

We assume that the stock of pollution evolves according to the stochastic differential equation

$$dX(t) = [c(t) - rX(t)]dt + \sigma X(t)dB(t), \quad X(0) = x > 0, \tag{10.1}$$

on a complete probability space $(\Omega, \mathcal{F}, P)$, carrying a standard Brownian motion $\{B(t)\}$, endowed with the natural filtration $\mathcal{F}_t$ generated by $\sigma(B(s), s \leq t)$. The consumption process $c = \{c(t)\}$ is said to be admissible if it is an $\{\mathcal{F}_t\}$-progressively measurable process such that

$$0 \leq c(t) \leq L, \qquad \text{for all } t \geq 0, \tag{10.2}$$

where L is the given quantitive restriction which represents the global standard that must be met in order to promote sustainable development. We denote the set of all admissible controls by $\mathcal{A}$. For $c \in \mathcal{A}$, we observe by the comparison theorem that (10.1) admits a unique positive solution $\{X(t)\}$.

The objective is to find an optimal policy $c^* = \{c^*(t)\}$ so as to maximize the long-run average welfare defined by

$$\bar{J}(c) = \liminf_{T\to\infty} \frac{1}{T} E[\int_0^T \{U(c(t)) - D(X(t))\}dt] \tag{10.3}$$

over the class $\mathcal{A}$. Here $U \in C^2(0,\infty) \cap C[0,\infty)$ and $D \in C(0,\infty)$ are assumed to have the following properties:

$$\begin{cases} U''(c) < 0, \\ U'(\infty) = U(0) = 0, \quad U'(0+) = U(\infty) = \infty, \end{cases} \tag{10.4}$$

$$\begin{cases} D(x) \geq 0 : \text{convex}, \\ D_0 x - D_1 \leq D(x) \leq D_1(1 + x^m), \\ |D(x) - D(y)| \leq D_1|x - y|(1 + x^{m-1} + y^{m-1}), \quad x, y > 0, \\ \quad \text{for some } m \in \mathbf{N} \text{ and } D_0, D_1 > 0. \end{cases} \tag{10.5}$$

The HJB equation associated with this pollution control problem is given by

$$\lambda = \frac{1}{2}\sigma^2 x^2 v'' - rxv' + \tilde{U}_L(-v') - D(x), \qquad x > 0, \tag{10.6}$$

where the unknown is a pair $(\lambda, v) \in [0,\infty) \times C^2(0,\infty)$ and $\tilde{U}_L(x) = \sup_{0\leq c\leq L}\{U(c) - xc\}$. Let u be a solution of the HJB equation (10.11) with discount rate $\beta > 0$:

$$\beta u = \frac{1}{2}\sigma^2 x^2 u'' - rxu' + \tilde{U}_L(-u') - D(x), \quad x > 0.$$

It will be shown that we obtain the solution (λ, v) of (10.6) by the limit of $(\beta u(0+), u(x) - u(0+))$ as $\beta \to 0$. This is called the vanishing discount method (cf. A. Bensoussan [17]).

Remark 10.1.1. *When $D(x) = x$, we can show that* (10.6) *admits a solution of the form*

$$\lambda = \tilde{U}_L(1/r), \quad v(x) = -x/r + \text{constant}$$

If $D(x) = x^m$, then (10.5) *is fulfilled.*

10.2 Moments

In this section, we present some bounds on the moment $E[X(t)^k]$ of the solution $\{X(t)\}$ to (10.1) for $k \in \mathbf{N}$. Let $\{Z(t)\}$ be the solution of

$$dZ(t) = [L - rZ(t)]dt + \sigma Z(t)dB(t), \quad Z(0) = x > 0. \tag{10.7}$$

Lemma 10.2.1. *Let $k \in \mathbf{N}$ be such that*

$$2r > (k-1)\sigma^2. \tag{10.8}$$

Then we have

$$\sup_{t \geq 0} E[Z(t)^k] \leq x^k + C \tag{10.9}$$

for some constant $C > 0$.

Proof.

(1) Let $k = 1$. By (10.7) and Itô's formula,

$$\begin{aligned} e^{rt}Z(t) &= x + \int_0^t re^{rs}Z(s)ds + \int_0^t e^{rs}dZ(s) \\ &= x + \int_0^t Le^{rs}ds + \int_0^t e^{rs}\sigma Z(s)dB(s), \quad \text{a.s.} \end{aligned}$$

Therefore, we have

$$\begin{aligned} E[Z(t)] = E[Z(t)e^{rt}]e^{-rt} &= (x + L\int_0^t e^{rs}ds)e^{-rt} \\ &\leq x + L/r. \end{aligned}$$

(2) Let $k \geq 2$. By (10.7) and Itô's formula,

$$\begin{aligned} dZ(t)^k &= kZ(t)^{k-1}dZ(t) + \tfrac{1}{2}k(k-1)\sigma^2 Z(t)^k dt \\ &= k(\tfrac{k-1}{2}\sigma^2 Z(t)^k + LZ(t)^{k-1} - rZ(t)^k)dt + k\sigma Z(t)^k dB(t). \end{aligned}$$

By (10.8), we take $\nu > 0$ such that

$$\mu := -\{\frac{k(k-1)}{2}\sigma^2 + (k-1)\nu^{k/(k-1)} - kr\} > 0.$$

By Young's inequality,

$$kZ(t)^{k-1}L = k(\nu Z(t)^{k-1} \cdot L/\nu) \leq (k-1)\nu^{k/(k-1)}Z(t)^k + (L/\nu)^k.$$

Then, setting $\theta := (L/\nu)^k$, we have

$$k(\frac{k-1}{2}\sigma^2 Z(t)^k + LZ(t)^{k-1} - rZ(t)^k) \le \theta - \mu Z(t)^k.$$

Let $\{\bar{Z}(t)\}$ be the solution of

$$d\bar{Z}(t) = [\theta - \mu\bar{Z}(t)]dt + k\sigma\bar{Z}(t)dB(t), \qquad \bar{Z}(0) = x^k.$$

We apply the comparison theorem to $\{Z(t)^k\}$ and $\{\bar{Z}(t)\}$ to obtain

$$Z(t)^k \le \bar{Z}(t), \quad t \ge 0, \quad \text{a.s.}$$

Therefore, by (1),

$$E[Z(t)^k] \le E[\bar{Z}(t)] \le x^k + \theta/\mu,$$

which implies (10.9). ■

Proposition 10.2.2. *We assume* (10.8). *Then*

$$\sup_{t\ge 0} E[X(t)^k] \le x^k + C_L, \tag{10.10}$$

where the constant $C_L > 0$ *depends only on* $L > 0$.

Proof. By (10.2), it is clear that

$$c(t) - rX(t) \le L - rX(t), \quad t \ge 0, \quad \text{a.s.}$$

By the comparison theorem, we have

$$X(t)^k \le Z(t)^k, \quad t \ge 0, \quad \text{a.s.}$$

Thus, by Lemma 10.2.1, we obtain (10.10). ■

10.3 The HJB Equations with Discount Rates

In this section, we study the HJB equation with discount rate $0 < \beta < 1$:

$$\beta u = \frac{1}{2}\sigma^2 x^2 u'' - rxu' + \tilde{U}_L(-u') - D(x), \quad x > 0. \tag{10.11}$$

We define

$$\begin{cases} u(x) = \sup_{c\in\mathcal{A}} J_x(c), \\ J_x(c) = E[\int_0^\infty e^{-\beta t}\{U(c(t)) - D(X(t))\}dt], \quad x > 0, \end{cases}$$

where the supremum is taken over all systems $(\Omega, \mathcal{F}, P, \{\mathcal{F}_t\}; \{B(t)\}, \{c(t)\})$. We will show that u is a viscosity solution of (10.11) and it is smooth.

10.3.1 Viscosity Solutions

Definition 10.3.1. *$u \in C(0,\infty)$ is a viscosity solution of* (10.11) *if*

(a) *u is a viscosity subsolution of* (10.11), *that is, for any $\phi \in C^2(0,\infty)$ and any local maximum point $z > 0$ of $u - \phi$,*

$$\beta u(z) \leq \frac{1}{2}\sigma^2 z^2 \phi''(z) - rz\phi'(z) + \tilde{U}_L(-\phi'(z)) - D(z),$$

and

(b) *u is a viscosity supersolution of* (10.11), *that is, for any $\phi \in C^2(0,\infty)$ and any local minimum point $z > 0$ of $u - \phi$,*

$$\beta u(z) \geq \frac{1}{2}\sigma^2 z^2 \phi''(z) - rz\phi'(z) + \tilde{U}_L(-\phi'(z)) - D(z).$$

Lemma 10.3.2. *Assume* (10.4) *and* (10.5) *and that*

$$2r > (m-1)\sigma^2. \tag{10.12}$$

Then u is concave and there exist $\gamma_i > 0$, $i = 0,1,2,3$, independent of β, such that

$$-\gamma_0(x^m + 1) - \gamma_1/\beta \leq u(x) \leq -\gamma_2 x + \gamma_3/\beta. \tag{10.13}$$

Proof.

(1) Let $x_i > 0$, $i = 1,2$, and $0 \leq \xi \leq 1$. For any $\varepsilon > 0$, there exists $c_i \in \mathcal{A}$ such that

$$u(x_i) - \varepsilon < E[\int_0^\infty e^{-\beta t}\{U(c_i(t)) - D(X_i(t))\}dt],$$

on the same probability space, where $\{X_i(t)\}$ is the solution of (10.1) corresponding to c_i with $X_i(0) = x_i$. We set $c_\xi(t) = \xi c_1(t) + (1-\xi)c_2(t)$ and $X_\xi(t) = \xi X_1(t) + (1-\xi)X_2(t)$. By (10.1), it is easy to see that

$$dX_\xi(t) = [c_\xi(t) - rX_\xi(t)]dt + \sigma X_\xi(t)dB(t), \quad X_\xi(0) = \xi x_1 + (1-\xi)x_2.$$

Furthermore, $\{c_\xi(t)\}$ belongs to $\mathcal{A}$. By (10.4) and (10.5), we note that $U(c)$ and $-D(x)$ are concave. Then

$$\begin{aligned} u(\xi x_1 + (1-\xi)x_2) &\geq E[\int_0^\infty e^{-\beta s}\{U(c_\xi(t)) - D(X_\xi(t))\}ds] \\ &\geq \xi E[\int_0^\infty e^{-\beta s}\{U(c_1(s)) - D(X_1(t))\}ds] \\ &\quad + (1-\xi)E[\int_0^\infty e^{-\beta s}\{U(c_2(s)) - D(X_2(t))\}ds] \\ &> \xi u(x_1) + (1-\xi)u(x_2) - \varepsilon. \end{aligned}$$

Therefore, letting $\varepsilon \to 0$, we deduce that u is concave.

(2) Let $\{\hat{X}(t)\}$ be the solution of (10.1) corresponding to $c = 0$. Then

$$\hat{X}(t) = x \exp(-rt + \sigma B(t) - \frac{1}{2}\sigma^2 t).$$

Since

$$E[\exp(\sigma m B(t) - \frac{1}{2}\sigma^2 m^2 t)] = 1,$$

we have by (10.5) and (10.12),

$$\begin{aligned} u(x) &\geq -E[\int_0^\infty e^{-\beta t} D(\hat{X}(t))dt] \geq -E[\int_0^\infty e^{-\beta t} D_1(1 + \hat{X}(t)^m)dt] \\ &= -\int_0^\infty e^{-\beta t} D_1(1 + x^m \exp\{(\frac{m-1}{2}\sigma^2 - r)mt\})dt \\ &\geq -\gamma_0 x^m - \gamma_1/\beta, \end{aligned}$$

where $\gamma_0 := D_1/\{r - \frac{(m-1)}{2}\sigma^2\}m$ and $\gamma_1 = D_1$.

(3) We consider the supersolution $\varphi(x)$ of (10.11), that is,

$$\beta\varphi(x) \geq \frac{1}{2}\sigma^2 x^2 \varphi''(x) - rx\varphi'(x) + \tilde{U}_L(-\varphi'(x)) - D(x), \quad x > 0. \tag{10.14}$$

Substituting $\varphi(x) := -\gamma_2 x + \gamma_3/\beta$ into (10.14), we have

$$D(x) - (r + \beta)\gamma_2 x + \gamma_3 \geq \tilde{U}_L(\gamma_2).$$

By (10.5), we choose $0 < \gamma_2 \leq \gamma_0$ and $\gamma_3 > \gamma_1 > 0$ such that

$$D_0 > (r + 1)\gamma_2, \quad -D_1 + \gamma_3 \geq \tilde{U}_L(\gamma_2).$$

Then

$$D(x) - (r + \beta)\gamma_2 x + \gamma_3 \geq \min_x\{D_0 x - (r + 1)\gamma_2 x\} - D_1 + \gamma_3 \geq \tilde{U}_L(\gamma_2).$$

Thus we see that $\varphi(x)$ satisfies (10.14) and

$$\varphi(x) \geq -\gamma_0(x^m + 1).$$

From (10.14) and Itô's formula it follows that

$$\begin{aligned} e^{-\beta t}\varphi(X(t)) &= \varphi(x) + \int_0^t e^{-\beta s}\{-\beta\varphi(X(s)) + \varphi'(X(s))(c(s) - rX(s)) \\ &\quad + \frac{1}{2}\sigma^2 X(s)^2 \varphi''(X(s))\}ds + \int_0^t e^{-\beta s}\varphi'(X(s))\sigma X(s)dB(s) \\ &\leq \varphi(x) - \int_0^t e^{-\beta s}\{U(c(s)) - D(X(s))\}ds \\ &\quad + \int_0^t e^{-\beta s}\varphi'(X(s))\sigma X(s)dB(s), \quad \text{a.s.} \end{aligned}$$

By the comparison theorem, we get

$$E[\int_0^t X(s)^2 ds] \le E[\int_0^t Z(s)^2 ds] < \infty.$$

Hence $\{\int_0^t e^{-\beta s}\varphi'(X(s))\sigma X(s) dB(s)\}$ is a martingale. Therefore,

$$E[e^{-\beta t}\varphi(X(t))] + E[\int_0^t e^{-\beta s}\{U(c(s)) - D(X(s))\}ds] \le \varphi(x).$$

Note by (10.10) that

$$\liminf_{t\to\infty} E[e^{-\beta t}\varphi(X(t))] \ge \lim_{t\to\infty} e^{-\beta t}\{-\gamma_0 E[X(t)^m + 1]\} = 0.$$

Thus, we deduce $u(x) \le \varphi(x)$, which implies (10.13). ■

Lemma 10.3.3. *Assume* (10.4), *and* (10.5) *and that*

$$2r > m(m-1)\sigma^2. \tag{10.15}$$

Then there exists $C_\rho > 0$, *for any* $\rho > 0$, *such that*

$$|u(x) - u(y)| \le C_\rho |x-y|^m + \rho(1 + x^m + y^m), \quad x,\ y > 0. \tag{10.16}$$

Proof. Let $\{Y(t)\}$ be the solution of (10.1) with $Y(0) = y > 0$. Clearly,

$$|X(t) - Y(t)| = |x-y|\exp(-rt + \sigma B(t) - \frac{1}{2}\sigma^2 t).$$

By definition,

$$\begin{aligned} u(x) - u(y) &\le \sup_{c\in\mathcal{A}} E[\int_0^\infty e^{-\beta t}\{\{U(c(t)) - D(X(t))\} \\ &\quad - \{U(c(t)) - D(Y(t))\}\}dt] \\ &\le \sup_{c\in\mathcal{A}} E[\int_0^\infty e^{-\beta t}|D(X(t)) - D(Y(t))|dt]. \end{aligned}$$

By (10.5) and (10.10), we get (10.16) with $C_\rho = 3D_1/r$ if $m = 1$. Let $m \ge 2$. By (10.5) and Young's inequality, we have for $\varrho > 0$

$$\begin{aligned} &E[|D(X(t)) - D(Y(t))|] \\ &\le D_1 E[e^{-rt}\cdot\frac{1}{\varrho}e^{rt}|X(t) - Y(t)|\varrho(1 + X(t)^{m-1} + Y(t)^{m-1})] \\ &\le D_1 E[e^{-rt}\{\frac{|x-y|^m}{m\varrho^m}\exp(\sigma m B(t) - \frac{m}{2}\sigma^2 t) \\ &\quad + \frac{m-1}{m}\varrho^{m/(m-1)}(1 + X(t)^{m-1} + Y(t)^{m-1})^{m/(m-1)}\}] \\ &\le D_1 e^{-rt}\{\frac{|x-y|^m}{m\varrho^m}\exp(\frac{m(m-1)}{2}\sigma^2 t) \\ &\quad + \frac{m-1}{m}(3\varrho)^{m/(m-1)}E[1 + X(t)^m + Y(t)^m]\}. \end{aligned} \tag{10.17}$$

By (10.15), we get

$$\int_0^\infty e^{-\beta t} e^{-rt} \exp(\frac{m(m-1)}{2}\sigma^2 t) dt \le 1/(r - \frac{m(m-1)}{2}\sigma^2) =: \bar{m}.$$

Also, by (10.10)

$$E[\int_0^\infty e^{-\beta t} e^{-rt} \, X(t)^m dt] \le (x^m + C_L)/r$$

for some constant $C_L > 0$. Therefore, by (10.17), we deduce

$$\begin{aligned} u(x) - u(y) \le D_1 \{ & \bar{m}|x-y|^m / m\varrho^m \\ & + \frac{m-1}{m}(3\varrho)^{\frac{m}{m-1}} \{1 + (x^m + C_L)/r + (y^m + C_L)/r\}. \end{aligned}$$

Interchanging x and y and taking sufficiently small $\varrho > 0$, we obtain (10.16). ■

Theorem 10.3.4. *We assume* (10.4), (10.5), *and* (10.15). *Then u is a viscosity solution of* (10.11).

Proof. In view of Theorem 4.3.1, the viscosity property of u follows from the dynamic programming principle for u, that is,

$$u(x) = \sup_{c \in \mathcal{A}} E[\int_0^\tau e^{-\beta s} \{U(c(s)) - D(X(s))\} ds + e^{-\beta\tau} u(X(\tau))], \quad x > 0, \tag{10.18}$$

for any $\tau \in \mathcal{S}_b$, where the supremum is taken over all systems $(\Omega, \mathcal{F}, P, \{\mathcal{F}_t\}; \{B(t)\}, \{c(t)\})$. In order to see the important roles of (10.10) and (10.16), by the same line as Theorem 3.2.5, we sketch the proof of (10.18) on the Polish space Ω when τ is nonrandom.

(1) Let $\tilde{u}(x)$ denote the right-hand side of (10.18). By Theorem 3.2.5 (1), for $c = \{c(t)\} \in \mathcal{A}$, we have

$$\begin{aligned} J_{X(\tau)}(\tilde{c}) &= E[\int_0^\infty e^{-\beta t} \{U(\tilde{c}(t)) - D(\tilde{X}(t))\} dt | \mathcal{F}_\tau] \\ &= e^{\beta\tau} E[\int_\tau^\infty e^{-\beta t} \{U(c(t)) - D(X(t))\} dt | \mathcal{F}_\tau], \quad \text{a.s.,} \end{aligned}$$

and hence

$$\begin{aligned} J_x(c) &= E[\int_0^\tau e^{-\beta t} \{U(c(t)) - D(X(t))\} dt \\ &\quad + \int_\tau^\infty e^{-\beta t} \{U(c(t)) - D(X(t))\} dt] \\ &= E[\int_0^\tau e^{-\beta t} \{U(c(t)) - D(X(t))\} dt + e^{-\beta\tau} J_{X(\tau)}(\tilde{c})], \end{aligned}$$

where $\{\tilde{X}(t)\}$ is the solution of (10.1) corresponding to the shifted process $\tilde{c}$ of c by τ with $\tilde{X}(0) = X(\tau)$. Therefore,

$$J_x(c) \leq E[\int_0^\tau e^{-\beta t}\{U(c(t)) - D(X(t))\}dt + e^{-\beta\tau}u(X(\tau))].$$

Taking the supremum, we deduce $u \leq \tilde{u}$.

(2) Let $\rho > 0$ be arbitrary. Let $\{S_j : j = 1, \ldots, n+1\}$ be a sequence of of disjoint subsets of $(0, \infty)$ such that

$$\operatorname{diam}(S_j) < \delta, \quad \bigcup_{j=1}^{n} S_j = (0, R) \quad \text{and} \quad S_{n+1} = [R, \infty),$$

for $\delta, R > 0$ chosen later. We take $x_j \in S_j$ and $c^{(j)} \in \mathcal{A}$ such that

$$u(x_j) - \rho \leq J_{x_j}(c^{(j)}), \quad j = 1, \ldots, n+1. \tag{10.19}$$

By the same calculation as the proof of Lemma 10.3.3, we find $C_\rho > 0$ such that for $c \in \mathcal{A}$,

$$|J_x(c) - J_y(c)| \leq C_\rho|x - y|^m + \rho(1 + x^m + y^m), \quad x, y > 0.$$

Also, by (10.16), we take $0 < \delta < 1$ as $C_\rho\delta^m < \rho$ to obtain for $|x - y| < \delta$,

$$\begin{aligned}|u(x) - u(y)| &\leq \rho(2 + x^m + y^m)\\ &< \Xi_\rho(x) := \rho(2 + x^m + 2^m(1 + x^m)), \quad x, y > 0.\end{aligned}$$

Hence

$$\begin{aligned}J_{X(\tau)}(c^{(j)}) &= \{J_{X(\tau)}(c^{(j)}) - J_{x_j}(c^{(j)})\} + J_{x_j}(c^{(j)})\\ &\geq -\Xi_\rho(X(\tau)) + J_{x_j}(c^{(j)})\\ &\geq -\Xi_\rho(X(\tau)) + u(x_j) - \rho\\ &\geq -2\Xi_\rho(X(\tau)) + u(X(\tau)) - \rho \quad \text{if } X(\tau) \in S_j, j = 1, \ldots, n.\end{aligned} \tag{10.20}$$

By definition, we can find $c \in \mathcal{A}$ such that

$$\tilde{u}(x) - \rho \leq E[\int_0^\tau e^{-\beta t}\{U(c(t)) - D(X(t))\}dt + e^{-\beta\tau}u(X(\tau))].$$

Define $c^\tau = \{c^\tau(t)\} \in \mathcal{A}$ by

$$c^\tau(t) = c(t)1_{\{t<\tau\}} + c^{(j)}(t - \tau)1_{\{t\geq\tau\}} \quad \text{if } X(\tau) \in S_j, j = 1, \ldots, n+1,$$

on the same probability space. Clearly, c^τ belongs to $\mathcal{A}$.

Now, taking into account Theorem 1.7.2, we choose $R > 0$ such that

$$\begin{aligned}\sup_{c\in\mathcal{A}} E[u(X(\tau))1_{\{X(\tau)\geq R\}}] &\leq \tfrac{U(L)}{\beta}\sup_{c\in\mathcal{A}} P(X(\tau) \geq R)\\ &\leq \tfrac{U(L)}{\beta}P(Z(\tau) \geq R) \leq \tfrac{U(L)}{\beta R^2}E[Z(\tau)^2] < \rho,\end{aligned}$$

and

$$\begin{aligned}E[e^{-\beta\tau}J_{X(\tau)}(c)1_{\{X(\tau)\geq R\}}] &\geq -\tilde{C}E[e^{-\beta\tau}(1+X(\tau)^m)1_{\{X(\tau)\geq R\}}]\\ &\geq -\tilde{C}E[(1+X(\tau)^m)1_{\{X(\tau)\geq R\}}]\\ &\geq -\tilde{C}(E[X(\tau)^2]+E[X(\tau)^{2m}]/R^2)\\ &\geq -\tilde{C}(E[Z(\tau)^2]+E[Z(\tau)^{2m}]/R^2) > -\rho,\end{aligned}$$

since

$$J_x(c) \geq -\sup_{c\in\mathcal{A}} E[\int_0^\infty e^{-\beta t}D_1(1+X(t)^m)dt] \geq -\tilde{C}(1+x^m), \quad \forall c\in\mathcal{A},$$

where $\tilde{C} = D_1(1+C_L)/\beta$. Also, by (10.10) and (10.15)

$$E[\Xi_\rho(X(\tau))] \leq \rho C(1+x^m),$$

for some constant $C > 0$, independent of ρ. Hence, by (10.19) and (10.20), we have

$$\begin{aligned}&E[\int_\tau^\infty e^{-\beta t}\{U(c^\tau(t)) - D(X^\tau(t))\}dt]\\ &= E[E[\int_\tau^\infty e^{-\beta t}\{U(c^\tau(t)) - D(X^\tau(t))\}dt|\mathcal{F}_\tau]]\\ &= E[e^{-\beta\tau}J_{X(\tau)}(\tilde{c}^\tau)]\\ &= E[e^{-\beta\tau}\sum_{j=1}^{n+1}J_{X(\tau)}(c^{(j)})1_{\{X(\tau)\in S_j\}}]\\ &\geq E[e^{-\beta\tau}\sum_{j=1}^{n}\{u(X(\tau)) - 2\Xi_\rho(X(\tau)) - \rho\}1_{\{X(\tau)\in S_j\}}] - \rho\\ &\geq E[e^{-\beta\tau}u(X(\tau))] - \rho - \{2\rho C(1+x^m)+\rho\} - \rho,\end{aligned}$$

where $\{X^\tau(t)\}$ denotes the solution of (10.1) corresponding to c^τ. Thus

$$\begin{aligned}u(x) &\geq E[\int_0^\tau e^{-\beta t}\{U(c^\tau(t)) - D(X^\tau(t))\}dt\\ &\quad + \int_\tau^\infty e^{-\beta t}\{U(c^\tau(t)) - D(X^\tau(t))\}dt]\\ &\geq E[\int_0^\tau e^{-\beta t}\{U(c(t)) - D(X(t))\}dt + e^{-\beta\tau}u(X(\tau))]\\ &\quad - \{2C(1+x^m)+3\}\rho\\ &\geq \tilde{u}(x) - \rho - \{2C(1+x^m)+3\}\rho.\end{aligned}$$

Therefore, letting $\rho \to 0$, we can obtain $\tilde{u} \leq u$. ■

10.3.2 Classical Solutions

Theorem 10.3.5. *We make the assumptions of Theorem 10.3.4. Then u is a classical solution of* (10.11) *in* $C^2(0,\infty)$.

Proof. Let $[a,b] \subset (0,\infty)$ be arbitrary and we consider the boundary value problem:

$$\beta w(x) = \frac{1}{2}\sigma^2 x^2 w''(x) - rxw'(x) + \tilde{U}_L(-w'(x)) - D(x) \text{ in } (a,b),$$
$$w(a) = u(a), \quad w(b) = u(b). \tag{10.21}$$

By (10.5), the uniform ellipticity, and the Lipshitz continuity of $\tilde{U}_L(\cdot)$, Theorem 5.3.7 yields that (10.21) has a smooth solution w. By Theorem 5.3.1, we obtain the uniqueness of the viscosity solutions of (10.21). Therefore, we deduce that $u = w \in C^2(a,b)$, and hence $u \in C^2(0,\infty)$. ■

10.4 Solution of the HJB Equation

In this section, we study the HJB equation associated with the pollution control problem:

$$\lambda = \frac{1}{2}\sigma^2 x^2 v'' - rxv' + \tilde{U}_L(-v') - D(x), \qquad x > 0. \tag{10.22}$$

We give a priori estimates of the solution of (10.11), denoted by u_β, for approximation.

Theorem 10.4.1. *We assume* (10.4), (10.5), *and* (10.15). *Then we have*

$$\sup_{0<\beta<1} \beta \|u_\beta\|_{C[a,b]} < \infty, \tag{10.23}$$

$$\sup_{0<\beta<1} \|u'_\beta\|_{C[a,b]} < \infty, \tag{10.24}$$

$$\sup_{0<\beta<1} \|u''_\beta\|_{C[a,b]} < \infty, \tag{10.25}$$

for each $0 < a < b$.

Proof. (10.23) follows from (10.13) immediately. We change the variable by $y = \log x$ and set $f_\beta(y) = -u_\beta(e^y)$. Since

$$\begin{cases} f'_\beta(y) = -u'_\beta(x)x, \\ f''_\beta(y) = -x^2 u''_\beta(x) + f'_\beta(y), \end{cases} \tag{10.26}$$

we can rewrite (10.11) as

$$\beta f_\beta = \frac{1}{2}\sigma^2 f''_\beta - (r + \frac{\sigma^2}{2}) f'_\beta - \tilde{U}_L(f'_\beta\, e^{-y}) + D(e^y), \quad y \in [c,d] \subset \mathbf{R}. \tag{10.27}$$

We will show that

$$\sup_{0<\beta<1} \|f''_\beta\|_{C[c,d]} < \infty. \tag{10.28}$$

We set $h(y) = f_\beta(y) - \min_{z\in[c,d]} f_\beta(z)$. By (10.27), we have

$$h'' = G(y, h, h') \quad \text{in } [c, d],$$

where

$$G(y, h, h') = \frac{2}{\sigma^2}\{(r + \frac{\sigma^2}{2})h' + \tilde{U}_L(h'e^{-y}) - D(e^y) + \beta \min_z f_\beta(z) + \beta h\}.$$

Also, by (10.16)

$$\sup_{0<\beta<1} \|h\|_{C[c,d]} < \infty.$$

Hence, taking into account (10.23), we see that

$$|G(y, h, h')| \le \bar{C}(1 + |h'|), \quad y \in [c, d]$$

for some constant $\bar{C} > 0$, independent of $\beta \in (0, 1)$. By the interpolation inequality (Theorem 5.1.2), there exists $C > 0$, for any $\varepsilon \in (0, (c + d)/2)$, such that

$$\|h'\|_{C[c,d]} \le \varepsilon\|h''\|_{C[c,d]} + \frac{C}{\varepsilon}\|h\|_{C[c,d]}. \tag{10.29}$$

Thus

$$\|h''\|_{C[c,d]} \le \bar{C}(1 + \varepsilon\|h''\|_{C[c,d]} + \frac{C}{\varepsilon}\|h\|_{C[c,d]}).$$

Taking $\varepsilon \le \min(1/2\bar{C}, (c + d)/2))$, we get

$$\|h''\|_{C[c,d]} \le 2\bar{C}(1 + \frac{C}{\varepsilon}\|h\|_{C[c,d]}),$$

which implies (10.28). By (10.29) again, we have

$$\sup_{0<\beta<1} \|h'\|_{C[c,d]} < \infty.$$

Therefore, by (10.26), we deduce (10.24) and (10.25). ∎

Theorem 10.4.2. *We make the assumptions of Theorem 10.4.1. Then there exists a subsequence $\beta \to 0$ such that*

$$\beta u_\beta(0+) \to \lambda \in \mathbf{R}_+, \tag{10.30}$$

$$v_\beta(x) := u_\beta(x) - u_\beta(0+) \to v(x) \in C^2(0, \infty) \quad \textit{locally uniformly}. \tag{10.31}$$

Furthermore, the limit (λ, v) satisfies (10.22).

Proof. By Lemma 10.3.2, $u_\beta(0+) := \lim_{x\to 0+} u_\beta(x)$ exists. By (10.13), $\beta u_\beta(0+)$ is bounded. Thus we obtain (10.30). By Theorem 10.4.1, we observe that $\{v_\beta\}$ and

$\{v'_\beta\}$ are uniformly bounded and equicontinuous on each $[a, b] \subset (0, \infty)$. Since

$$\begin{aligned}|h''(y_1) - h''(y_2)| &= |G(y_1, h, h') - G(y_2, h, h')| \\ &\leq C_1|y_1 - y_2|, \quad y_1, y_2 \in [c, d],\end{aligned}$$

we see by (10.26) that

$$|u''_\beta(x_1) - u''_\beta(x_2)| \leq C_2|x_1 - x_2|, \quad x_1, x_2 \in [a, b],$$

where the constants $C_1, C_2 > 0$ depend only on a, b. Hence, $\{v''_\beta\}$ is also uniformly bounded and equicontinuous on $[a, b]$. By the Ascoli–Arzelà Theorem, extracting a subsequence, we have

$$\begin{cases} v_\beta & \to \quad v, \\ v'_\beta & \to \quad v', \\ v''_\beta & \to \quad v'' \quad \text{uniformly on } [a, b], \end{cases}$$

which implies (10.31). Furthermore, we rewrite (10.11) as

$$\beta u_\beta(0+) + \beta v_\beta(x) = \frac{1}{2}\sigma^2 x^2 v''_\beta - r x v'_\beta + \tilde{U}_L(-v'_\beta) - D(x), \quad x > 0.$$

Passing to the limit as $\beta \to 0$, we deduce (10.22). ■

10.5 Optimal Policies

In this section, we present a synthesis of optimal policies for the optimization problem. We consider the stochastic differential equation,

$$dX^*(t) = \{I_L \circ (-v')(X^*(t)) - rX^*(t)\}dt + \sigma X^*(t)dB(t), \quad X^*(0) = x > 0, \tag{10.32}$$

where $I_L(x)$ denotes the maximizer of $\tilde{U}_L(x)$, that is,

$$I_L(x) = \begin{cases} (U')^{-1}(x) & \text{if} \quad (U')^{-1}(x) \leq L, \ x > 0, \\ L & \text{otherwise.} \end{cases}$$

Theorem 10.5.1. *We assume* (10.4), (10.5), *and* (10.15). *Then the optimal policy* $\{c^*(t)\}$ *and the maximum value of* (10.3) *are given by*

$$\begin{aligned}&c^*(t) = I_L \circ (-v')(X^*(t)), \\ &\sup_{c \in \mathcal{A}} \bar{J}(c) = \lambda, \quad \textit{for all } x > 0.\end{aligned}$$

Proof.

(1) By Lemma 10.3.2 and Theorem 10.4.2, we note that v is concave, $g(x) := I_L \circ (-v')(x) - rx$ is nonincreasing and locally Lipschitz on $(0, \infty)$. By analogy with Theorem 1.9.1, there exists a solution $\{X^*(t)\}$ of (10.32) up to the explosion time e (cf. Theorem 12.4.1 (1) below). By the comparison theorem, we have

$$0 < X^*(t) \leq Z(t), \quad t \geq 0, \quad a.s.,$$

and $\mathsf{e} = \infty$ a.s. Let $\tilde{X}(t)$ be another solution of (10.32). By Proposition 1.8.3, we see that

$$E[\sup_{0\le t\le T} \tilde{X}(t)^{2k}] \le E[\sup_{0\le t\le T} Z(t)^{2k}] < \infty, \quad \forall k \in \mathbf{N}, \quad T > 0.$$

Hence, by monotonicity,

$$\begin{aligned} E[(X^*(t) - \tilde{X}(t))^2] &= E[\int_0^t 2(g(X^*(s)) - g(\tilde{X}(s)))(X^*(s) - \tilde{X}(s)) \\ &\quad + \sigma^2(X^*(s) - \tilde{X}(s))^2 ds \\ &\quad + \int_0^t 2\sigma(X^*(s) - \tilde{X}(s))^2 dB(s)] \\ &\le \sigma^2 E[\int_0^t (X^*(s) - \tilde{X}(s))^2 ds], \quad 0 \le t \le T, \end{aligned}$$

which implies $X^*(t) = \tilde{X}(t)$ for all $t \ge 0$ a.s. Therefore, (10.32) admits a unique positive solution $\{X^*(t)\}$.

(2) Applying Itô's formula to (10.22), we have

$$\begin{aligned} v(X^*(T \wedge \tau_n)) &= v(x) + \int_0^{T\wedge\tau_n} \{v'(X^*(t))(c^*(t) - rX^*(t)) \\ &\quad + \frac{1}{2}\sigma^2 X^*(t)^2 v''(X^*(t))\}dt \\ &\quad + \int_0^{T\wedge\tau_n} v'(X^*(t))\sigma X^*(t) dB(t) \\ &= v(x) + \int_0^{T\wedge\tau_n} \{\lambda - U(c^*(t)) + D(X^*(t))\}dt \\ &\quad + \int_0^{T\wedge\tau_n} v'(X^*(t))\sigma X^*(t) dB(t), \quad \text{a.s.,} \end{aligned}$$

where $\{\tau_n\}$ is a sequence of localizing stopping times $\tau_n \uparrow \infty$ of the local martingale. Hence

$$E[v(X^*(T \wedge \tau_n))] = v(x) + E[\int_0^{T\wedge\tau_n} \{\lambda - U(c^*(t)) + D(X^*(t))\}dt].$$

By (10.16) and (10.31), we see that

$$|v(x)| \le C(1 + x^m),$$

so that

$$\sup_n E[|v(X^*(T \wedge \tau_n)|^{1+1/m}] \le C(1 + E[\sup_{0\le t\le T} X^*(t)^{m+1}]) < \infty$$

for some constant $C > 0$. Thus $\{v(X^*(T \wedge \tau_n)) : n \geq 1\}$ is uniformly integrable. Letting $n \to \infty$, we get

$$E[v(X^*(T))] = v(x) + E[\int_0^T \{\lambda - U(c^*(t)) + D(X^*(t))\}dt].$$

Dividing both sides by $T > 0$ and letting $T \to \infty$, by (10.10) and (10.5), we get

$$\lambda = \liminf_{T\to\infty} \frac{1}{T} E[\int_0^T \{U(c^*(t)) - D(X^*(t))\}dt].$$

By the same calculation as above, we have

$$\lambda \geq \liminf_{T\to\infty} \frac{1}{T} E[\int_0^T \{U(c(t)) - D(X(t))\}dt], \qquad \text{for } c \in \mathcal{A}.$$

Thus we obtain the desired result. ∎

11

Optimal Stopping Problems

In this chapter, we study the theory of optimal stopping problems in mathematical finance. We consider an investor, that is, an economic agent who can trade continuously in the stock market.

The stock price $X(t)$ is governed by a stochastic differential equation with an expected return r. An American call or put option gives the right to buy or to sell the underlying asset with the reward function $g(X(t))$ at any time t. The objective of the investor is to decide the exercise time τ when he buys or sells the risky asset for the maximal expected reward $E[e^{-r\tau} g(X(\tau))]$.

By the method of penalization, we solve the variational inequality associated with this problem. The value function coincides with its solution v and, by using v, the optimal stopping time is shown to exist.

11.1 The Model

Consider the optimal stopping problem for the stock price in mathematical finance. Define the following quantities:

$X(t)$ = stock price at time t.
r = expected return of the stock, $r > 0$.
$B(t)$ = the standard Brownian motion.
σ = the nonzero diffusion constant.
τ = exercise time or stopping time.
$g(x)$ = reward function of stock x.

We assume that the stock price $X = \{X(t)\}$ evolves according to the stochastic differential equation,

$$dX(t) = rX(t)dt + \sigma X(t)dB(t), \quad X(0) = x \geq 0, \tag{11.1}$$

on a complete probability space $(\Omega, \mathcal{F}, P)$, carrying a standard Brownian motion $\{B(t)\}$, endowed with the natural filtration $\mathcal{F}_t$ generated by $\sigma(B(s), s \leq t)$. The reward function $g(x)$ is assumed to have the following property:

$$g \geq 0, \quad g \in \mathcal{C}, \tag{11.2}$$

where $\mathcal{C}$ denotes the Banach space $C_0[0, \infty)$ of all continuous functions h on $[0, \infty)$ vanishing at infinity with norm $\|h\| = \sup_{x \geq 0} |h(x)|$.

The objective is to find an optimal stopping time τ^* so as to maximize the expected reward function:

$$J(\tau) = E[e^{-r\tau} g(X(\tau))] \tag{11.3}$$

over the class $\mathcal{S}$ of all stopping times τ, where $e^{-r\tau} g(X_\tau)$ at $\tau = \infty$ is interpreted as zero. Instead of HJB equations, we consider the variational inequality:

$$\begin{cases} -rv + \frac{1}{2}\sigma^2 x^2 v'' + rxv' \leq 0, & x > 0, \\ v(x) \geq g(x), & x \geq 0, \\ (-rv + \frac{1}{2}\sigma^2 x^2 v'' + rxv')(v - g)^+ = 0, \\ v(0) = g(0). \end{cases} \tag{11.4}$$

To solve (11.4), we need to study the penalty equation for $\varepsilon > 0$:

$$\begin{cases} ru = \frac{1}{2}\sigma^2 x^2 u'' + rxu' + \frac{1}{\varepsilon}(u - g)^- \quad \text{in} \quad (0, \infty), \\ u(0) = g(0)/(r\varepsilon + 1), \end{cases} \tag{11.5}$$

originated by A. Bensoussan and J. L. Lions [22]. We will show that $v = \lim_{\varepsilon \to 0} u$, and give a synthesis of the optimal stopping time τ^*.

Remark 11.1.1. *The condition* (11.2) *is fulfilled if the reward function is given by*

$$g(x) = (K - x)^+$$

for the strike price $K > 0$ of a put option. (cf. I. Karatzas and S. E. Shreve [89]).

11.2 Remarks on Variational Inequalities

In this section, we explain why the variational inequality (11.4) solves the optimal stopping problem (11.3). Suppose that the variational inequality (11.4) admits a solution $v \in C^2$. Then the optimal stopping time $\hat{\tau}$ is given by

$$\hat{\tau} = \inf\{t : \ v(X(t)) = g(X(t))\}.$$

From (11.4) it follows that

$$-rv + \frac{1}{2}\sigma^2 x^2 v'' + rxv' = 0 \quad \text{if} \quad v > g.$$

Hence

$$-rv + \frac{1}{2}\sigma^2 x^2 v'' + rxv' \,|_{x=X(t)} = 0 \quad \text{for } t < \hat{\tau}.$$

By Itô's formula, under some additional assumptions on v, we obtain

$$\begin{aligned} E[e^{-r\hat{\tau}}v(X(\hat{\tau}))] &= v(x) + E[\int_0^{\hat{\tau}} e^{-rt}\{-rv + \frac{1}{2}\sigma^2 x^2 v'' + rxv'\}\,|_{x=X(t)}\,dt] \\ &\quad + E[\int_0^{\hat{\tau}} e^{-rt}v'(X(t))\sigma X(t)dB(t)] \\ &= v(x). \end{aligned}$$

Thus

$$E[e^{-r\hat{\tau}}g(X(\hat{\tau}))] = E[e^{-r\hat{\tau}}v(X(\hat{\tau}))] = v(x).$$

Since

$$-rv + \frac{1}{2}\sigma^2 x^2 v'' + rxv' \le 0,$$

Itô's formula gives

$$E[e^{-r\tau}g(X(\tau))] \le E[e^{-r\tau}v(X(\tau))] \le v(x), \qquad \tau \in \mathcal{S}.$$

Therefore we seem to obtain the optimality of $\hat{\tau}$.

But, we remark that $v \in C^2$ fails, because v is connected to g at some point x. Thus we make the crucial improvements in Sections 11.3–11.6.

11.3 Penalized Problem

In this section, we show the existence of a unique solution u of the penalty equation (11.5).

11.3.1 Probabilistic Penalization

We begin with the probabilistic penalty equation,

$$u(x) = E[\int_0^{\infty} e^{-(r+\frac{1}{\varepsilon})t}\frac{1}{\varepsilon}(u \vee g)(X(t))dt], \quad x \ge 0. \tag{11.6}$$

Theorem 11.3.1. *We assume* (11.2). *Then there exists a unique nonnegative solution $u = u_\varepsilon \in \mathcal{C}$ of* (11.6).

Proof. Define

$$\mathcal{T}h(x) = E[\int_0^{\infty} e^{-(r+\frac{1}{\varepsilon})t}\frac{1}{\varepsilon}(h \vee g)(X(t))dt], \quad \text{for } h \in \mathcal{C}_+, \tag{11.7}$$

where $\mathcal{C}_+ = \{h \in \mathcal{C} : h \ge 0\}$. Clearly, $\mathcal{C}_+$ is a closed subset of $\mathcal{C}$. By (11.7), we have

$$\begin{aligned} 0 \le \mathcal{T}h(x) &\le \|h \vee g\| \int_0^{\infty} e^{-(r+\frac{1}{\varepsilon})t}\frac{1}{\varepsilon}dt \\ &= \frac{\|h \vee g\|}{r\varepsilon + 1}. \end{aligned}$$

We note that

$$X(t) = x \exp\{rt + \sigma B(t) - \frac{1}{2}\sigma^2 t\}.$$

Then, by the dominated convergence theorem,

$$\begin{aligned}|\mathcal{T}h(x) - \mathcal{T}h(y)| &\le E[\int_0^\infty e^{-(r+\frac{1}{\varepsilon})t}\frac{1}{\varepsilon}\{|(h \vee g)(X(t)) - (h \vee g)(Y(t))|\}dt] \\ &\to 0 \quad \text{as } y \to x,\end{aligned}$$

where $\{Y(t)\}$ be the solution of (11.1) with $Y(0) = y > 0$. Moreover,

$$\mathcal{T}h(x) = E[\int_0^\infty e^{-(r+\frac{1}{\varepsilon})t}\frac{1}{\varepsilon}(h \vee g)(X(t))dt] \to 0 \quad \text{as } x \to \infty.$$

Thus $\mathcal{T}$ maps $\mathcal{C}_+$ into $\mathcal{C}_+$.

Now, by (11.7), we have

$$\begin{aligned}|\mathcal{T}h_1(x) - \mathcal{T}h_2(x)| &\le E[\int_0^\infty e^{-(r+\frac{1}{\varepsilon})t}\frac{1}{\varepsilon}|h_1(X(t)) - h_2(X(t))|dt] \\ &\le E[\int_0^\infty e^{-(r+\frac{1}{\varepsilon})t}\frac{1}{\varepsilon}\|h_1 - h_2\|dt] \\ &= \frac{1}{r\varepsilon + 1}\|h_1 - h_2\|, \quad h_1, h_2 \in \mathcal{C}_+.\end{aligned}$$

This yields that $\mathcal{T}$ is a contraction mapping. Thus $\mathcal{T}$ has a fixed point u, which solves (11.6). ∎

11.3.2 Viscosity Solutions of the Penalty Equation

Consider the penalty equation,

$$ru = \frac{1}{2}\sigma^2 x^2 u'' + rxu' + \frac{1}{\varepsilon}(u - g)^- \quad \text{in } (0, \infty), \tag{11.8}$$

with the boundary condition $u(0) = g(0)/(r\varepsilon + 1)$. Since

$$u \vee g = u + (u - g)^-,$$

we rewrite (11.8) as

$$(r + \frac{1}{\varepsilon})u = \frac{1}{2}\sigma^2 x^2 u'' + rxu' + \frac{1}{\varepsilon}(u \vee g) \quad \text{in } (0, \infty). \tag{11.9}$$

Definition 11.3.2. *Let $w \in C[0, \infty)$ and $w(0) = g(0)/(r\varepsilon + 1)$. Then w is a viscosity solution of* (11.8) *if*

(a) *w is a viscosity subsolution of* (11.8)*, that is, for any $\phi \in C^2(0, \infty)$ and any local maximum point $z > 0$ of $w - \phi$,*

$$rw(z) \le \frac{1}{2}\sigma^2 z^2 \phi''(z) + rz\phi'(z) + \frac{1}{\varepsilon}(w - g)^-(z),$$

and

(b) *w is a viscosity supersolution of* (11.8)*, that is, for any $\phi \in C^2(0,\infty)$ and any local minimum point $\bar{z} > 0$ of $w - \phi$,*

$$r w(\bar{z}) \geq \frac{1}{2}\sigma^2 \bar{z}^2 \phi''(\bar{z}) + r\bar{z}\phi'(\bar{z}) + \frac{1}{\varepsilon}(w-g)^-(\bar{z}).$$

Theorem 11.3.3. *We make the assumption of Theorem 11.3.1. Then u is a viscosity solution of* (11.8).

Proof. By Theorem 2.6.2, we see that $(\Omega, \mathcal{F}, P, \{\mathcal{F}_t^X\}; X)$ is a strong Markov process, that is,

$$P_x(X(t+\tau) \in A | \mathcal{F}_\tau^X) = P_{X_\tau}(X(t) \in A), \quad P_x\text{-a.s.}, \quad t \geq 0,$$

for any Borel set A of $\mathbf{R}$ and $\tau \in \mathcal{S}_b$, where P_x denotes the probability measure P with $X(0) = x$. By (11.6), we get

$$u(x) = E[\int_0^\tau e^{-(r+\frac{1}{\varepsilon})t} \frac{1}{\varepsilon}(u \vee g)(X(t))dt + \int_\tau^\infty e^{-(r+\frac{1}{\varepsilon})t} \frac{1}{\varepsilon}(u \vee g)(X(t))dt].$$

By the strong Markov property,

$$\begin{aligned} &E[\int_\tau^\infty e^{-(r+\frac{1}{\varepsilon})t} \frac{1}{\varepsilon}(u \vee g)(X(t))dt | \mathcal{F}_\tau^X] \\ &= e^{-(r+\frac{1}{\varepsilon})\tau} E[\int_0^\infty e^{-(r+\frac{1}{\varepsilon})t} \frac{1}{\varepsilon}(u \vee g)(X(t+\tau))dt | \mathcal{F}_\tau^X] \\ &= e^{-(r+\frac{1}{\varepsilon})\tau} u(X(\tau)), \quad \text{a.s.} \end{aligned}$$

Thus

$$u(x) = E[\int_0^\tau e^{-(r+\frac{1}{\varepsilon})t} \frac{1}{\varepsilon}(u \vee g)(X(t))dt + e^{-(r+\frac{1}{\varepsilon})\tau} u(X(\tau))].$$

This corresponds to the dynamic programming principle for u. Also, we have $u(0) = (u \vee g)(0)/(r\varepsilon + 1)$, so that

$$u(0) = g(0)/(r\varepsilon + 1).$$

By the same line as the proof of Theorem 4.3.1, we deduce that u is a viscosity solution of (11.9), and hence (11.8). ∎

11.3.3 Classical Solutions of the Penalty Equation

We study the smoothness of the solution u to (11.8).

Theorem 11.3.4. *We assume* (11.2) *and that*

$$g: \text{ locally Lipschitz on } (0,\infty). \tag{11.10}$$

Then there exists a unique classical solution u of (11.8) *in $C^2(0,\infty)$. Furthermore, for any $\tau \in \mathcal{S}$, we have*

$$u(x) = E[\int_0^\tau e^{-rt} \frac{1}{\varepsilon}(u-g)^-(X(t))dt + e^{-r\tau} u(X(\tau))]. \tag{11.11}$$

In particular,

$$u(x) = E[\int_0^\infty e^{-rt}\frac{1}{\varepsilon}(u-g)^-(X(t))dt]. \tag{11.12}$$

Proof.

(1) Let $[a, b] \subset (0, \infty)$ be arbitrary and we consider the boundary value problem:

$$\begin{aligned}(r+\frac{1}{\varepsilon})\chi(x) &= \tfrac{1}{2}\sigma^2x^2\chi''(x) + rx\chi'(x) + \tfrac{1}{\varepsilon}(\chi \vee g)(x) \ \text{ in } (a, b),\\ \chi(a) &= u(a), \quad \chi(b) = u(b),\end{aligned} \tag{11.13}$$

By (11.10), we see that

$$\begin{aligned}&|\chi \vee g(x_1) - \chi \vee g(x_2)|\\ &\leq |\chi(x_1) - \chi(x_2)| + C_0|x_1 - x_2|, \quad x_1, x_2 \in [a, b],\end{aligned}$$

for some constant $C_0 > 0$, depending only on a, b. Hence, by uniform ellipticity, Theorem 5.3.7 yields that (13) has a smooth solution χ. By Theorem 5.3.1, we obtain the uniqueness of the viscosity solutions of (11.13). Therefore we deduce that $u = \chi \in C^2(a, b)$, and hence $u \in C^2(0, \infty)$.

(2) We set

$$\tau_R = \inf\{t \geq 0 : X(t) > R \text{ or } X(t) < 1/R\}, \tag{11.14}$$

for $R > 1$ and $\varrho = \tau \wedge \tau_R$. By Itô's formula and (11.8), we get

$$\begin{aligned}e^{-r(\varrho\wedge n)}u(X(\varrho \wedge n)) &= u(x) + \int_0^{\varrho\wedge n} e^{-rt}(-ru(x)\\ &\quad + \frac{1}{2}\sigma^2x^2u''(x) + rxu'(x))|_{x=X(t)}dt\\ &\quad + \int_0^{\varrho\wedge n} e^{-rt}u'(X(t))\sigma X(t)dB(t)\\ &= u(x) - \int_0^{\varrho\wedge n} e^{-rt}\frac{1}{\varepsilon}(u-g)^-(X(t))dt\\ &\quad + \int_0^{\varrho\wedge n} e^{-rt}u'(X(t))\sigma X(t)dB(t), \quad \text{a.s.,} \quad \forall n \in \mathbf{N}.\end{aligned}$$

Since u' is bounded on $[1/R, R]$, we see that

$$\begin{aligned}&E[\int_0^{\varrho\wedge n} e^{-rt}u'(X(t))\sigma X(t)dB(t)]\\ &= E[\int_0^{n} e^{-rt}u'(X(t))\sigma X(t)1_{\{t\leq\varrho\}}dB(t)] = 0.\end{aligned}$$

Hence

$$u(x) = E[\int_0^{\varrho\wedge n} e^{-rt}\frac{1}{\varepsilon}(u-g)^-(X(t))dt + e^{-r(\varrho\wedge n)}u(X(\varrho \wedge n))].$$

Letting $n \to \infty$, by the dominated convergence theorem, we have

$$u(x) = E[\int_0^{\tau\wedge\tau_R} e^{-rt}\frac{1}{\varepsilon}(u-g)^-(X(t))dt + e^{-r(\tau\wedge\tau_R)}u(X(\tau\wedge\tau_R))].$$

Note that $\tau_R \nearrow \infty$ as $R \nearrow \infty$. Passing to the limit, we deduce (11.11). Equation (11.12) is immediate from (11.11) with $\tau = \infty$.

(3) Let $u_i, i = 1, 2$, be two solutions of (11.8) in $\mathcal{C}_+$. By the same line as (11.12), we have

$$u_i(x) = E[\int_0^{\infty} e^{-(r+\frac{1}{\varepsilon})t}\frac{1}{\varepsilon}(u_i \vee g)(X(t))dt].$$

Thus

$$\|u_1 - u_2\| \le \frac{1}{r\varepsilon+1}\|u_1 - u_2\|,$$

which implies $u_1 = u_2$. ■

11.4 Passage to the Limit as $\varepsilon \to 0$

We study the convergence of $u = u_\varepsilon \in \mathcal{C}_+$ as $\varepsilon \to 0$. Define

$$G_\beta h(x) = E[\int_0^{\infty} e^{-\beta t}h(X(t))dt], \quad \beta > 0,$$

and

$$\mathcal{D} = \{G_\beta(\beta h) : \ h \in \mathcal{C}, \beta > r\}.$$

Our objective is to prove the following theorem.

Theorem 11.4.1. *We assume* (11.2) *and* (11.10). *Then we have*

$$u_{\varepsilon_n} \to v \in \mathcal{C}, \tag{11.15}$$

where $\varepsilon_n = 2^{-n}$.

Lemma 11.4.2. *The class $\mathcal{D}$ is dense in $\mathcal{C}$.*

Proof.

(1) Let $h \in \mathcal{C}$ be arbitrary. It is clear that $\|G_\beta(\beta h)\| \le \|h\|$. By the dominated convergence theorem, we get

$$\begin{aligned}|G_\beta(\beta h)(y) - G_\beta(\beta h)(x)| &\le E[\int_0^{\infty} e^{-\beta t}\beta|h(Y(t)) - h(X(t))|dt] \\ &\to 0 \quad \text{as} \quad y \to x,\end{aligned}$$

and

$$G_\beta(\beta h)(x) = E[\int_0^{\infty} e^{-\beta t}\beta h(X(t))dt] \quad \to \quad 0 \quad \text{as} \quad x \to \infty.$$

Thus

$$\mathcal{D} \subset \mathcal{C}.$$

(2) For each $x \geq 0$, it is easy to see that

$$\begin{aligned} |G_\beta(\beta h)(x) - h(x)| &\leq E[\int_0^\infty e^{-\beta t}\beta|h(X(t)) - h(x)|dt] \\ &= E[\int_0^\infty e^{-s}|h(X(s/\beta)) - h(x)|ds] \\ &\to 0 \quad \text{as} \quad \beta \to \infty. \end{aligned}$$

We extend $h \in \mathcal{C}$ on the compactification $[0, \infty) \cup \{\infty\}$ as $h(\infty) = 0$. Hence

$$G_\beta(\beta h)(x) \to h(x) \quad \text{as } \beta \to \infty, \quad x \in [0, \infty) \cup \{\infty\}.$$

Let Λ be in the dual space $\mathcal{C}^*$. By the Riesz representation theorem, there exists a measure μ on $[0, \infty) \cup \{\infty\}$ such that

$$\Lambda(h) = \int h d\mu,$$

for all $h \in \mathcal{C}$. Let $\Lambda(f) = 0$ for all $f = G_\beta(\beta h) \in \mathcal{D}$. By the dominated convergence theorem, we have

$$\Lambda(h) = \lim_{\beta \to \infty} \Lambda(\beta G_\beta h) = 0, \quad \text{for all } h \in \mathcal{C}.$$

Now, suppose that $\bar{\mathcal{D}} \neq \mathcal{C}$, where $\bar{\mathcal{D}}$ is the closure of $\mathcal{D}$. Then, there is $f_0 \notin \bar{\mathcal{D}}$ such that

$$\delta := \inf_{f \in \bar{\mathcal{D}}} \|f - f_0\| > 0.$$

By the Hahn–Banach theorem (cf. H. L. Royden [139]), we find $\Lambda^* \in \mathcal{C}^*$ such that

$$\Lambda^*(f_0) = \delta, \quad \Lambda^*(f) = 0 \quad \text{for all } f \in \bar{\mathcal{D}}.$$

This is a contradiction. Therefore, we deduce $\bar{\mathcal{D}} = \mathcal{C}$. ■

Lemma 11.4.3. *Let $\tilde{u} \in \mathcal{C}_+$ be the solution of* (11.8) *with $\tilde{g} \in \mathcal{C}_+$ replacing g. Then we have*

$$\|u - \tilde{u}\| \leq \|g - \tilde{g}\|. \tag{11.16}$$

Proof. Let $z, \tilde{z} \in \mathcal{C}_+$ satisfy

$$\|z - \tilde{z}\| \leq \|g - \tilde{g}\|.$$

It is easy to check that

$$|a \vee b - c \vee d| \leq |b - d| \quad \text{if } |a - c| \leq |b - d|, \quad a, b, c, d \in \mathbf{R}.$$

Hence, by a contradiction argument, we have

$$\begin{aligned}\|z \vee g - \tilde{z} \vee \tilde{g}\| &\leq \max(\|z - \tilde{z}\|, \|g - \tilde{g}\|) \\ &= \|g - \tilde{g}\|.\end{aligned}$$

Let $\tilde{\mathcal{T}}$ denote $\mathcal{T}$ with $\tilde{g}$ repalcing g. By (11.7), we have

$$\begin{aligned}|\mathcal{T}z(x) - \tilde{\mathcal{T}}\tilde{z}(x)| &\leq E[\int_0^\infty e^{-(r+\frac{1}{\varepsilon})t}\frac{1}{\varepsilon}|z \vee g - \tilde{z} \vee \tilde{g}|(X(t))dt] \\ &\leq \|g - \tilde{g}\|.\end{aligned}$$

Clearly,

$$\begin{aligned}|\mathcal{T}0(x) - \tilde{\mathcal{T}}0(x)| &\leq E[\int_0^\infty e^{-(r+\frac{1}{\varepsilon})t}\frac{1}{\varepsilon}|0 \vee g - 0 \vee \tilde{g}|(X(t))dt] \\ &\leq \|g - \tilde{g}\|.\end{aligned}$$

Taking $z = \mathcal{T}0, \tilde{z} = \tilde{\mathcal{T}}0$, we get

$$\|\mathcal{T}^2 0 - \tilde{\mathcal{T}}^2 0\| \leq \|g - \tilde{g}\|.$$

By induction,

$$\|\mathcal{T}^n 0 - \tilde{\mathcal{T}}^n 0\| \leq \|g - \tilde{g}\|, \quad n = 1, 2, \ldots.$$

Letting $n \to \infty$, by Theorem 11.3.1, we obtain (11.16). ■

Lemma 11.4.4. *Under* (11.2) *and* (11.10)*, we have*

$$u_\varepsilon(x) = \sup_{\tau \in \mathcal{S}} E[e^{-r\tau}\{g - (u_\varepsilon - g)^-\}(X(\tau))]. \tag{11.17}$$

Proof. By (11.11), we have

$$\begin{aligned}u_\varepsilon(x) &= E[\int_0^\tau e^{-rt}\frac{1}{\varepsilon}(u_\varepsilon - g)^-(X(t))dt + e^{-r\tau}u_\varepsilon(X(\tau))] \\ &\geq E[\int_0^\tau e^{-rt}\frac{1}{\varepsilon}(u_\varepsilon - g)^-(X(t))dt + e^{-r\tau}(u_\varepsilon \wedge g)(X(\tau))] \\ &\geq E[e^{-r\tau}(u_\varepsilon \wedge g)(X(\tau))] \\ &= E[e^{-r\tau}\{g - (u_\varepsilon - g)^-\}(X(\tau))].\end{aligned}$$

On the other hand, we take $\tau = \tau_\varepsilon := \inf\{t : u_\varepsilon(X(t)) \leq g(X(t))\}$. Since

$$e^{-r\tau_\varepsilon}u_\varepsilon(X(\tau_\varepsilon)) = e^{-r\tau_\varepsilon}(u_\varepsilon \wedge g)(X(\tau_\varepsilon)) = e^{-r\tau_\varepsilon}[g - (u_\varepsilon - g)^-](X(\tau_\varepsilon)),$$

we get

$$\begin{aligned}u_\varepsilon(x) &= E[\int_0^{\tau_\varepsilon} e^{-rt}\frac{1}{\varepsilon}(u_\varepsilon - g)^-(X(t))dt + e^{-r\tau_\varepsilon}u_\varepsilon(X(\tau_\varepsilon))] \\ &= E[\int_0^{\tau_\varepsilon} e^{-rt}\frac{1}{\varepsilon}(u_\varepsilon - g)^-(X(t))dt + e^{-r\tau_\varepsilon}(u_\varepsilon \wedge g)(X(\tau_\varepsilon))] \\ &= E[e^{-r\tau_\varepsilon}(u_\varepsilon \wedge g)(X(\tau_\varepsilon))],\end{aligned}$$

which implies (11.17). ■

Proof of Theorem 11.4.1.

(1) We claim that

$$(u_\varepsilon - g)^- \le \varepsilon \|\beta h + (r-\beta)g\|, \tag{11.18}$$

if $g = G_\beta(\beta h) \in \mathcal{D}$ for some $h \in \mathcal{C}$. By the same line as the proof of Theorems 4.4.9 and 4.6.2, we observe that g is a unique viscosity solution of

$$\begin{cases} \beta g = \frac{1}{2}\sigma^2 x^2 g'' + rxg' + \beta h & \text{in} \quad (0,\infty), \\ g(0) = h(0), \end{cases}$$

or, equivalently,

$$\begin{cases} (r+\frac{1}{\varepsilon})g = \frac{1}{2}\sigma^2 x^2 g'' + rxg' + \hat{h} + \frac{1}{\varepsilon}g & \text{in} \quad (0,\infty), \\ g(0) = \frac{\varepsilon}{r\varepsilon+1}\{\hat{h}(0) + \frac{1}{\varepsilon}g(0)\}, \end{cases}$$

where $\hat{h} = \beta h + (r-\beta)g$. Hence we have $g = G_{r+\frac{1}{\varepsilon}}(\hat{h} + \frac{1}{\varepsilon}g)$. Therefore, by (11.6),

$$\begin{aligned} u_\varepsilon(x) - g(x) &= E[\int_0^\infty e^{-(r+\frac{1}{\varepsilon})t}\{\frac{1}{\varepsilon}(u_\varepsilon \vee g)(X(t)) - (\hat{h}(X(t)) \\ &\qquad + \frac{1}{\varepsilon}g(X(t)))\}dt] \\ &\ge -E[\int_0^\infty e^{-(r+\frac{1}{\varepsilon})t}\ \hat{h}(X(t))dt] \\ &\ge -\varepsilon\|\hat{h}\|, \quad x > 0, \end{aligned}$$

which implies (11.18).

(2) Let $g = G_\beta(\beta h) \in \mathcal{D}$. By Lemma 11.4.4, applying (11.18) to $u_{\varepsilon_{n+1}}(x)$ and $u_{\varepsilon_n}(x)$, we have

$$\begin{aligned} |u_{\varepsilon_{n+1}}(x) - u_{\varepsilon_n}(x)| &\le \sup_{\tau\in\mathcal{S}} E[e^{-r\tau}|(u_{\varepsilon_{n+1}} - g)^- - (u_{\varepsilon_n} - g)^-|(X(\tau))] \\ &\le (\varepsilon_{n+1} + \varepsilon_n)\|\beta h + (r-\beta)g\|. \end{aligned}$$

Thus

$$\sum_{n=1}^\infty \|u_{\varepsilon_{n+1}} - u_{\varepsilon_n}\| \le \sum_{n=1}^\infty (\varepsilon_{n+1} + \varepsilon_n)\|\beta h + (r-\beta)g\| < \infty.$$

This implies that $\{u_{\varepsilon_n}\}$ is a Cauchy sequence in $\mathcal{C}$, and we get (11.15).

(3) Let g satisfy (11.2). By Lemma 11.4.2, there exists a sequence $\{g_m\} \subset \mathcal{D}$ such that $g_m \to g$. Let u_ε^m be the solution of (11.8) corresponding to g_m. By (2), we see that

$$u_{\varepsilon_n}^m \to v^m \in \mathcal{C} \quad \text{as} \quad n \to \infty. \tag{11.19}$$

By Lemma 11.4.3,

$$\|u_{\varepsilon_n}^m - u_{\varepsilon_n}^{m'}\| \le \|g_m - g_{m'}\|.$$

Letting $n \to \infty$, we have

$$\|v^m - v^{m'}\| \le \|g_m - g_{m'}\|.$$

Hence $\{v^m\}$ is a Cauchy sequence, and

$$v^m \to v \in \mathcal{C}. \tag{11.20}$$

Thus

$$\begin{aligned}\|u_{\varepsilon_n} - v\| &\le \|u_{\varepsilon_n} - u^m_{\varepsilon_n}\| + \|u^m_{\varepsilon_n} - v^m\| + \|v^m - v\| \\ &\le \|g - g_m\| + \|u^m_{\varepsilon_n} - v^m\| + \|v^m - v\|.\end{aligned}$$

Letting $n \to \infty$ and then $m \to \infty$, we obtain (11.15). ■

11.5 Viscosity Solutions of Variational Inequalities

In this section, we study the viscosity solution of the variational inequality:

$$\begin{cases} -rv + \frac{1}{2}\sigma^2 x^2 v'' + rxv' \le 0, & x > 0, \\ v(x) \ge g(x), & x \ge 0, \\ (-rv + \frac{1}{2}\sigma^2 x^2 v'' + rxv')(v - g)^+ = 0, \\ v(0) = g(0). \end{cases} \tag{11.21}$$

Definition 11.5.1. *Let $v \in C[0,\infty)$ and $v(0) = g(0)$. Then v is a viscosity solution of (11.21), if the following assertions are satisfied:*

(a) *For any $\phi \in C^2$ and any local minimum point $\bar{z} > 0$ of $v - \phi$,*

$$-rv(\bar{z}) + \frac{1}{2}\sigma^2 \bar{z}^2 \phi''(\bar{z}) + r\bar{z}\phi'(\bar{z}) \le 0,$$

(b) $v(x) \ge g(x)$, *for all $x \ge 0$,*

(c) *For any $\phi \in C^2$ and any local maximum point $z > 0$ of $v - \phi$,*

$$(-rv + \frac{1}{2}\sigma^2 x^2 \phi'' + rx\phi')(v - g)^+ |_{x=z} \ge 0.$$

Theorem 11.5.2. *We assume* (11.2) *and* (11.10). *Then the limit v in Theorem 11.4.1 is a viscosity solution of* (11.21).

Proof. By (11.8) and Theorem 11.4.1, it is obvious that $v(0) = g(0)$. Let $\phi \in C^2$ and $z > 0$ be the maximizer of $v - \phi$ such that

$$v(z) - \phi(z) > v(x) - \phi(x), \quad x \in \bar{B}_\delta(z), \quad z \ne x,$$

for some $\delta > 0$. By the uniform convergence in Theorem 11.4.1, $u_{\varepsilon_n} - \phi$ attains a local maximum at $x_n \in \bar{B}_\delta(z)$. By the same argument as Theorem 4.5.1 (1), we deduce

$$x_n \to z \quad \text{as} \quad n \to \infty.$$

Now, by Theorem 11.3.3, we have

$$-ru_{\varepsilon_n}(x) + \frac{1}{2}\sigma^2 x^2 \phi''(x) + rx\phi'(x) + \frac{1}{\varepsilon_n}(u_{\varepsilon_n} - g)^-(x)|_{x=x_n} \geq 0.$$

Multiply both sides by $(u_{\varepsilon_n} - g)^+$ to obtain

$$(-ru_{\varepsilon_n}(x_n) + \frac{1}{2}\sigma^2 x_n^2 \phi''(x_n) + rx_n\phi'(x_n))(u_{\varepsilon_n} - g)^+(x_n) \geq 0.$$

Letting $n \to \infty$, we get

$$(-rv(z) + \frac{1}{2}\sigma^2 z^2 \phi''(z) + rz\phi'(z))(v - g)^+(z) \geq 0.$$

Next, by (11.18), we have

$$(u_{\varepsilon_n}^m - g_m)^- \leq \varepsilon_n \|\beta h_m + (r - \beta)g_m\|,$$

where $g_m = G_\beta(\beta h_m)$ for some $h_m \in \mathcal{C}$ and $u_{\varepsilon_n}^m$ is as in the proof of Theorem 11.4.1. Letting $n \to \infty$, by (11.19), we have

$$v^m(x) \geq g_m(x), \quad x \geq 0,$$

and then, by (11.20)

$$v(x) \geq g(x), \qquad \text{for all} \quad x \geq 0.$$

Finally, let $\bar{z}$ be the minimizer of $v - \phi$, and $\bar{x}_n$ be the local minimizer of $u_{\varepsilon_n} - \phi$ with $\bar{x}_n \to \bar{z}$. Then, by Theorem 11.3.3

$$-ru_{\varepsilon_n}(x) + \frac{1}{2}\sigma^2 x^2 \phi''(x) + rx\phi'(x) + \frac{1}{\varepsilon_n}(u_{\varepsilon_n} - g)^-(x)|_{x=\bar{x}_n} \leq 0,$$

from which

$$-ru_{\varepsilon_n}(\bar{x}_n) + \frac{1}{2}\sigma^2 \bar{x}_n^2 \phi''(\bar{x}_n) + r\bar{x}_n\phi'(\bar{x}_n) \leq 0.$$

Letting $n \to \infty$, we deduce

$$-rv(\bar{z}) + \frac{1}{2}\sigma^2 \bar{z}^2 \phi''(\bar{z}) + r\bar{z}\phi'(\bar{z}) \leq 0.$$

Thus we get the assertion of the theorem. ∎

Theorem 11.5.3. *We make the assumption of Theorem 11.5.2. Let $v_i \in \mathcal{C}, i = 1, 2,$ be two viscosity solutions of* (11.21). *Then we have*

$$v_1 = v_2.$$

Proof.

(1) Let $\phi \in C^2$ and $z > 0$ be the local maximizer of $v_1 - \phi$. If $v_1 \leq v_2$, then $(v_1 - v_2)^+ = 0$. If $v_1 > v_2$, then $v_1 > v_2 \geq g$, and thus $(v_1 - g)^+ > 0$. By Definition 11.5.1(c), we have

$$(-rv_1(z) + \frac{1}{2}\sigma^2 z^2 \phi''(z) + rz\phi'(z))(v_1 - g)^+(z) \geq 0.$$

Hence

$$-rv_1(z) + \frac{1}{2}\sigma^2 z^2 \phi''(z) + rz\phi'(z) \geq 0.$$

Thus

$$(-rv_1(z) + \frac{1}{2}\sigma^2 z^2 \phi''(z) + rz\phi'(z))(v_1 - v_2)^+(z) \geq 0.$$

Therefore, by Proposition 4.4.2, we obtain

$$(-rv_1(x) + \frac{1}{2}\sigma^2 x^2 q + rxp)(v_1 - v_2)^+(x) \geq 0,$$
$$\forall (p, q) \in \bar{J}^{2,+} v_1(x), \quad \forall x > 0.$$

(2) We prove this theorem by the same line as Theorem 4.4.9. Suppose there exists $x_0 > 0$ such that $v_1(x_0) - v_2(x_0) > 0$. Then we find $\eta > 0$ such that

$$M_0 := \sup_{x \geq 0} [v_1(x) - v_2(x) - 2\eta\psi(x)] > 0,$$

where $\psi(x) = (1 + x^2)^{1/2} + \psi_0$, for some constant $\psi_0 > 0$ chosen later. Define

$$\Psi_k(x, y) = v_1(x) - v_2(y) - \frac{k}{2}|x - y|^2 - \eta\{\psi(x) + \psi(y)\},$$

for $k \in \mathbf{N}$. Then there exists the maximizer $(x_k, y_k) \in [0, \infty)^2$ of $\Psi_k(x, y)$ such that

$$\Psi_k(x_k, y_k) = \sup \Psi_k(x, y) \geq \max(\Psi_k(\bar{x}, \bar{x}), \Psi_k(x_k, x_k)),$$

where $\bar{x} > 0$ is the maximum point of M_0. Hence

$$\frac{k}{2}|x_k - y_k|^2 < v_1(x_k) - v_2(y_k) - \eta\{\psi(x_k) + \psi(y_k)\},$$

and

$$\frac{k}{2}|x_k - y_k|^2 \leq v_2(x_k) - v_2(y_k) + \eta\{\psi(x_k) - \psi(y_k)\} \to 0 \quad \text{as } k \to \infty.$$

By extracting a subsequence, we obtain that $x_k \to \tilde{x}$, $y_k \to \tilde{x}$. Therefore,

$$v_1(\tilde{x}) - v_2(\tilde{x}) - 2\eta\psi(\tilde{x}) > 0, \quad \tilde{x} > 0. \tag{11.22}$$

Now, applying Ishii's lemma to $\Psi_k(x, y)$, we obtain $q_1, q_2 \in \mathbf{R}$ such that

$$(k(x_k - y_k), q_1) \in \bar{J}^{2,+} \tilde{v}_1(x_k),$$
$$(k(x_k - y_k), q_2) \in \bar{J}^{2,-} \tilde{v}_2(y_k),$$
$$\begin{pmatrix} q_1 & 0 \\ 0 & -q_2 \end{pmatrix} \leq 3k \begin{pmatrix} 1 & -1 \\ -1 & 1 \end{pmatrix},$$

where $\tilde{v}_1(x) = v_1(x) - \eta\psi(x)$, $\tilde{v}_2(y) = v_2(y) + \eta\psi(y)$. Hence

$$(\hat{p}_1, \hat{q}_1) := (k(x_k - y_k), q_1) + \eta(\psi'(x_k), \psi''(x_k)) \in \bar{J}^{2,+}v_1(x_k),$$
$$(\hat{p}_2, \hat{q}_2) := (k(x_k - y_k), q_2) - \eta(\psi'(y_k), \psi''(y_k)) \in \bar{J}^{2,-}v_2(y_k).$$

By Definition 11.5.1(a), we have

$$-rv_2(y_k) + \frac{1}{2}\sigma^2 y_k^2 \hat{q}_2 + ry_k\hat{p}_2 \le 0.$$

Also, by (1),

$$(-rv_1(x_k) + \frac{1}{2}\sigma^2 x_k^2 \hat{q}_1 + rx_k\hat{p}_1)(v_1 - v_2)^+(x_k) \ge 0.$$

Hence, by (11.22),

$$-rv_1(x_k) + \frac{1}{2}\sigma^2 x_k^2 \hat{q}_1 + rx_k\hat{p}_1 \ge 0$$

for sufficiently large k. Putting these inequalities together, we get

$$\begin{aligned} r\{v_1(x_k) - v_2(y_k)\} &\le \frac{1}{2}\sigma^2(x_k^2\hat{q}_1 - y_k^2\hat{q}_2) + r(x_k\hat{p}_1 - y_k\hat{p}_2) \\ &\equiv I_1 + I_2. \end{aligned}$$

Letting $k \to \infty$, we have

$$I_1 \le \frac{\sigma^2}{2}\{3k|x_k - y_k|^2 + \eta\{x_k^2\psi''(x_k) + y_k^2\psi''(y_k)\}\} \to \sigma^2\eta\tilde{x}^2\psi''(\tilde{x}),$$

and

$$I_2 = r\{k|x_k - y_k|^2 + \eta\{x_k\psi'(x_k) + y_k\psi'(y_k)\}\} \to 2r\eta\tilde{x}\psi'(\tilde{x}).$$

Thus, choosing $\psi_0 > 0$ such that $\frac{1}{2}\sigma^2 \le r\psi_0$, we obtain

$$\begin{aligned} r\{v_1(\tilde{x}) - v_2(\tilde{x})\} &\le 2\eta\{\tfrac{1}{2}\sigma^2\tilde{x}^2\psi''(\tilde{x}) + r\tilde{x}\psi'(\tilde{x})\} \\ &\le 2\eta(\tfrac{1}{2}\sigma^2 + r\tilde{x}^2)(1 + \tilde{x}^2)^{-1/2} \\ &\le 2\eta\{r\psi_0(1 + \tilde{x}^2)^{1/2} + r(1 + \tilde{x}^2)\}(1 + \tilde{x}^2)^{-1/2} \\ &= 2r\eta\psi(\tilde{x}). \end{aligned}$$

This is contrary to (11.22). ■

Example 11.5.4. *Let $g(x)$ be the reward function of the put option in Remark 11.1.1. We set*

$$p(x) = \begin{cases} K - x, & 0 \le x \le c, \\ (K - c)(\frac{x}{c})^{-\alpha}, & c < x. \end{cases}$$

Choosing $\alpha = 2r/\sigma^2$ *and* $0 < c = \frac{\alpha}{\alpha+1}K < K$, *we have a solution* $p \in C^1(0,\infty) \cap C^2((0,\infty) \setminus \{c\})$ *of the variational inequality,*

$$\begin{cases} -rp + \frac{1}{2}\sigma^2 x^2 p'' + rxp' = -rK < 0, & 0 \le x < c, \\ p(x) = g(x), & 0 \le x \le c, \\ -rp + \frac{1}{2}\sigma^2 x^2 p'' + rxp' = 0, & x > c, \\ p(x) > g(x), & x > c. \end{cases}$$

Let $\bar{z} > 0$ *be the minimum point of* $p - \phi$ *for* $\phi \in C^2$. *Then*

$$p'(\bar{z}) = \phi'(\bar{z}) \quad \text{and} \qquad p''(\bar{z}) \ge \phi''(\bar{z}) \quad \text{if } \bar{z} \ne c,$$

If $\bar{z} = c$, *then*

$$\begin{aligned} g(c+h) = p(c+h) &\ge p(c) + \phi(c+h) - \phi(c) \\ &= g(c) + \phi'(c)h + \frac{1}{2}\phi''(c)h^2 + o(h^2) \ \text{for } h < 0, \end{aligned}$$

so that

$$\frac{1}{2}\phi''(c) \le \lim_{h \to 0-} \frac{g(c+h) - g(c) - g'(c)h}{h^2} = 0.$$

Hence

$$-rp(c) + \frac{1}{2}\sigma^2 c^2 \phi''(c) + rc\phi'(c) \le -r(K - c) + rc(-1) = -rK < 0.$$

Thus we easily see that p *is a viscosity solution of* (11.21). *By Theorem 11.5.3, we obtain* $v = p$. *(cf. I. Karatzas and S. E. Shreve [89, Theorem 2.7.2, pp. 69–70])*

11.6 Solution of the Optimal Stopping Problem

In this section, we give a synthesis of the optimal stopping time.

Theorem 11.6.1. *We assume* (11.2) *and* (11.10). *Then we have*

$$v(x) = \sup_{\tau \in \mathcal{S}} E[e^{-r\tau} g(X(\tau))].$$

Proof.

(1) Let $\tau \in \mathcal{S}$. By (11.11), we get

$$\begin{aligned} u_{\varepsilon_n}(x) &= E[\int_0^\tau e^{-rt} \frac{1}{\varepsilon_n}(u_{\varepsilon_n} - g)^-(X(t))dt + e^{-r\tau} u_{\varepsilon_n}(X(\tau))] \\ &\ge E[e^{-r\tau} u_{\varepsilon_n}(X(\tau))]. \end{aligned}$$

Letting $n \to \infty$, by Theorems 11.4.1 and 11.5.2, we have

$$v(x) \ge E[e^{-r\tau} v(X(\tau))] \ge E[e^{-r\tau} g(X(\tau))].$$

(2) For any $m \in \mathbf{N}$, we set

$$\varrho_m = \inf\{t \geq 0 : v(X(t)) - \frac{1}{m} \leq g(X(t))\}. \tag{11.23}$$

Since

$$v(X(t)) - \frac{1}{m} > g(X(t)) \quad \text{on} \quad \{t < \varrho_m\},$$

we have

$$\begin{aligned} E[\int_0^{\varrho_m} e^{-rt}(u_{\varepsilon_n} - g)^-(X(t))dt] &\leq E[\int_0^{\varrho_m} e^{-rt}(u_{\varepsilon_n} - (v - \frac{1}{m}))^-(X(t))dt] \\ &\leq E[\int_0^{\varrho_m} e^{-rt}(\frac{1}{m} - \|u_{\varepsilon_n} - v\|)^-(X(t))dt] \\ &= 0 \end{aligned} \tag{11.24}$$

for sufficiently large n. Hence, by (11.11),

$$\begin{aligned} u_{\varepsilon_n}(x) &= E[\int_0^{\varrho_m} e^{-rt}\frac{1}{\varepsilon_n}(u_{\varepsilon_n} - g)^-(X(t))dt + e^{-r\varrho_m}u_{\varepsilon_n}(X(\varrho_m))] \\ &= E[e^{-r\varrho_m}u_{\varepsilon_n}(X(\varrho_m))]. \end{aligned} \tag{11.25}$$

Letting $n \to \infty$, by (11.23), we get

$$\begin{aligned} v(x) &= E[e^{-r\varrho_m}v(X(\varrho_m))] = E[e^{-r\varrho_m}\{g(X(\varrho_m)) + \frac{1}{m}\}] \\ &\leq \sup_{\tau \in \mathcal{S}} E[e^{-r\tau}g(X(\tau))] + \frac{1}{m}. \end{aligned}$$

Passing to the limit, we deduce

$$v(x) \leq \sup_{\tau \in \mathcal{S}} E[e^{-r\tau}g(X(\tau))]. \qquad \blacksquare$$

Theorem 11.6.2. *We make the assumptions of Theorem 11.6.1. Then the optimal stopping time τ^* is given by*

$$\tau^* = \inf\{t \geq 0 : v(X(t)) = g(X(t))\}.$$

Proof. We set $\bar{\tau} = \tau_R \wedge \varrho_m$, for ϱ_m of (11.23) and τ_R of (11.14). By (11.24), it is clear that

$$E[\int_0^{\bar{\tau}} e^{-rt}(u_{\varepsilon_n} - g)^-(X(t))dt] = 0$$

for sufficiently large n. By the same calculation as (11.25), we have

$$E[e^{-r\bar{\tau}}u_{\varepsilon_n}(X(\bar{\tau}))] = u_{\varepsilon_n}(x).$$

Letting $n \to \infty$ and then $R \to \infty$, we get

$$E[e^{-r\varrho_m} v(X(\varrho_m))] = v(x).$$

Note that $\varrho_m \nearrow \tau^*$ as $m \nearrow \infty$. Passing to the limit, we deduce

$$E[e^{-r\tau^*} g(X(\tau^*))] = E[e^{-r\tau^*} v(X(\tau^*))] = v(x).$$

Thus, we obtain the assertion. ■

12

Investment and Exit Decisions

In this chapter, we study the decision problem of investment rates and exit times for a private business in a stochastic environment. We consider a competitive firm that produces a capital good with the capital stock $K(t)$.

The market price $P(t)$ of the capital good is governed by a stochastic differential equation, and the market value of the capital given by $X(t) := P(t)K(t)$. The net cash flow of the firm coincides with the profit $rX(t)$ for the profit rate r net the total cost $\varphi(q(t))X(t)$ of investment $q(t)K(t)$. The firm sells its whole business at any time τ, given a resale value function $g(\cdot)$, when the integration of capital goods is costly. The firm's fundamental value is the total of the net cash flow and its abandonment payoff $g(X(\tau))$. The objective of the firm is to find the investment rate $q^*(t)$ and the exit time τ^* simultaneously, which maximize the expected fundamental value.

We solve the nonlinear variational inequality associated with the firm's value maximizing problem. The optimal policies are shown to exist from the optimality conditions.

12.1 The Model

Consider the decision problem of investment rates and exit times. Define the following quantities:

$P(t) =$ market price of the capital good at time t.
$\nu =$ expected decline rate of the price.
$B(t) =$ the standard Brownian motion.
$\sigma =$ the standard deviation of the decline ratc, $\sigma > 0$.
$K(t) =$ capital stock at time t.
$\lambda =$ depreciation rate of the capital stock, $\lambda > 0$.
$q(t)K(t) =$ investment for the gross investment capital rate $q(t)$ at time t.
$X(t) = P(t)K(t)$, i.e., market value of the capital at time t.
$rX(t) =$ profit for the profit rate $r > 0$.

$\varphi(q) =$ adjustment cost function.
$\bar{\varphi} =$ exogeneous financial constraint of $\varphi(q)$.
$\varphi(q(t))X(t) =$ total cost of investment at time t.
$\tau =$ exit time or stopping time.
$g(x) =$ resale value of the firm for the market value x of the capital.

We assume that the market price $\{P(t)\}$ of the capital good evolves according to the stochastic differential equation,

$$dP(t) = \nu P(t)dt + \sigma P(t)dB(t), \quad P(0) = p > 0, \tag{12.1}$$

where $\nu \in \mathbf{R}$ is the expected decline rate of the price, $\sigma > 0$ the standard deviation of the decline rate, and $\{B(t)\}$ denotes the standard Brownian motion on a complete probability space $(\Omega, \mathcal{F}, P)$, endowed with the natural filtration $\mathcal{F}_t$ generated by $\sigma(B(s), s \leq t)$. The total change in the capital stock $\{K(t)\}$ is given by

$$\frac{d}{dt}K(t) = q(t)K(t) - \lambda K(t), \quad K(0) = k > 0, \tag{12.2}$$

where $\{q(t)K(t)\}$ is the investment, that is, $q(t)$ denotes the gross investment-capital ratio, and $\lambda > 0$ is the depreciation rate. Let $\{X(t)\}$ be the market value of the capital $K(t)$ defined by $X(t) = P(t)K(t)$. By (12.1) and (12.2), we observe that $\{X(t)\}$ satisfies

$$dX(t) = (q(t) - \mu)X(t)dt + \sigma X(t)dB(t), \quad X(0) = x := pk > 0, \tag{12.3}$$

where $\mu = \lambda - \nu$.

We assume that the total cost of investing at time t is of the form:

$$P(t)\varphi(q(t))K(t) = \varphi(q(t))X(t)$$

for the adjustment cost function $\varphi \in C[0,\infty) \cap C^2(0,\infty)$, called the Penrose–Uzawa function, such that

$$\begin{aligned} &\varphi(q) \geq 0, \quad \varphi'(q) > 0, \quad \varphi''(q) > 0, \\ &\varphi(0) = 0, \quad \varphi'(0) = 1, \quad \varphi'(\infty) = \infty. \end{aligned}$$

This is fulfilled if $\varphi(q) = Aq^2 + q$ for $A > 0$. In addition, we impose an exogenous financial restriction on the firm's investing behavior such that

$$\varphi(q(t))X(t) \leq \bar{\varphi}X(t),$$

that is,

$$0 \leq q(t) \leq \zeta := \varphi^{-1}(\bar{\varphi}) \tag{12.4}$$

for some constant $\bar{\varphi} > 0$. The control $q = \{q(t)\}$ is said to be admissible if it is an $\{\mathcal{F}_t\}$-progressively measurable process satisfying (12.4). We denote by $\mathcal{A}$ the class of all admissible controls.

The objective of the firm is to find an optimal policy $(q^*, \tau^*) \in \mathcal{A} \times \mathcal{S}$ so as to maximize the expected fundamental value $J(q, \tau)$ with discount rate $\alpha > 0$:

$$J(q,\tau) = E[\int_0^\tau e^{-\alpha t}(r - \varphi(q(t)))X(t)dt + e^{-\alpha\tau}g(X(\tau))], \quad (q,\tau) \in \mathcal{A} \times \mathcal{S}, \tag{12.5}$$

subject to (12.3), where $\mathcal{S}$ is the class of all stopping times, the $rX(t)$ means the profit with profit rate $r > 0$, and g denotes the resale value or the abandonment payoff of the firm. We make the following assumptions on α, φ, μ and g:

$$\zeta - \mu < \alpha, \tag{12.6}$$

$$g \in \mathcal{B}, \quad g \geq 0, \quad g(0) = 0, \tag{12.7}$$

where $\mathcal{B}$ is the Banach space of all continuous functions h on $[0, \infty)$ such that

$$\begin{cases} \| h \| := \sup_{x \geq 0} |h(x)|/(1 + x) < \infty, \\ \text{There exists } C_\rho > 0, \text{ for any } \rho > 0, \text{ such that} \\ \qquad |h(x) - h(y)| \leq C_\rho|x - y| + \rho(1 + x + y), \quad x, y \geq 0. \end{cases} \tag{12.8}$$

We investigate the nonlinear variational inequality associated with the firm's value maximization problem:

$$\begin{cases} H(x, v, v', v'') := -\alpha v + \frac{1}{2}\sigma^2 x^2 v'' - \mu x v' + rx + \Gamma(v')x \leq 0, \quad x > 0, \\ v(x) \geq g(x), \quad x \geq 0, \\ H(x, v, v', v'')(v - g)^+ = 0, \\ v(0) = 0, \end{cases} \tag{12.9}$$

where $\Gamma(y) := \sup_{0 \leq q \leq \zeta}(qy - \varphi(q))$ and the supremum is attained by

$$\gamma(y) = \begin{cases} \zeta & \text{if } y \geq \varphi'(\zeta), \\ (\varphi)'^{-1}(y) & \text{if } 1 < y < \varphi'(\zeta), \\ 0 & \text{if } y \leq 1. \end{cases} \tag{12.10}$$

By using the classical solution v of (12.9), the optimal policy is shown to exist.

12.2 Penalized Problem

In this section, we show the existence of a unique smooth solution of the penalty equation.

12.2.1 Probabilistic Penalization

We begin with

$$u(x) = \sup_{q \in \mathcal{A}} E[\int_0^\infty e^{-(\alpha + \frac{1}{\varepsilon})t}\{(r - \varphi(q(t)))X(t) + \frac{1}{\varepsilon}(u \vee g)(X(t))\}dt], \tag{12.11}$$

where the supremum is taken over all systems $(\Omega, \mathcal{F}, P, \{\mathcal{F}_t\}; \{B(t)\}, \{q(t)\})$.

Lemma 12.2.1. *We assume* (12.6). *Then there exist* $\kappa > 1$ *and* $\kappa_0 > 0$ *such that*

$$E[e^{-\alpha\tau} X(\tau)^\kappa] + \kappa_0 E[\int_0^\tau e^{-\alpha t} X(t)^\kappa dt] \le x^\kappa, \qquad (12.12)$$

for any $(q, \tau) \in \mathcal{A} \times \mathcal{S}$.

Proof. By (12.6), we choose $\kappa > 1$ such that

$$-\alpha + \kappa(\zeta - \mu) + \frac{1}{2}\sigma^2\kappa(\kappa - 1) < 0, \qquad (12.13)$$

and set

$$\kappa_0 = \alpha - \kappa(\zeta - \mu) - \frac{1}{2}\sigma^2\kappa(\kappa - 1).$$

By (12.3), (12.6), and Itô's formula, we have

$$\begin{aligned} e^{-\alpha t} X(t)^\kappa &= x^\kappa + \int_0^t e^{-\alpha s}(-\alpha + \kappa(q(s) - \mu) + \frac{1}{2}\sigma^2\kappa(\kappa - 1))X(s)^\kappa ds \\ &\quad + \int_0^t e^{-\alpha s}\kappa\sigma X(s)^\kappa dB(s) \\ &\le x^\kappa - \int_0^t e^{-\alpha s}\kappa_0 X(s)^\kappa ds + \int_0^t e^{-\alpha s}\kappa\sigma X(s)^\kappa dB(s), \quad \text{a.s.} \end{aligned}$$

By Proposition 1.8.3, $\{\int_0^t e^{-\alpha s} X(s)^\kappa dB(s)\}$ is a martingale. Hence, for $T > 0$,

$$E[e^{-\alpha(\tau\wedge T)} X(\tau \wedge T)^\kappa] + \kappa_0 E[\int_0^{\tau\wedge T} e^{-\alpha s} X(s)^\kappa ds] \le x^\kappa.$$

Letting $T \to \infty$, we obtain (12.12), where $e^{-\alpha\tau} X(\tau)^\kappa$ at $\tau = \infty$ is interpreted as $\liminf_{t\to\infty} e^{-\alpha t} X(t)^\kappa = 0$ a.s. ∎

Theorem 12.2.2. *We assume* (12.6) *and* (12.7). *Then there exists a unique solution* $u = u_\varepsilon \in \mathcal{B}$ *of* (12.11) *such that* $u \ge 0$ *and* $u(0) = 0$.

Proof. We define

$$\begin{cases} Th(x) = \sup_{q\in\mathcal{A}} T_q h(x), \\ T_q h(x) = E[\int_0^\infty e^{-(\alpha+\frac{1}{\varepsilon})t}\{(r - \varphi(q(t)))X(t) + \frac{1}{\varepsilon}(h \vee g)(X(t))\}dt], \end{cases} \qquad (12.14)$$

for any $h \in \mathcal{B}$, where the supremum is taken over all systems $(\Omega, \mathcal{F}, P, \{\mathcal{F}_t\};$ $\{B(t)\}, \{q(t)\})$. Let $\{\bar{X}(t)\}$ be the solution of (12.3) corresponding to the constant control ζ, that is,

$$d\bar{X}(t) = (\zeta - \mu)\bar{X}(t)dt + \sigma\bar{X}(t)dB(t), \quad \bar{X}(0) = x > 0.$$

Applying the comparison theorem to $\{X(t)\}$ and $\{\bar{X}(t)\}$, we get

$$X(t) \le \bar{X}(t) = x\exp((\zeta - \mu)t + \sigma B(t) - \frac{1}{2}\sigma^2 t), \quad \text{a.s.} \qquad (12.15)$$

and thus

$$E[X(t)] \le E[\bar{X}(t)] = xe^{(\zeta-\mu)t}.$$

Hence, taking into account $\varphi(q) \ge 0$, we have

$$\begin{aligned} 0 \le \mathcal{T}h(x) &\le \int_0^\infty e^{-(\alpha+\frac{1}{\varepsilon})t}\{rxe^{(\zeta-\mu)t} + \frac{\|h\|+\|g\|}{\varepsilon}(1+xe^{(\zeta-\mu)t})\}dt \\ &\le C(1+x) \end{aligned}$$

for some constant $C > 0$. This implies $\| \mathcal{T}h \| < \infty$. Furthermore,

$$E[|X(t) - Y(t)|] \le |x-y|e^{(\zeta-\mu)t}, \quad x, y \ge 0, \tag{12.16}$$

where $\{Y(t)\}$ is the solution of (12.3) with $Y(0) = y$. By (12.7) and (12.8), it is clear that

$$\begin{aligned} |(h \vee g)(x) - (h \vee g)(y)| &\le |h(x) - h(y)| + |g(x) - g(y)| \\ &\le C_\rho |x-y| + \rho(1+x+y), \quad x, y \ge 0. \end{aligned}$$

Then, by (12.6), (12.15), and (12.16), we find $C'_\rho > 0$, for any $\rho > 0$, such that

$$\begin{aligned} |\mathcal{T}h(x) - \mathcal{T}h(y)| &\le \sup_{q\in\mathcal{A}} E[\int_0^\infty e^{-(\alpha+\frac{1}{\varepsilon})t}\{|r - \varphi(q(s))||X(t) - Y(t)| \\ &\quad + \frac{1}{\varepsilon}|(h \vee g)(X(t)) - (h \vee g)(Y(t))|\}dt] \\ &\le \sup_{q\in\mathcal{A}} E[\int_0^\infty e^{-(\alpha+\frac{1}{\varepsilon})t}\{(r + \varphi(\zeta))|X(t) - Y(t)| \\ &\quad + \frac{1}{\varepsilon}(C_\rho|X(t) - Y(t)| + \rho(1 + X(t) + Y(t)))\}dt] \\ &\le C'_\rho|x-y| + \rho(1+x+y), \quad x, y \ge 0. \end{aligned} \tag{12.17}$$

Therefore $\mathcal{T}$ maps $\mathcal{B}$ into itself.

Now, by (12.8) and (12.15), we have

$$\begin{aligned} |\mathcal{T}h_1(x) - \mathcal{T}h_2(x)| &\le \sup_{q\in\mathcal{A}} E[\int_0^\infty e^{-(\alpha+\frac{1}{\varepsilon})t}\frac{1}{\varepsilon}|(h_1 \vee g)(X(t)) - (h_2 \vee g)(X(t))|dt] \\ &\le \sup_{q\in\mathcal{A}} E\left[\int_0^\infty e^{-(\alpha+\frac{1}{\varepsilon})t}\frac{1}{\varepsilon}(1 + X(t))dt\right] \|h_1 - h_2\| \\ &\le \max\{\frac{1}{(\alpha-(\zeta-\mu))\varepsilon+1}, \frac{1}{\alpha\varepsilon+1}\}\|h_1 - h_2\|(1+x), \\ &\qquad\qquad h_1, h_2 \in \mathcal{B}. \end{aligned}$$

Thus $\mathcal{T}$ is a contraction mapping, which has a fixed point $u \in \mathcal{B}$. Therefore, u is a unique solution to (12.11). By (12.11), we have $u = \mathcal{T}u \geq 0$. Furthermore, by (12.11)

$$u(0) \leq \frac{1}{\alpha\varepsilon + 1}(u \vee g)(0),$$

which implies $u(0) = 0$. ∎

12.2.2 Viscosity Solutions of the Penalty Equation

Consider the nonlinear penalty equation of the form:

$$\begin{aligned} H_\varepsilon(x, u, u', u'') := -\alpha u + \tfrac{1}{2}\sigma^2 x^2 u'' - \mu x u' + rx + \Gamma(u')x \\ + \frac{1}{\varepsilon}(u - g)^- = 0, \quad x > 0, \end{aligned} \tag{12.18}$$

for $\varepsilon > 0$, with the boundary condition

$$u(0) = 0.$$

Definition 12.2.3. *Let $w \in C[0, \infty)$ and $w(0) = 0$. Then w is a viscosity solution of* (12.18) *if*

(a) *w is a viscosity subsolution of* (12.18)*, that is, for any $\phi \in C^2(0, \infty)$ and any local maximum point $z > 0$ of $w - \phi$,*

$$H_\varepsilon(z, u, \phi', \phi'') \geq 0,$$

and

(b) *w is a viscosity supersolution of* (12.18)*, that is, for any $\phi \in C^2(0, \infty)$ and any local minimum point $\bar{z} > 0$ of $w - \phi$,*

$$H_\varepsilon(\bar{z}, u, \phi', \phi'') \leq 0.$$

Theorem 12.2.4. *We make the assumptions of Theorem 12.2.2. Then u of* (12.11) *is a viscosity solution of* (12.18).

Proof. In view of Theorem 4.3.1, the viscosity property of u follows from the dynamic programming principle for u, that is,

$$u(x) = \sup_{q \in \mathcal{A}} E\Big[\int_0^\tau e^{-\beta t}\{(r - \varphi(q(t)))X(t) + \frac{1}{\varepsilon}(u \vee g)(X(t))\}dt + e^{-\beta\tau}u(X(\tau))\Big], \tag{12.19}$$

for any $\tau \in \mathcal{S}_b$, where the supremum is taken over all systems $(\Omega, \mathcal{F}, P, \{\mathcal{F}_t\}; \{B(t)\}, \{q(t)\})$, and $\beta := \alpha + 1/\varepsilon$. In order to see the roles of Lemma 12.2.1, by the same line as Theorem 3.2.5, we sketch the proof of (12.19) on the Polish space Ω when τ is nonrandom.

(1) Let $\tilde{u}(x)$ denote the right-hand side of (12.19) and $J_x(q) = T_q u(x)$ for $q = \{q(t)\} \in \mathcal{A}$. By Theorem 3.2.5 (1), we have

$$J_{X(\tau)}(\tilde{q}) = E[\int_0^\infty e^{-\beta t}\{(r - \varphi(\tilde{q}(t)))\tilde{X}(t) + \frac{1}{\varepsilon}(u \vee g)(\tilde{X}(t))\}dt|\mathcal{F}_\tau]$$
$$= e^{\beta\tau} E[\int_\tau^\infty e^{-\beta t}\{(r - \varphi(q(t)))X(t) + \frac{1}{\varepsilon}(u \vee g)(X(t))\}dt|\mathcal{F}_\tau],$$
a.s.,

and hence

$$J_x(q) = E[\int_0^\tau + \int_\tau^\infty e^{-\beta t}\{(r - \varphi(q(t)))X(t) + \frac{1}{\varepsilon}(u \vee g)(X(t))\}dt]$$
$$= E[\int_0^\tau e^{-\beta t}\{(r - \varphi(q(t)))X(t) + \frac{1}{\varepsilon}(u \vee g)(X(t))\}dt$$
$$+ e^{-\beta\tau} J_{X(\tau)}(\tilde{q})],$$

where $\{\tilde{X}(t)\}$ is the solution of (12.3) corresponding to the shifted process $\tilde{q}$ of q by τ with $\tilde{X}(0) = X(\tau)$. Therefore,

$$J_x(q) \le E[\int_0^\tau e^{-\beta t}\{(r - \varphi(q(t)))X(t) + \frac{1}{\varepsilon}(u \vee g)(X(t))\}dt$$
$$+ e^{-\beta\tau} u(X(\tau))].$$

Taking the supremum, we deduce $u \le \tilde{u}$.

(2) Let $\rho > 0$ be arbitrary. Let $\{S_j : j = 1, \ldots, n+1\}$ be a sequence of disjoint subsets of $(0, \infty)$ such that

$$\mathrm{diam}(S_j) < \delta, \quad \bigcup_{j=1}^{n} S_j = (0, R) \quad \text{and} \quad S_{n+1} = [R, \infty),$$

for $\delta, R > 0$ chosen later. We take $x_j \in S_j$ and $q^{(j)} \in \mathcal{A}$ such that

$$u(x_j) - \rho \le J_{x_j}(q^{(j)}), \quad j = 1, \ldots, n+1. \tag{12.20}$$

By the same calculation as (12.17), we find $C_\rho > 0$ such that for $q \in \mathcal{A}$,

$$|J_x(q) - J_y(q)| \le C_\rho |x - y| + \rho(1 + x + y), \quad x, y > 0.$$

Also, by (12.17), we take $0 < \delta < 1$ as $C_\rho \delta < \rho$. Then, for $|x - y| < \delta$

$$|u(x) - u(y)| \le \rho(2 + x + y)$$
$$< \Xi_\rho(x) := \rho(2 + x + (1 + x)), \quad x, y > 0.$$

Hence

$$\begin{aligned} J_{X(\tau)}(q^{(j)}) &= \{J_{X(\tau)}(q^{(j)}) - J_{x_j}(q^{(j)})\} + J_{x_j}(q^{(j)}) \\ &\ge -\Xi_\rho(X(\tau)) + J_{x_j}(q^{(j)}) \\ &\ge -\Xi_\rho(X(\tau)) + u(x_j) - \rho \\ &\ge -2\Xi_\rho(X(\tau)) + u(X(\tau)) - \rho \quad \text{if } X(\tau) \in S_j, j = 1, \ldots, n. \end{aligned} \tag{12.21}$$

By definition, we can find $q \in \mathcal{A}$ such that

$$\tilde{u}(x) - \rho \le E[\int_0^\tau e^{-\beta t}\{(r - \varphi(q(t)))X(t) + \frac{1}{\varepsilon}(u \vee g)(X(t))\}dt + e^{-\beta\tau}u(X(\tau))].$$

Define $q^\tau = \{q^\tau(t)\} \in \mathcal{A}$ by

$$q^\tau(t) = q(t)1_{\{t<\tau\}} + q^{(j)}(t-\tau)1_{\{t\ge\tau\}} \quad \text{if } X(\tau) \in S_j, j = 1, \dots, n+1,$$

on the same probability space. Clearly, q^τ belongs to $\mathcal{A}$.

Now, taking into account Lemmas 12.2.1, we choose $R > 0$ such that

$$\begin{aligned}\sup_{q\in\mathcal{A}} E[e^{-\beta\tau}u(X(\tau))1_{\{X(\tau)\ge R\}}] &\le \sup_{q\in\mathcal{A}} E[e^{-\beta\tau}(1 + X(\tau))1_{\{X(\tau)\ge R\}}]\|u\| \\ &\le \{\frac{1}{R}E[e^{-\beta\tau}X(\tau)] \\ &\quad + \frac{1}{R^{\kappa-1}}E[e^{-\beta\tau}X(\tau)^\kappa]\}\|u\| < \rho,\end{aligned}$$

and

$$\begin{aligned}E[e^{-\beta\tau}J_{X(\tau)}(q)1_{\{X(\tau)\ge R\}}] &\ge -\tilde{C}E[e^{-\beta\tau}X(\tau)1_{\{X(\tau)\ge R\}}] \\ &\ge -\tilde{C}E[e^{-\beta\tau}X(\tau)^\kappa]/R^{\kappa-1} \\ &> -\rho,\end{aligned}$$

since

$$J_x(q) \ge -\varphi(\zeta)E[\int_0^\infty e^{-\beta t}X(t)dt] \ge -\tilde{C}x, \quad \forall q \in \mathcal{A},$$

where $\tilde{C} = \varphi(\zeta)/(\alpha - (\zeta - \mu))$. Also, by (12.15),

$$E[\Xi_\rho(X(\tau))] \le \rho C(1 + x)$$

for some constant $C > 0$, independent of ρ. Hence, by (12.20) and (12.21), we have

$$\begin{aligned}&E[\int_\tau^\infty e^{-\beta t}\{(r - \varphi(q(t)))X^\tau(t) + \frac{1}{\varepsilon}(u \vee g)(X^\tau(t))\}dt] \\ &= E[e^{-\beta\tau}J_{X(\tau)}(\tilde{q}^\tau)] \\ &= E[e^{-\beta\tau}\sum_{j=1}^{n+1} J_{X(\tau)}(q^{(j)})1_{\{X(\tau)\in S_j\}}] \\ &\ge E[e^{-\beta\tau}\sum_{j=1}^{n}\{u(X(\tau)) - (2\Xi_\rho(X(\tau)) + \rho)\}1_{\{X(\tau)\in S_j\}}] - \rho \\ &\ge E[e^{-\beta\tau}u(X(\tau))] - \rho - \{2\rho C(1 + x) + \rho\} - \rho,\end{aligned}$$

where $\{X^\tau(t)\}$ denotes the solution of (12.3) corresponding to q^τ. Thus

$$\begin{aligned}
u(x) &\geq E[\int_0^\tau + \int_\tau^\infty e^{-\beta t}\{(r - \varphi(q(t)))X^\tau(t) + \frac{1}{\varepsilon}(u \vee g)(X^\tau(t))\}dt] \\
&\geq E[\int_0^\tau e^{-\beta t}\{(r - \varphi(q(t)))X(t) + \frac{1}{\varepsilon}(u \vee g)(X(t))\}dt \\
&\quad + e^{-\beta\tau}u(X(\tau))] - \{2C(1+x)+3\}\rho \\
&\geq \tilde{u}(x) - \rho - \{2C(1+x)+3\}\rho.
\end{aligned}$$

Therefore, letting $\rho \to 0$, we can obtain $\tilde{u} \leq u$. ∎

Theorem 12.2.5. *We make the assumptions of Theorem 12.2.2. Let u_1 and u_2 be two viscosity solutions of* (12.18) *in $\mathcal{B}$. Then we have $u_1 = u_2$.*

Proof. We will show that

$$u_1 \leq u_2.$$

Suppose that $u_1(x_0) - u_2(x_0) > 0$ for some $x_0 > 0$. We take a constant $\eta > 0$ such that

$$\sup_{x \geq 0}\{u_1(x) - u_2(x) - 2\eta\psi(x)\} > 0,$$

where $\psi(x) := x^\kappa$ for $\kappa > 1$ of (12.13). Since $u_1(x) - u_2(x) - 2\eta\psi(x) \to -\infty$ as $x \to \infty$, there exists $\bar{x} > 0$ such that

$$\sup_{x \geq 0}\{u_1(x) - u_2(x) - 2\eta\psi(x)\} = u_1(\bar{x}) - u_2(\bar{x}) - 2\eta\psi(\bar{x}) > 0.$$

Define

$$\Psi_n(x, y) := u_1(x) - u_2(y) - \frac{n}{2}|x - y|^2 - \eta(\psi(x) + \psi(y)),$$

for $n \in \mathbf{N}$. It is clear that

$$\begin{aligned}
\Psi_n(x, y) &\leq (\|u_1\| + \|u_2\|)(2 + x + y) - \eta(\psi(x) + \psi(y)) \\
&\to -\infty \quad \text{as} \quad x + y \to \infty.
\end{aligned}$$

Hence we find $(x_n, y_n) \in [0, \infty)^2$ such that

$$\begin{aligned}
\sup_{x,y \geq 0} \Psi_n(x, y) &= \Psi_n(x_n, y_n) \\
&= u_1(x_n) - u_2(y_n) - \frac{n}{2}|x_n - y_n|^2 - \eta(\psi(x_n) + \psi(y_n)) \\
&\geq u_1(\bar{x}) - u_2(\bar{x}) - 2\eta\psi(\bar{x}) > 0, \qquad (12.22)
\end{aligned}$$

from which

$$\frac{n}{2}|x_n - y_n|^2 < u_1(x_n) - u_2(y_n) - \eta(\psi(x_n) + \psi(y_n)).$$

Thus the sequences $\{x_n\}$, $\{y_n\}$, and $\{n|x_n - y_n|^2\}$ are bounded by some constant $C > 0$. Therefore, extracting a subsequence, we deduce

$$\begin{aligned}
&|x_n - y_n| \leq C/\sqrt{n} \to 0,\\
&x_n, y_n \;\to\; \tilde{x} \;\in [0, \infty) \quad \text{as } n \to \infty.
\end{aligned}$$

Furthermore, we recall $\Psi_n(x_n, y_n) \geq \Psi_n(x_n, x_n)$ to obtain

$$\frac{n}{2}|x_n - y_n|^2 \leq u_2(x_n) - u_2(y_n) + \eta(\psi(x_n) - \psi(y_n)) \to 0.$$

Passing to the limit in (12.22), we get

$$u_1(\tilde{x}) - u_2(\tilde{x}) - 2\eta\psi(\tilde{x}) > 0, \quad \tilde{x} \neq 0. \tag{12.23}$$

Now we apply Ishii's lemma to

$$\Psi_n(x, y) = U_1(x) - U_2(y) - \frac{n}{2}|x - y|^2,$$

where $U_1(x) := u_1(x) - \eta\psi(x)$ and $U_2(y) := u_2(y) + \eta\psi(y)$. Then we obtain $X_1, X_2 \in \mathbf{R}$ such that

$$\begin{aligned}
&(n(x_n - y_n), X_1) \in \bar{J}^{2,+}U_1(x_n),\\
&(n(x_n - y_n), X_2) \in \bar{J}^{2,-}U_2(y_n),\\
&\begin{pmatrix} X_1 & 0 \\ 0 & -X_2 \end{pmatrix} \leq 3n \begin{pmatrix} 1 & -1 \\ -1 & 1 \end{pmatrix}.
\end{aligned}$$

It is easy to see that

$$\begin{aligned}
J^{2,+}u_1(x) &= \{(p + \eta\psi'(x), X + \eta\psi''(x) : (p, X) \in J^{2,+}U_1(x)\},\\
J^{2,-}u_2(x) &= \{(p - \eta\psi'(x), X - \eta\psi''(x) : (p, X) \in J^{2,-}U_2(x)\}.
\end{aligned}$$

Hence

$$\begin{aligned}
&(p_1, \bar{X}_1) := (n(x_n - y_n) + \eta\psi'(x_n), X_1 + \eta\psi''(x_n)) \in \bar{J}^{2,+}u_1(x_n)\\
&(p_2, \bar{X}_2) := (n(x_n - y_n) - \eta\psi'(y_n), X_2 - \eta\psi''(y_n)) \in \bar{J}^{2,-}u_2(y_n),\\
&x_n^2 X_1 - y_n^2 X_2 \leq 3n(x_n - y_n)^2 \to 0 \quad \text{as } n \to \infty.
\end{aligned}$$

By the viscosity properties of u_1 and u_2, we have

$$\begin{aligned}
&-\alpha u_1(x) + \frac{1}{2}\sigma^2 x^2 \bar{X}_1 - \mu x p_1 + r x + \Gamma(p_1)x + \frac{1}{\varepsilon}(u_1 - g)^-|_{x=x_n} \geq 0,\\
&-\alpha u_2(y) + \frac{1}{2}\sigma^2 y^2 \bar{X}_2 - \mu y p_2 + r y + \Gamma(p_2)y + \frac{1}{\varepsilon}(u_2 - g)^-|_{y=y_n} \leq 0.
\end{aligned}$$

Putting these inequalities together, we have

$$\begin{aligned}\alpha(u_1(x_n) - u_2(y_n)) &\le \frac{1}{2}\sigma^2(x_n^2\bar{X}_1 - y_n^2\bar{X}_2)\\ &\quad - \mu(p_1x_n - p_2y_n) + r(x_n - y_n)\\ &\quad + \Gamma(p_1)x_n - \Gamma(p_2)y_n\\ &\quad + \frac{1}{\varepsilon}\{(u_1 - g)^-(x_n) - (u_2 - g)^-(y_n)\}\\ &\equiv I_1 + I_2 + I_3 + I_4.\end{aligned}$$

We note that

$$\begin{aligned}I_1 &= \frac{\sigma^2}{2}(x_n^2X_1 - y_n^2X_2 + x_n^2\eta\psi''(x_n) + y_n^2\eta\psi''(y_n)) \to \sigma^2\tilde{x}^2\eta\psi''(\tilde{x}),\\ I_2 &= -\mu n(x_n - y_n)^2 - \mu x_n\eta\psi'(x_n) - \mu y_n\eta\psi'(y_n) + r(x_n - y_n)\\ &\to -2\mu\tilde{x}\eta\psi'(\tilde{x}),\\ I_3 &\le \zeta|p_1x_n - p_2y_n| + \varphi(\zeta)|x_n - y_n| \to 2\zeta\tilde{x}\eta\psi'(\tilde{x}),\\ I_4 &\to \frac{1}{\varepsilon}\{(u_1 - g)^-(\tilde{x}) - (u_2 - g)^-(\tilde{x})\} \le 0 \qquad \text{by (12.23)}.\end{aligned}$$

Thus, by (12.13), we deduce

$$\begin{aligned}\alpha(u_1(\tilde{x}) - u_2(\tilde{x})) &\le 2\eta\{\frac{1}{2}\sigma^2\tilde{x}^2\psi''(\tilde{x}) + (\zeta - \mu)\tilde{x}\psi'(\tilde{x})\}\\ &= 2\eta\psi(\tilde{x})\{\frac{1}{2}\sigma^2\kappa(\kappa - 1) + \kappa(\zeta - \mu)\}\\ &< 2\alpha\eta\psi(\tilde{x}),\end{aligned}$$

which is contrary to (12.23). ■

12.2.3 Classical Solutions of the Penalty Equation

Theorem 12.2.6. *We assume* (12.6) *and* (12.7) *and that g is locally Lipschitz on* $(0, \infty)$. *Then the unique viscosity solution u of* (12.18) *belongs to* $C^2(0, \infty)$.

Proof. Let $0 < a < b$ be arbitrary, and we consider the boundary value problem:

$$\begin{aligned}&-(\alpha + \frac{1}{\varepsilon})\chi + \frac{1}{2}\sigma^2x^2\chi'' - \mu x\chi' + rx + \Gamma(\chi')x + \frac{1}{\varepsilon}\chi \vee g = 0 \text{ in } (a, b),\\ &\chi(a) = u(a), \quad \chi(b) = u(b).\end{aligned} \tag{12.24}$$

By the uniform ellipticity and the Lipschitz continuity of Γ and g, Theorem 5.3.7 yields that (12.24) has a smooth solution χ. By Theorem 5.3.1, we see that the viscosity solution of (12.24) is unique. Therefore we conclude that $u = \chi \in C^2(a, b)$, and thus $u \in C^2(0, \infty)$. ■

12.3 Nonlinear Variational Inequalities

In this section, we show the existence of a classical solution v of the nonlinear variational inequality:

$$\begin{cases} H(x, v, v', v'') := -\alpha v + \frac{1}{2}\sigma^2 x^2 v'' - \mu x v' + rx + \Gamma(v')x \le 0, & x > 0, \\ v(x) \ge g(x), \quad x \ge 0, \\ H(x, v, v', v'')(v - g)^+ = 0, \\ v(0) = 0. \end{cases} \tag{12.25}$$

Definition 12.3.1. *Let $v \in C[0, \infty)$ and $v(0) = 0$. Then v is a viscosity solution of* (12.25) *if*

(a) *For any $\phi \in C^2(0, \infty)$ and any local maximum point $z > 0$ of $v - \phi$,*

$$H(z, v, \phi', \phi'')(v - g)^+ \ge 0,$$

(b) *$v(x) \ge g(x)$, for all $x \in [0, \infty)$, and*
(c) *For any $\phi \in C^2(0, \infty)$ and any local minimum point $\bar{z} > 0$ of $v - \phi$,*

$$H(\bar{z}, v, \phi', \phi'') \le 0.$$

To prove the convergence of $u = u_\varepsilon$, we make the following assumption on g:

$$\begin{cases} g \in C^2(0, \infty) \cap C[0, \infty), \\ -\alpha g + \frac{1}{2}\sigma^2 x^2 g'' + (\zeta - \mu)xg' + h_0 = 0, \quad \text{for some } h_0 \in \mathcal{B}. \end{cases} \tag{12.26}$$

We can show that if $g(x) = Ax^\theta$ for $0 < \theta \le 1$ and $A > 0$, then g fulfills (12.26) and also, (12.7) by Young's inequality.

Lemma 12.3.2. *Under* (12.6), (12.7)*, and* (12.26)*, we have*

$$u(x) = \sup_{q \in \mathcal{A}, \tau \in \mathcal{S}} E[\int_0^\tau e^{-\alpha t}(r - \varphi(q(t)))X(t)dt + e^{-\alpha\tau}\{g - (u - g)^-\}(X(\tau))].$$

Proof. By Lemma 12.2.1, we have

$$\begin{aligned} \sup_{\tau \in \mathcal{S}} E[(e^{-\alpha\tau}u(X(\tau)))^\kappa] &\le \sup_{\tau \in \mathcal{S}} E[e^{-\alpha\tau}C^\kappa(1 + X(\tau))^\kappa] \\ &\le C^\kappa 2^\kappa(1 + \sup_{\tau \in \mathcal{S}} E[e^{-\alpha\tau}X(\tau)^\kappa]) < \infty. \end{aligned}$$

This yields that $\{e^{-\alpha\tau_m}u(X(\tau_m)) : m \in \mathbf{N}\}$ is uniformly integrable for any sequence $\{\tau_m\}$ of stopping times. Applying Itô's formula to (12.18), we get

$$\begin{aligned} u(x) &= E[\int_0^\tau e^{-\alpha t}\{(r - \varphi(q(t)))X(t) + \frac{1}{\varepsilon}(u - g)^-(X(t))\}dt + e^{-\alpha\tau}u(X(\tau))] \\ &\ge E[\int_0^\tau e^{-\alpha t}(r - \varphi(q(t)))X(t)dt + e^{-\alpha\tau}(u \wedge g)(X(\tau))], \end{aligned}$$

for any $\tau \in \mathcal{S}$. We note by Theorem 12.2.6 that $\gamma \circ u'(x)$ is locally Lipschitz, and

$$|(\gamma \circ u'(x) - \mu)x| \leq C(1+x), \quad x > 0$$

for some constant $C > 0$, where γ is as in (12.10). By analogy with Theorem 1.9.1, we see that there exists a unique solution $\{\hat{X}(t)\}$ of

$$d\hat{X}(t) = (\gamma \circ u'(\hat{X}(t)) - \mu)\hat{X}(t)dt + \sigma \hat{X}(t)dB(t), \quad \hat{X}(0) = x > 0.$$

Define

$$\hat{\tau} = \inf\{t \geq 0 : u(\hat{X}(t)) \leq g(\hat{X}(t))\}.$$

Then, by Lemma 12.2.1 and Itô's formula again, we deduce

$$\begin{aligned} u(x) &= E[\int_0^{\hat{\tau}} e^{-\alpha t}\{(r - \varphi(\gamma \circ u'(\hat{X}(t)))\hat{X}(t) \\ &\quad + \frac{1}{\varepsilon}(u-g)^-(\hat{X}(t))\}dt + e^{-\alpha\hat{\tau}}u(\hat{X}(\hat{\tau}))] \\ &= E[\int_0^{\hat{\tau}} e^{-\alpha t}(r - \varphi(\gamma \circ u'(\hat{X}(t)))\hat{X}(t)dt + e^{-\alpha\hat{\tau}}\{g - (u-g)^-\}(\hat{X}(\hat{\tau}))]. \end{aligned}$$

Thus we get the assertion. ■

Theorem 12.3.3. *We assume* (12.6), (12.7), *and* (12.26). *Then we have*

$$u_{\varepsilon_n} \to v \in \mathcal{B},$$

as $\varepsilon_n := 2^{-n} \to 0$, *and* v *is a viscosity solution of* (12.25).

Proof.

(1) We claim that

$$(u_\varepsilon - g)^-(x) \leq \varepsilon C(1+x), \quad x \geq 0, \tag{12.27}$$

for some constant $C > 0$, independent of $\varepsilon > 0$. By (12.6), (12.7), and (12.15), we have

$$0 \leq E[e^{-(\alpha+\frac{1}{\varepsilon})t}g(\bar{X}(t))] \leq \| g \| e^{-(\alpha+\frac{1}{\varepsilon})t}(1 + xe^{(\zeta-\mu)t}) \to 0 \quad \text{as } t \to \infty.$$

Then, by (12.26) and Itô's formula, g can be rewritten as

$$g(x) = E[\int_0^\infty e^{-(\alpha+\frac{1}{\varepsilon})t}(h_0 + \frac{1}{\varepsilon}g)(\bar{X}(t))dt].$$

Also, by (12.11),

$$u_\varepsilon(x) \geq E[\int_0^\infty e^{-(\alpha+\frac{1}{\varepsilon})t}\{(r - \varphi(\zeta))\bar{X}(t) + \frac{1}{\varepsilon}(u_\varepsilon \vee g)(\bar{X}(t))\}dt].$$

Hence, by (12.15),

$$\begin{aligned}(u_\varepsilon - g)(x) &\geq E[\int_0^\infty e^{-(\alpha+\frac{1}{\varepsilon})t}\{(r-\varphi(\zeta))\bar{X}(t) - h_0(\bar{X}(t)) \\ &\quad + \frac{1}{\varepsilon}(u_\varepsilon \vee g - g)(\bar{X}(t))\}dt] \\ &\geq -\frac{|r-\varphi(\zeta)|\varepsilon x}{(\alpha-(\zeta-\mu))\varepsilon+1} - \frac{\|h_0\|\varepsilon(1+x)}{(\alpha-(\zeta-\mu))\varepsilon+1} \\ &\geq \frac{-\varepsilon C(1+x)}{(\alpha-(\zeta-\mu))\varepsilon+1},\end{aligned}$$

which implies (12.27).

(2) By (12.15) and (12.6), taking into account the optional sampling theorem and Fatou's lemma, we have

$$E[e^{-\alpha\tau}X(\tau)] \leq E[e^{-\alpha\tau}\bar{X}(\tau)] \leq x, \quad \tau \in \mathcal{S}.$$

Hence, by (12.27) and Lemma 12.3.2,

$$\begin{aligned}|u_{\varepsilon_{n+1}}(x) - u_{\varepsilon_n}(x)| &\leq \sup_{q\in\mathcal{A},\tau\in\mathcal{S}} E[e^{-\alpha\tau}|(u_{\varepsilon_{n+1}} - g)^- - (u_{\varepsilon_n} - g)^-|(X(\tau))] \\ &\leq (\varepsilon_{n+1} + \varepsilon_n)C \sup_{\tau\in\mathcal{S}} E[e^{-\alpha\tau}(1+\bar{X}(\tau))] \\ &\leq (\varepsilon_{n+1} + \varepsilon_n)C(1+x).\end{aligned}$$

Therefore,

$$\sum_{n=1}^\infty \| u_{\varepsilon_{n+1}} - u_{\varepsilon_n} \| \leq C\sum_{n=1}^\infty(\varepsilon_{n+1} + \varepsilon_n) < \infty.$$

This yields that $\{u_{\varepsilon_n}\}$ is a Cauchy sequence and u_{ε_n} converges to v in $\mathcal{B}$.

(3) Let $\phi \in C^2(0,\infty)$ and $z > 0$ be the maximizer of $v - \phi$ such that

$$v(z) - \phi(z) > v(x) - \phi(x), \quad x \in \bar{B}_\delta(z), \ z \neq x$$

for some $\delta > 0$. By (2), we see that u_{ε_n} converges to v uniformly on $\bar{B}_\delta(z)$, and $u_{\varepsilon_n} - \phi$ attains a local maximum at $x_n \in \bar{B}_\delta(z)$. By the same argument as Theorem 4.5.1 (1), we deduce

$$x_n \to z \quad \text{as} \quad n \to \infty.$$

Then, by Theorem 12.2.4

$$(-\alpha u_{\varepsilon_n} + \frac{1}{2}\sigma^2x^2\phi'' - \mu x\phi' + rx + \Gamma(\phi')x)(u_{\varepsilon_n} - g)^+|_{x=x_n} \geq 0.$$

Passing to the limit, we get

$$(-\alpha v + \frac{1}{2}\sigma^2x^2\phi'' - \mu x\phi' + rx + \Gamma(\phi')x)(v - g)^+|_{x=z} \geq 0.$$

Next, letting $\varepsilon_n \to 0$ in (12.27), we have

$$v(x) \geq g(x), \quad \text{for all } x \in [0, \infty).$$

Finally, let $\bar{z}$ be the minimizer of $v - \phi$, and $\bar{x}_n$ be the minimum point of $u_{\varepsilon_n} - \phi$ such that $\bar{x}_n \to \bar{z}$. Then, by Theorem 12.2.4,

$$-\alpha u_{\varepsilon_n} + \frac{1}{2}\sigma^2 x^2 \phi'' - \mu x \phi' + rx + \Gamma(\phi')x|_{x=\bar{x}_n} \leq 0.$$

Passing to the limit, we have

$$-\alpha v + \frac{1}{2}\sigma^2 x^2 \phi'' - \mu x \phi' + rx + \Gamma(\phi')x|_{x=\bar{z}} \leq 0.$$

Thus we obtain the assertion of the theorem. ∎

Theorem 12.3.4. *We make the assumptions of Theorem 3.3. Then we have*

$$v \in C^{1,1}([a, b])$$

for any compact interval $[a, b] \subset (0, \infty)$, *where* $C^{1,1}$ *denotes the Hölder space in Chapter 5.*

Proof.

(1) We claim that

$$\sup_n \|u''_{\varepsilon_n}\|_{C[a,b]} < \infty. \tag{12.28}$$

We change the variable by $y = \log x$ and set $f_\varepsilon(y) = u_\varepsilon(e^y)$, $\tilde{g}(y) = g(e^y)$. Since

$$\begin{cases} f'_\varepsilon(y) = u'_\varepsilon(x)x, \\ f''_\varepsilon(y) = x^2 u''_\varepsilon(x) + f'_\varepsilon(y), \end{cases} \tag{12.29}$$

we can rewrite (12.18) as

$$-\alpha f_\varepsilon + \frac{1}{2}\sigma^2 f''_\varepsilon - (\frac{1}{2}\sigma^2 + \mu) f'_\varepsilon + re^y + \Gamma(f'_\varepsilon e^{-y})e^y + \frac{1}{\varepsilon}(f_\varepsilon - \tilde{g})^- = 0,$$
$$y \in [c, d] \subset \mathbf{R},$$

or, equivalently,

$$f'' = G(y, f, f') \quad \text{in } [c, d]$$

where we supress ε and

$$G(y, f, f') = \frac{2}{\sigma^2}\{\alpha f + (\frac{1}{2}\sigma^2 + \mu) f' - re^y - \Gamma(f' e^{-y})e^y - \frac{1}{\varepsilon}(f - \tilde{g})^-\}.$$

By Theorem 12.3.3, we have $\sup_n \|u_{\varepsilon_n} - v\| < \infty$, so that

$$0 \leq f_{\varepsilon_n}(y) \leq C(1 + e^y)$$

for some constant $C > 0$, independent of ε_n. Hence

$$\sup_n \|f_{\varepsilon_n}\|_{C[c,d]} < \infty. \tag{12.30}$$

Taking into account (12.27), we see that

$$|G(y, f, f')| \le \bar{C}(1 + |f'|), \quad y \in [c, d]$$

for some constant $\bar{C} > 0$, independent of $\varepsilon > 0$. By the interpolation inequality (Theorem 5.1.2), there exists $C > 0$, for any $\rho \in (0, (c+d)/2)$, such that

$$\|f'\|_{C[c,d]} \le \rho\|f''\|_{C[c,d]} + \frac{C}{\rho}\|f\|_{C[c,d]}. \tag{12.31}$$

Thus

$$\|f''\|_{C[c,d]} \le \bar{C}(1 + \rho\|f''\|_{C[c,d]} + \frac{C}{\rho}\|f\|_{C[c,d]}).$$

Taking $\rho \le \min(1/2\bar{C}, (c+d)/2))$, we get

$$\|f''\|_{C[c,d]} \le 2\bar{C}(1 + \frac{C}{\rho}\|f\|_{C[c,d]}),$$

so that

$$\sup_n \|f''_{\varepsilon_n}\|_{C[c,d]} < \infty.$$

By (12.31) again, we have

$$\sup_n \|f'_{\varepsilon_n}\|_{C[c,d]} < \infty.$$

Therefore, by (12.29), we deduce (12.28).

(2) By (1), we observe that $\{u_{\varepsilon_n}\}$ and $\{u'_{\varepsilon_n}\}$ are uniformly bounded and equicontinuous on $[a, b]$. By the Ascoli–Arzelà Theorem, extracting a subsequence, we have

$$\begin{cases} u_{\varepsilon_n} \to v, \\ u'_{\varepsilon_n} \to v' \quad \text{uniformly on } [a, b]. \end{cases} \tag{12.32}$$

Also,

$$|v'(x_1) - v'(x_2)| \le \hat{C}|x_1 - x_2|, \quad x_1, x_2 \in [a, b],$$

since

$$|u'_{\varepsilon_n}(x_1) - u'_{\varepsilon_n}(x_2)| \le \hat{C}|x_1 - x_2|, \quad x_1, x_2 \in [a, b],$$

where the constant $\hat{C} > 0$ depends only on a, b. Thus we obtain the desired result. ∎

12.4 Optimal Policies

In this section, we give a synthesis of optimal policies for the firm's value maximization problem.

Theorem 12.4.1. *We assume* (12.6), (12.7), *and* (12.26). *Then the optimal policy* $(q^*, \tau^*) \in \mathcal{A} \times \mathcal{S}$ *is given by*

$$q^*(t) = \gamma \circ v'(P(t)K^*(t)),$$
$$\tau^* = \inf\{t \geq 0 : v(P(t)K^*(t)) = g(P(t)K^*(t))\},$$

where

$$\frac{d}{dt}K^*(t) = \{\gamma \circ v'(P(t)K^*(t)) - \lambda\}K^*(t), \quad K^*(0) = k > 0. \tag{12.33}$$

Proof.

(1) We show the existence of a unique positive solution $\{X^*(t)\}$ of

$$dX^*(t) = \{\bar{\gamma}(X^*(t)) - \mu X^*(t)\}dt + \sigma X^*(t)dB(t), \quad X^*(0) = x > 0, \tag{12.34}$$

where $\bar{\gamma}(x) = \gamma \circ v'(x)x$. By (12.10) and Theorem 12.3.4, we observe that

$$\begin{cases} |\bar{\gamma}(x) - \bar{\gamma}(y)| \leq C|x - y|, & x, y \in [a, b], \\ 0 \leq \bar{\gamma}(x) \leq C(1 + x), \end{cases}$$

for some constant $C > 0$, depending only on each $0 < a < b$. By the same line as Theorem 1.7.2, for $n \in \mathbf{N}$, there exists a unique solution $\{X_n(t)\}$ of

$$dX_n(t) = \{\bar{\gamma} \circ \xi_n(X_n(t)) - \mu X_n(t)\}dt + \sigma X_n(t)dB(t), \quad X_n(0) = x > 0,$$

where the bounded and Lipschitz function $\xi_n(x)$ on $\mathbf{R}$ is defined by

$$\xi_n(x) = \begin{cases} n & \text{if } x > n, \\ x & \text{if } 1/n \leq x \leq n, \\ 1/n & \text{if } x < 1/n. \end{cases}$$

By analogy with the proof of Theorem 1.9.1, we can show the following:

① If $n < m$, $X_n(t \wedge \varrho_n) = X_m(t \wedge \varrho_n)$ for all $t \geq 0$ a.s., where $\varrho_n = \inf\{t \geq 0 : X_n(t) < 1/n \text{ or } X_n(t) > n\}$.

② ϱ_n is nondecreasing since $1/m < X_n(\varrho_n) < m$.

③ $X_n(t) = X_m(t)$ for $t \leq \varrho_n$, a.s.

④ Define $\mathsf{e} = \lim_{n\to\infty} \varrho_n$ and $X^*(t) = \lim_{n\to\infty} X_n(t)$, a.s. Clearly, $\xi_n \circ X_n(t) = X_n(t)$ for $t \le \varrho_n$.

⑤ The process $\{X^*(t)\}$ is $\mathcal{F}_t$-adapted and satisfies

$$dX^*(t) = [\bar{\gamma}(X^*(t)) - \mu X^*(t)]dt + \sigma X^*(t)dB(t),$$
$$X^*(0) = x > 0, \qquad \text{on } \{t < \mathsf{e}\}.$$

⑥ By the comparison theorem, $X^*(t) > 0$ on $\{t < \mathsf{e}\}$.

⑦ By Gronwall's inequality, $P(\mathsf{e} < \infty) = 0$ and the uniqueness holds.

Therefore, we obtain (12.34). It is obvious that $K^*(t) = X^*(t)/P(t)$ solves (12.33).

(2) By Lemma 12.2.1, we recall that $\{e^{-\alpha\varrho_m} u_{\varepsilon_n}(X(\varrho_m)) : m \in \mathbf{N}\}$ is uniformly integrable for any sequence $\{\varrho_m\}$ of stopping times. Then, by (12.18), Theorem 12.2.6, and Itô's formula,

$$\begin{aligned} E[e^{-\alpha\tau} u_{\varepsilon_n}(X(\tau))] &= u_{\varepsilon_n}(x) + E[\int_0^\tau e^{-\alpha t}\{-\alpha u_{\varepsilon_n} + \frac{1}{2}\sigma^2 x^2 u''_{\varepsilon_n} \\ &\quad + (q(t) - \mu)x u'_{\varepsilon_n}\}|_{x=X(t)} dt] \\ &\le u_{\varepsilon_n}(x) - E[\int_0^\tau e^{-\alpha t}(r - \varphi(q(t)))X(t)dt]. \end{aligned}$$

Letting $\varepsilon_n \to 0$, by Lemma 12.2.1 and Theorem 12.3.3, we deduce $v(x) \ge J(q, \tau)$ for all $(q, \tau) \in \mathcal{A} \times \mathcal{S}$.

(3) To prove the optimality, let

$$\tau_m = \inf\{t \ge 0 : v(X^*(t)) - \frac{1}{m}(1 + X^*(t)) \le g(X^*(t))\}, \quad m \in \mathbf{N}.$$

By Theorem 12.3.3, we get

$$\begin{aligned} &E[\int_0^{\tau_m} e^{-\alpha t}(u_{\varepsilon_n} - g)^-(X^*(t))dt] \\ &\le E[\int_0^{\tau_m} e^{-\alpha t}(u_{\varepsilon_n} - (v - \frac{1}{m}(1 + x)))^-|_{x=X^*(t)} dt] \\ &\le E[\int_0^{\tau_m} e^{-\alpha t}(\frac{1}{m} - \|u_{\varepsilon_n} - v\|)^-(1 + X^*(t))dt] \\ &= 0 \end{aligned}$$

for sufficiently large n. We set

$$\tau_R = \inf\{t \ge 0 : X^*(t) > R \text{ or } X^*(t) < 1/R\}$$

for $R > 1$ and $\bar{\tau} = \tau_R \wedge \tau_m$. Applying Itô's formula to (12.18), by (12.10) and (12.34), we have

$$\begin{aligned}
E[e^{-\alpha\bar{\tau}} u_{\varepsilon_n}(X^*(\bar{\tau}))] &= u_{\varepsilon_n}(x) + E[\int_0^{\bar{\tau}} e^{-\alpha t}\{-\alpha u_{\varepsilon_n} + \frac{1}{2}\sigma^2 x^2 u''_{\varepsilon_n} \\
&\quad + (\gamma \circ v' - \mu) x u'_{\varepsilon_n}\}|_{x=X^*(t)} dt] \\
&= u_{\varepsilon_n}(x) - E[\int_0^{\bar{\tau}} e^{-\alpha t}\{(r - \varphi(\gamma \circ u'_{\varepsilon_n}))x \\
&\quad + \frac{1}{\varepsilon_n}(u_{\varepsilon_n} - g)^-\}|_{x=X^*(t)} dt] \\
&\quad + E[\int_0^{\bar{\tau}} e^{-\alpha t}\{\gamma \circ v' - \gamma \circ u'_{\varepsilon_n}\} x u'_{\varepsilon_n}|_{x=X^*(t)} dt] \\
&= u_{\varepsilon_n}(x) - E[\int_0^{\bar{\tau}} e^{-\alpha t}\,(r - \varphi(q^*(t)))X^*(t) dt] \\
&\quad + E[\int_0^{\bar{\tau}} e^{-\alpha t}\{\gamma \circ v' - \gamma \circ u'_{\varepsilon_n}\} x u'_{\varepsilon_n}|_{x=X^*(t)} dt] \\
&\quad - E[\int_0^{\bar{\tau}} e^{-\alpha t}\{\varphi(\gamma \circ v'(X^*(t))) \\
&\quad - \varphi(\gamma \circ u'_{\varepsilon_n}(X^*(t)))\} X^*(t) dt]
\end{aligned}$$

for sufficiently large n. Letting $n \to \infty$, by (12.32), we see that

$$E[e^{-\alpha\bar{\tau}} v(X^*(\bar{\tau}))] = v(x) - E[\int_0^{\bar{\tau}} e^{-\alpha t}\,(r - \varphi(q^*(t)))X^*(t) dt].$$

Note that $\bar{\tau} \nearrow \tau_m$ as $R \nearrow \infty$ and $\tau_m \nearrow \tau^*$ as $m \nearrow \infty$. Passing to the limit, we have

$$E[e^{-\alpha\tau^*} v(X^*(\tau^*))] = v(x) - E[\int_0^{\tau^*} e^{-\alpha t}\,(r - \varphi(q^*(t)))X^*(t) dt].$$

Thus, $J(q^*, \tau^*) = v(x)$. Therefore, we obtain the assertion. ■

Part III

Appendices

Appendix A

Dini's Theorem

Theorem A.1 (Dini). *Let $C(\Omega)$ be the space of real-valued continuous functions on a compact metric space Ω. Let $\{f_n, n \in \mathbf{N}\}$ be a nonincreasing sequence in $C(\Omega)$ such that $f_n(\omega) \to f(\omega)$ as $n \to \infty$, for every $\omega \in \Omega$. Suppose that $f \in C(\Omega)$. Then $f_n \to f$ uniformly in Ω.*

Proof. We set $\varphi_n(\omega) = f_n(\omega) - f(\omega)$. Then $\varphi_n(\omega)$ is nonnegative and $\varphi_n(\omega) \searrow 0$. Since $\varphi_n \in C(\Omega)$, the maximum of φ_n is attained at $\omega_n \in \Omega$ for every $n \in \mathbf{N}$. Also, by extracting a subsequence, we may consider that $\omega_n \to \omega_0 \in \Omega$ as $n \to \infty$. For $n \geq m$, we have

$$\varphi_n(\omega_n) \leq \varphi_m(\omega_n) = \{\varphi_m(\omega_n) - \varphi_m(\omega_0)\} + \varphi_m(\omega_0).$$

Letting $n \to \infty$ and then $m \to \infty$, we deduce $\limsup_{n\to\infty} \varphi_n(\omega_n) = 0$. This shows that the whole sequence $\{\varphi_n\}$ converges to zero uniformly in Ω. ∎

Appendix B

The Stone–Weierstrass Theorem

Theorem B.1 (Stone–Weierstrass). *Let Ω be a compact metric space and let $\mathcal{D}$ be a a subset of the metric space $C(\Omega)$ of real-valued continuous functions on Ω. Assume that $\mathcal{D}$ has the following properties:*

(i) *if $f, g \in \mathcal{D}$ and $a, b \in \mathbf{R}$, then $f \times g \in \mathcal{D}$, $af + bg \in \mathcal{D}$,*
(ii) $1 \in \mathcal{D}$,
(iii) *if $x, y \in \Omega$ and $x \neq y$, there exists $h \in \mathcal{D}$ such that $h(x) \neq h(y)$.*

Then $\mathcal{D}$ is dense in $C(\Omega)$.

Proof.

(1) We show that there exists a nondecreasing sequence $\{u_n\}$ of real polynomials on $[0, 1]$ such that u_n converges uniformly to $\sqrt{t}$. Define

$$\begin{cases} u_{n+1}(t) = u_n(t) + \frac{1}{2}(t - u_n^2(t)), & n \geq 1, \\ u_1(t) = 0. \end{cases}$$

By induction, we have

$$\begin{aligned} \sqrt{t} - u_{n+1}(t) &= \sqrt{t} - u_n(t) - \frac{1}{2}(t - u_n^2(t)) \\ &= (\sqrt{t} - u_n(t))(1 - \frac{1}{2}(\sqrt{t} + u_n(t))) \geq 0, \end{aligned}$$

and

$$u_{n+1} \geq u_n \geq 0.$$

Hence $u_n(t)$ converges to some $v(t)$. Letting $n \to \infty$, we get $t - v^2(t) = 0$ and thus $v(t) = \sqrt{t}$. Dini's theorem yields that u_n converges uniformly to v.

(2) Claim: if $f \in \mathcal{D}$, then $|f|$ belongs to the closure $\bar{\mathcal{D}}$ of $\mathcal{D}$ in $C(\Omega)$. Let $\alpha = \sup_{x\in\Omega} |f(x)| > 0$. By (i), we see that $\frac{f^2}{\alpha^2} \in \mathcal{D}$ and $u_n(\frac{f^2}{\alpha^2}) \in \mathcal{D}$. By (1), we have

$$u_n(\frac{f^2}{\alpha^2}) \to \sqrt{\frac{f^2}{\alpha^2}} = |f/\alpha| \quad \text{uniformly},$$

which implies $|f/\alpha| \in \bar{\mathcal{D}}$.

(3) Let $f, g \in \mathcal{D}$. It is easy to see that

$$\min(f, g) = \frac{1}{2}(f + g - |f - g|), \quad \max(f, g) = \frac{1}{2}(f + g + |f - g|).$$

By (2), $\min(f, g)$ and $\max(f, g)$ belong to $\bar{\mathcal{D}}$.

(4) Let $f \in C(\Omega)$. Let $x, y \in \Omega$ with $x \neq y$. By (ii) and (iii), we find $a, b \in \mathbf{R}$ such that

$$f_{xy}(z) := ah(z) + b \cdot 1$$

satisfies $f_{xy}(x) = f(x)$, $f_{xy}(y) = f(y)$.

(5) Let $\varepsilon > 0$ be arbitrary. Then, by (4), there is a ball $B_r(y)$ such that $f_{xy}(z) < f(z) + \varepsilon$, for all $z \in B_r(y)$. Since Ω is compact, it is covered by a finite number of such balls $B_{r_1}(y_1), \ldots, B_{r_m}(y_m)$. We set

$$f_x(z) = \min(f_{xy_1}(z), \ldots, f_{xy_m}(z)), \quad z \in \Omega.$$

Since $f_{xy} \in \mathcal{D}$, we see by (3) that $f_x \in \bar{\mathcal{D}}$. By (4), we have $f_x(x) = f(x)$ and, also, $f_x(z) < f(z) + \varepsilon$, for all $z \in \Omega$.

(6) By (5), there is a ball $B_s(x)$ such that $f_x(z) > f(z) - \varepsilon$, for all $z \in B_s(x)$. By compactness, Ω is covered by a finite number of such balls, $B_{s_1}(x_1), \ldots, B_{s_k}(x_k)$. Define

$$g(z) = \max(f_{x_1}(z), \ldots, f_{x_k}(z)), \quad z \in \Omega.$$

By (3) and (5), we deduce that $g \in \bar{\mathcal{D}}$ and $f(z) - \varepsilon < g(z) < f(z) + \varepsilon$, for all $z \in \Omega$, that is, $\sup_{x\in\Omega} |f(x) - g(x)| < \varepsilon$. Thus we conclude that $f \in \bar{\mathcal{D}}$, which implies $\bar{\mathcal{D}} = C(\Omega)$. ■

Appendix C

The Riesz Representation Theorem

Theorem C.1 (Riesz). *Let $C(\Omega)$ be the Banach space of real-valued continuous functions h on a compact metric space Ω with norm $\|h\| = \sup_\omega |h(\omega)|$. Let Λ be a continuous linear functional on $C(\Omega)$. Then there exists a unique, finite signed measure μ on $\sigma(C(\Omega))$ such that*

$$\Lambda(f) = \int_\Omega f\, d\mu, \qquad f \in C(\Omega).$$

Proof. We apply the Daniell theorem to the space $C(\Omega)$. Note that $\sigma(C(\Omega))$ is the σ-algebra $\mathcal{B}(\Omega)$ of Borel sets in Ω. Let $\mathcal{L}^+$ denote the collection of linear functionals L on $C(\Omega)$ such that L is positive, that is, $L(f) \geq 0$ whenever $f \geq 0$, and $C_+ = \{f \in C(\Omega) : f \geq 0\}$.

(1) Define

$$\Lambda^+(f) = \sup\{\Lambda(g) : g \leq f, g \in C_+\}, \quad \text{for} \quad f \in C_+.$$

By the continuity of Λ,

$$\Lambda^+(f) \leq \sup\{\|\Lambda\|\|g\| : g \leq f, g \in C_+\} \leq \|\Lambda\|\|f\| < \infty.$$

If $f_i, g_i \in C_+$ and $g_i \leq f_i$, $i = 1, 2$, then

$$\Lambda(g_1) + \Lambda(g_2) = \Lambda(g_1 + g_2) \leq \Lambda^+(f_1 + f_2).$$

Take the suprenum over g_1 and g_2 to obtain $\Lambda^+(f_1) + \Lambda^+(f_2) \leq \Lambda^+ (f_1 + f_2)$.

Conversely, let $g, f_1, f_2 \in C_+$ and $g \leq f_1 + f_2$. Define

$$g_1 = g \wedge f_1, \quad g_2 = (g - f_1)^+ \in C_+.$$

Then $g_1 \leq f_1, g_2 \leq f_2$ and $g = g_1 + g_2$. Thus $\Lambda(g) = \Lambda(g_1) + \Lambda(g_2) \leq \Lambda^+(f_1) + \Lambda^+(f_2)$. Hence $\Lambda^+(f_1 + f_2) \leq \Lambda^+(f_1) + \Lambda^+(f_2)$, and, consequently,

$$\Lambda^+(f_1) + \Lambda^+(f_2) = \Lambda^+(f_1 + f_2), \quad f_1, f_2 \in C_+.$$

By definition, it is clear that $\Lambda^+(af) = a\Lambda^+(f)$ if $a \geq 0$ and $f \in C_+$.

(2) Define

$$\Lambda^+(f) = \Lambda^+(f^+) - \Lambda^+(f^-), \quad \text{for } f \in C(\Omega), \quad \text{and } f = f^+ - f^-,$$

and

$$\Lambda^-(f) = (-\Lambda)^+(f).$$

By (1), we have $\Lambda^+(f^+) - \Lambda^+(f^-) = \Lambda^+(f_1) - \Lambda^+(f_2)$ if $f = f_1 - f_2$, $f_1, f_2 \in C_+$. Hence

$$\begin{aligned}
&\Lambda^+(f_1 + f_2) = \Lambda^+(f_1) + \Lambda^+(f_2), \quad f_1, f_2 \in C(\Omega),\\
&\Lambda^+(af) = \Lambda^+(-(-a)f^+ + (-a)f^-) = (-a)\Lambda^+(f^-) - (-a)\Lambda^+(f^+)\\
&\qquad = a\Lambda^+(f), \quad a < 0, \ f \in C(\Omega).
\end{aligned}$$

Also, $\Lambda^+(f) \geq \Lambda(0) = 0$ for $f \in C_+$. Thus Λ^+, $\Lambda^- \in \mathcal{L}^+$. Note that

$$\begin{aligned}
\Lambda^-(f) &= -\inf\{\Lambda(g) : g \leq f, g \in C_+\}\\
&= -\inf\{\Lambda(f) - \Lambda(f - g) : g \leq f, g \in C_+\}\\
&= -\Lambda(f) + \sup\{\Lambda(h) : h \leq f, h \in C_+\}\\
&= -\Lambda(f) + \Lambda^+(f), \quad \text{for } f \in C_+.
\end{aligned}$$

Therefore, we get $\Lambda(f) = \Lambda^+(f) - \Lambda^-(f)$, for all $f \in C(\Omega)$.

(3) Let $\|f\| \leq 1$. Then $-\Lambda^+(1) \leq \Lambda^+(f) \leq \Lambda^+(1)$. We may consider $\Lambda^+ \neq 0$. Hence

$$0 < \|\Lambda^+\| = \sup_{\|f\| \leq 1} |\Lambda^+(f)| \leq \Lambda^+(1).$$

Define $\tilde{\Lambda} \in \mathcal{L}^+$ by

$$\tilde{\Lambda}(f) = \Lambda^+(f)/\Lambda^+(1).$$

Clearly, $\tilde{\Lambda}(1) = 1$.

(4) Let $f_n \in C(\Omega) \searrow 0$ as $n \to \infty$. By Dini's theorem, we have

$$\|f_n\| = \sup_{\omega \in \Omega} f_n(\omega) \to 0.$$

By positivity, we deduce

$$0 \leq \tilde{\Lambda}(f_n) \leq \tilde{\Lambda}(\|f_n\|) = \|f_n\| \to 0.$$

(5) By the Daniell theorem, there exist probability measures P^+, P^- on $\sigma(C(\Omega))$ such that

$$\tilde{\Lambda}(f) = \int_\Omega f \, dP^+, \quad f \in C(\Omega),$$

and, similarly,

$$\Lambda^-(f) = \Lambda^-(1) \int_\Omega f \, dP^-, \quad f \in C(\Omega).$$

Setting $d\mu = \Lambda^+(1)dP^+ - \Lambda^-(1)dP^-$, we have

$$\Lambda(f) = \int_\Omega f\, d\mu, \quad f \in C(\Omega).$$

(6) To prove uniqueness, let $\nu = \nu^+ - \nu^-$ be another such measure for two positive measures ν^+, ν^-. Then

$$\int_\Omega f\, d(\mu^+ - \mu^-) = \int_\Omega f\, d(\nu^+ - \nu^-), \quad f \in C(\Omega),$$

and hence

$$\int_\Omega f\, d\psi^+ = \int_\Omega f\, d\psi^-,$$

where $\psi^+ = \mu^+ + \nu^-$ and $\psi^- = \mu^- + \nu^+$. Since $\psi^+(\Omega) = \psi^-(\Omega)$, we have

$$\int_\Omega f\, \frac{1}{\psi^+(\Omega)} d\psi^+ = \int_\Omega f\, \frac{1}{\psi^-(\Omega)} d\psi^-.$$

By the Daniell theorem, we get $\frac{1}{\psi^+(\Omega)} d\psi^+ = \frac{1}{\psi^-(\Omega)} d\psi^-$ and thus $\mu = \nu$. ■

Appendix D

Rademacher's Theorem

Theorem D.1 (Rademacher). *Let $f : \mathbf{R}^N \to \mathbf{R}$ be Lipschitz. Then f is differentiable almost everywhere on $\mathbf{R}^N$, i.e.,*

$$\lim_{y \to x} \frac{f(y) - f(x) - (Df(x), y - x)}{|y - x|} = 0, \quad \textit{for almost all } x \in \mathbf{R}^N.$$

Proof.

(1) For $v \in \mathbf{R}^N$ with $|v| = 1$, $t \in \mathbf{R}$, and $x \in \mathbf{R}^N$, let $df(x, v)$ denote the directional derivative of f at x, and set

$$N_v = \{x \in \mathbf{R}^N : df(x, v) \text{ does not exist}\}.$$

Then

$$N_v = \{x \in \mathbf{R}^N : \liminf_{t \to 0} \frac{f(x + tv) - f(x)}{t} < \limsup_{t \to 0} \frac{f(x + tv) - f(x)}{t}\}$$

and is a Borel measurable set. Note that $\xi(t) := f(x + tv)$ is Lipschitz, absolutely continuous, and thus differentiable a.e. on $\mathbf{R}$. Hence

$$y_1 v \notin N_v, \quad a.e. \ y_1 > 0.$$

By Fubini's theorem

$$\text{meas}(N_v) = \int_{\mathbf{R}^{N-1}} (\int_{\mathbf{R}} 1_{N_v}(y_1, y_2, \ldots, y_N) dy_1) dy_2 \cdots dy_N = 0,$$

where $y_1, \ldots, y_N$ is the standard coordinate system of $\mathbf{R}^N$. In particular, we see that $Df(x) := (\frac{\partial f}{\partial x_1}(x), \cdots, \frac{\partial f}{\partial x_N}(x))$ exists a.e. on $\mathbf{R}^N$.

(2) Let ϕ be any infinitely differentiable function with compact support in $\mathbf{R}^N$, that is, $\phi \in C_c^\infty(\mathbf{R}^N)$. Then, it is clear that

$$\int_{\mathbf{R}^N} \frac{f(x + tv) - f(x)}{t} \phi(x) dx = -\int_{\mathbf{R}^N} f(x) \frac{\phi(x) - \phi(x - tv)}{t} dx.$$

By the dominated convergence thorem

$$\begin{aligned}\int_{\mathbf{R}^N} df(x,v)\phi(x)dx &= -\int_{\mathbf{R}^N} f(x)d\phi(x,v)dx\\ &= -\int_{\mathbf{R}^N} f(x)(D\phi(x),v)dx\\ &= \sum_i \int_{\mathbf{R}^N} v_i \frac{\partial f}{\partial x_i}(x)\phi(x)dx\\ &= \int_{\mathbf{R}^N} (Df(x),v)\phi(x)dx.\end{aligned}$$

This yields that

$$df(x,v) = (Df(x),v), \quad \text{a.e.} \quad x \in \mathbf{R}^N.$$

(3) Choose a countable dense subset $\{v_k\}$ of the boundary $\partial B_1(0)$ of $B_1(0)$. Set

$$\begin{aligned}M_k = \{x \in \mathbf{R}^N &: df(x,v_k), Df(x) \text{ exist and } df(x,v_k)\\ &= (Df(x),v_k)\}, \quad k = 1,2,\ldots,\end{aligned}$$

and define $M = \cap_{k=1}^{\infty} M_k$. Clearly, M^c is null. Fix any $x \in M$ and let

$$Q(x,v,t) := \frac{f(x+tv)-f(x)}{t} - (Df(x),v),$$

for each $v \in \partial B_1(0),\ t \neq 0$. By Lipschitz continuity,

$$\begin{aligned}|Q(x,v,t) - Q(x,v_k,t)| &\le \sup_{z\neq z'} \frac{|f(z)-f(z')|}{|z-z'|}|v-v_k| + |Df(x)||v-v_k|\\ &\le L(1+\sqrt{N})|v-v_k|\end{aligned}$$

for some constant $L > 0$, and hence

$$|Q(x,v,t)| \le |Q(x,v_k,t)| + L(1+\sqrt{N})|v-v_k|.$$

We take v_k such that $|v-v_k| \to 0$ as $k \to \infty$. Clearly, for any k,

$$\lim_{t\to 0} Q(x,v_k,t) = 0.$$

Therefore, letting $t \to 0$ and then $k \to \infty$, we deduce

$$\lim_{t\to 0} Q(x,v,t) = 0 \quad \text{for } x \in M.$$

(4) Let $x \in M, y \neq x$, and set $v = (y-x)/|y-x| \in \partial B_1(0)$, so that $y = x + tv$ for $t = |x-y|$. Then, by (3), we have

$$\begin{aligned}f(y) - f(x) - (Df(x), y-x) &= f(x+tv) - f(x) - (Df(x),tv)\\ &= Q(x,v,t)t\\ &= o(|y-x|) \quad \text{as } y \to x.\end{aligned}$$

■

Appendix E

Vitali's Covering Theorem

Lemma E.1. *Let $\mathcal{F}$ be a family of subsets of $\mathbf{R}^N$. Let Γ be a collection of $\mathcal{G} \subset \mathcal{F}$. We assume that for any family $\{\mathcal{G}_\lambda : \lambda \in \Lambda\} \subset \Gamma$ with the property that*

$$\mathcal{G}_\lambda \subset \mathcal{G}_{\bar{\lambda}} \text{ or } \mathcal{G}_{\bar{\lambda}} \subset \mathcal{G}_\lambda \quad \text{whenever} \quad \lambda, \bar{\lambda} \in \Lambda,$$

we have $\cup_{\lambda \in \Lambda} \mathcal{G}_\lambda \in \Gamma$. Then Γ has a maximal element $\mathcal{G}^*$, that is, $\mathcal{G}^* \subset \mathcal{G}^\square$, $\mathcal{G}^\square \in \Gamma$ implies $\mathcal{G}^* = \mathcal{G}^\square$.

Proof. Define the partial ordering $\leq$ on Γ by inclusion, that is, $\mathcal{G}_1 \leq \mathcal{G}_2$ if $\mathcal{G}_1 \subset \mathcal{G}_2$, for $\mathcal{G}_1, \mathcal{G}_2 \in \Gamma$. By assumption, every chain $\{\mathcal{G}_\lambda : \lambda \in \Lambda\}$ has an upper bound $\cup_{\lambda \in \Lambda} \mathcal{G}_\lambda \in \Gamma$. By Zorn's lemma, we conclude that there exists a maximal element $\mathcal{G}^*$ in Γ. ■

Theorem E.2. *Let $\mathcal{F}$ be a collection of nondegenerate closed balls in $\mathbf{R}^N$ with*

$$R := \sup\{\text{diam}(B) : B \in \mathcal{F}\} < \infty.$$

Then there exists a countable family $\mathcal{G} \subset \mathcal{F}$ of disjoint balls such that

$$\bigcup_{B \in \mathcal{F}} B \subset \bigcup_{B \in \mathcal{G}} \hat{B},$$

where $\hat{B} := \bar{B}_{5r}(x)$ if B is a ball $\bar{B}_r(x) \subset \mathbf{R}^N$.

Proof.

(1) For $j = 1, 2, \ldots,$ let $\mathcal{F}_j := \{B \in \mathcal{F} : R/2^j < \text{diam}(B) \leq R/2^{j-1}\}$. We define $\mathcal{G}_j \subset \mathcal{F}_j$ as follows:

① Let $\mathcal{G}$ be any family of disjoint balls in $\mathcal{F}_1$ and Γ_1 a collection of such $\mathcal{G}$. For any chain $\{\mathcal{G}_\lambda : \lambda \in \Lambda\} \subset \Gamma_1$, let $B_1, B_2 \in \mathcal{E} := \cup_{\lambda \in \Lambda} \mathcal{G}_\lambda$. Clearly, $B_1 \in \mathcal{G}_\lambda$ and $B_2 \in \mathcal{G}_{\bar{\lambda}}$ for some $\lambda, \bar{\lambda} \in \Lambda$. If $\mathcal{G}_\lambda \subset \mathcal{G}_{\bar{\lambda}}$, then $B_1, B_2 \in \mathcal{G}_{\bar{\lambda}}$, so that $B_1 \cap B_2 = \phi$. Thus $\mathcal{E} \in \Gamma_1$. By Lemma E.1, Γ_1 has a maximal element $\mathcal{G}_1$.

② Suppose that $\mathcal{G}_1, \mathcal{G}_2, \ldots, \mathcal{G}_{j-1}$ have been chosen. Let $\hat{\mathcal{G}}$ be any family of disjoint balls in $\hat{\mathcal{F}}_j$ and Γ_j a collection of such $\hat{\mathcal{G}}$, where

$$\hat{\mathcal{F}}_j = \{B \in \mathcal{F}_j : B \cap B' = \phi \text{ for all } B' \in \bigcup_{i=1}^{j-1} \mathcal{G}_i\}.$$

By the same argument as ①, using Lemma E.1, we deduce that Γ_j has a maximal element $\mathcal{G}_j$.

Now we define $\mathcal{G} = \cup_{j=1}^{\infty} \mathcal{G}_j$. It is clear that $\mathcal{G}$ is a collection of disjoint balls in $\mathcal{F}$ and $\mathcal{G} \subset \cup_{j=1}^{\infty} \mathcal{F}_j = \mathcal{F}$.

(2) Let $B \in \mathcal{F}$. Clearly, $B \in \mathcal{F}_k$ for some $k \geq 1$. Then there exists $B_0 \in \cup_{i=1}^{k} \mathcal{G}_i$ such that $B \cap B_0 \neq \phi$. If not, we have $B \in \hat{\mathcal{F}}_k$ and $B \cap B' = \phi$ for any $B' \in \mathcal{G}_k$. Therefore, $\hat{\mathcal{G}}_k := \{B\} \cup \mathcal{G}_k$ belongs to Γ_k. By the maximality of $\mathcal{G}_k$, we get $\hat{\mathcal{G}}_k = \mathcal{G}_k$, which is a contradiction. Furthermore,

$$\operatorname{diam}(B) \leq \frac{R}{2^{k-1}} = 2\frac{R}{2^k} < 2 \operatorname{diam}(B_0).$$

Hence, in view of geometry, $B \subset \hat{B}_0$. Thus we get the assertion. ∎

Appendix F

The Area Formula

Theorem F.1 (Sard). *Let $G \subset \mathbf{R}^N$ be open and $f : G \to \mathbf{R}^N$ be of class C^1. Let $F = \{x \in G : \det Df(x) = 0\}$. Then $f(F)$ has measure zero.*

Proof.

(1) We claim that $f(F \cap Q)$ has measure zero, where Q is a closed cube in G. For any $\varepsilon > 0$, there exists $\delta > 0$ such that for $x, y \in Q$, $|x - y| < \delta$,

$$|f(y) - f(x) - Df(x)(y - x)| < \varepsilon|x - y|.$$

Taking $M = \sup_{x \in Q} \|Df(x)\| < \infty$, we have

$$|f(y) - f(x)| \leq M|y - x|.$$

Let Q have sides of length l. We divide Q into k^N cubes $S = S_i$ of side length $l/k, (i = 1, \ldots, k^N)$, choosing $k \in \mathbf{N}$ as $\delta > \eta := \sqrt{N}l/k$. Then $|x - y| \leq diam(S) = \eta < \delta$ for $x, y \in S$. Suppose $F \cap S \neq \phi$. Then, for $x \in F \cap S$, the linear transform $Df(x)$ is degenerate. Hence $\{Df(x)y : y \in S\}$ is contained in an $(N-1)$-dimensional parallel polyhedron of side length $\leq M\eta$. Also,

$$\sup_{y \in S} |f(y) - f(x) - Df(x)(y - x)| \leq \varepsilon\eta.$$

Thus, the set $\{z \in \mathbf{R}^N : z = f(y) - f(x) + Df(x)x,\ y \in S\}$ is contained in a cylinder with height $2\varepsilon\eta$; the base of this cylinder is an $(N-1)$-dimensional parallel polyhedron of side length $\leq M\eta + 2\varepsilon\eta$. Since the Lebesgue measure is translation invariant, we get

$$\text{meas}(f(S)) \leq (M\eta + 2\varepsilon\eta)^{N-1}(2\varepsilon\eta) = (M + 2\varepsilon)^{N-1}2\varepsilon(\sqrt{N}l/k)^N.$$

Therefore

$$\text{meas}(f(F \cap Q)) \leq (M + 2\varepsilon)^{N-1}2\varepsilon(\sqrt{N}l)^N.$$

Letting $\varepsilon \to 0$, we obtain $\text{meas}(f(F \cap Q)) = 0$.

(2) To prove the theorem, it suffices to show that G can be expressed as a countable union of nonoverlapping closed cubes $Q_n, n \in \mathbf{N}$, in G. Consider the countable set of point of $\mathbf{R}^N$ with integral coordinate and unit closed cubes at each of these points. Then we have nonoverlapping (= disjoint interiors) cubes $\{Q_n^{(0)} : n \in \mathbf{N}\}$ of side length 1 such that $\mathbf{R}^N = \cup_n Q_n^{(0)}$. We split each side of each $Q_n^{(0)}$ into equal parts, so that we obtain nonoverlapping cubes $\{Q_{n'}^{(1)} : n' \in \mathbf{N}\}$ of side length $1/2$ such that $\mathbf{R}^N = \cup_{n'} Q_{n'}^{(1)}$. Repeating this procedure, we get at k th stage nonoverlapping cubes $\{Q_m^{(k)} : m \in \mathbf{N}\}$ of side length $1/2^k$ such that $\mathbf{R}^N = \cup_m Q_m^{(k)}$.

Now, let $\mathcal{C}_1$ be the set of $Q_n^{(1)}, n \in \mathbf{N}$, such that $Q_n^{(1)} \subset G$. Next, let $\mathcal{C}_2$ be the set of $Q_{n'}^{(2)}, n' \in \mathbf{N}$, such that $Q_{n'}^{(2)} \subset G$ and $Q_{n'}^{(2)} \not\subset Q$ for any $Q \in \mathcal{C}_1$. By induction, let $\mathcal{C}_k$ be the set of $Q_m^{(k)}, m \in \mathbf{N}$, such that $Q_m^{(k)} \subset G$ and $Q_m^{(k)} \not\subset Q$ for any $Q \in \mathcal{C}_1 \cup \cdots \cup \mathcal{C}_{k-1}$. Since each point of G is in one of the elements of $\mathcal{C}_r$ for some $r \in \mathbf{N}$, we deduce that

$$G = \cup\{Q : Q \in \bigcup_{k=1}^{\infty} \mathcal{C}_k\},$$

where the right-hand side is at most a countable union. ∎

Theorem F.2 (Area formula). *Let $A \subset \mathbf{R}^N$ be Lebesgue measurable and $g \in L^1(\mathbf{R}^N)$ nonnegative. Let $f : \mathbf{R}^N \to \mathbf{R}^N$ be of class C^1. Then*

$$\int_{f(A)} g(y)dy \leq \int_A g(f(x))|\det Df(x)|dx.$$

Proof.

(1) Let $g \in C_c(\mathbf{R}^N)$ and $g \geq 0$. Let G be open and set $F = \{x \in G : \det Df(x) = 0\}$. Then $G \setminus F$ is open. Thus, by Theorem F.1 (2), $G \setminus F$ can be expressed as a countable union of nonoverlapping closed cubes $Q_n, n = 1, 2, \ldots$. By the change of variables formula

$$\int_{f(Q_n)} g(y)dy = \int_{Q_n} g(f(x))|\det Df(x)|dx.$$

Hence

$$\begin{aligned}\int_{f(G\setminus F)} g(y)dy &= \int_{f(\cup_n Q_n)} g(y)dy = \int_{\cup_n f(Q_n)} g(y)dy \\ &\leq \sum_n \int_{f(Q_n)} g(y)dy = \sum_n \int_{Q_n} g(f(x))|\det Df(x)|dx \\ &= \int_{G\setminus F} g(f(x))|\det Df(x)|dx.\end{aligned}$$

By Sard's theorem,

$$\int_{f(F)} g(y)dy = 0.$$

Therefore,

$$\int_{f(G)} g(y)dy \le \int_{f(G\setminus F)} g(y)dy + \int_{f(F)} g(y)dy$$
$$\le \int_G g(f(x))|\det Df(x)|dx.$$

(2) We claim that $f(A)$ is Lebesgue measurable. Without loss of generality, we assume that A is bounded. Since the Lebesgue measure is regular, there exist compact sets K_n and open sets G_n such that $K_n \subset A \subset G_n$ and

$$\text{meas}(G_n \setminus K_n) < 1/n,$$

for $n \in \mathbf{N}$ (cf. W. Rudin [140]). We may consider that G_n is bounded and nonincreasing and K_n is nondecreasing. Since f is continuous, $f(K_n)$ is compact and, thus, is Lebesgue measurable. Hence $f(\cup_{n=1}^{\infty} K_n) = \cup_{n=1}^{\infty} f(K_n)$ is Lebesgue measurable, and so is $f(G_n)$. By (1), there exists $C > 0$, for any bounded open set G, such that

$$\int_{f(G)} dy \le \int_G |\det Df(x)|dx \le C\int_G dx.$$

Thus

$$\text{meas}(f(A) \setminus f(\overset{\infty}{\underset{n=1}{\cup}} K_n))$$
$$\le \text{meas}(f(\overset{\infty}{\underset{n=1}{\cap}} (G_n \setminus K_n))) \le \text{meas}(\overset{\infty}{\underset{n=1}{\cap}} f(G_n \setminus K_n))$$
$$= \lim_{n\to\infty} \text{meas}(f(G_n \setminus K_n)) \le \lim_{n\to\infty} C\,\text{meas}(G_n \setminus K_n) = 0.$$

Therefore, we get the assertion.

(3) Let A be Lebesgue measurable and bounded, and $g \in C_c(\mathbf{R}^N)$, $g \ge 0$. We set $G = \{x \in \mathbf{R}^N : |\det Df(x)| > 0\}$ and $F = \{x \in \mathbf{R}^N : \det Df(x) = 0\}$. Then G is open and $A \setminus F \subset G$. Furthermore, we note that

$$\int_G g(f(x))|\det Df(x)|dx < \infty.$$

As in (2), there exist open sets $\{G_n, n = 1, 2, \ldots\}$ such that $A \setminus F \subset G_n$ and $\text{meas}(G_n \setminus (A \setminus F)) \to 0$ as $n \to \infty$. Hence, by (1), we have

$$\int_{f(A\setminus F)} g(y)dy \le \int_{f(G\cap G_n)} g(y)dy \le \int_{G\cap G_n} g(f(x))|\det Df(x)|dx$$
$$\to \int_{A\setminus F} g(f(x))|\det Df(x)|dx \quad \text{as } n \to \infty.$$

Thus, by the same line as (1), we obtain the inequality.

(4) Let $g \in L^1(\mathbf{R}^N)$ and $g \ge 0$. Let $\hat{g}(x) = g(x)1_{\{x\in B_r(0):\, g(x)\le m\}}$ for $m \in \mathbf{N}, r > 0$. For $h > 0$, let $\hat{g}_h$ denote the mollification of $\hat{g}$, that is,

$$\hat{g}_h(x) = h^{-N}\int_{\mathbf{R}^N} \rho(\frac{x-y}{h})\hat{g}(y)dy = \int_{B_1(0)} \rho(z)\hat{g}(x-hz)dz,$$

as in Chapter 5. We take $\phi^{(k)} \in C_c(\mathbf{R}^N)$ such that $\|\hat{g} - \phi^{(k)}\|_{L^1} \to 0$ as $k \to \infty$. Hence, we can show that

$$\begin{aligned}\|\hat{g}_h - \hat{g}\|_{L^1} &\leq \|\hat{g}_h - \phi_h^{(k)}\|_{L^1} + \|\phi_h^{(k)} - \phi^{(k)}\|_{L^1} + \|\phi^{(k)} - \hat{g}\|_{L^1} \\ &\leq 2\|\hat{g} - \phi^{(k)}\|_{L^1} + \|\phi_h^{(k)} - \phi^{(k)}\|_{L^1} \\ &\to 0 \text{ as } h \to 0 \text{ and } k \to \infty.\end{aligned}$$

We set $g_n = \hat{g}_h$ for $h = 1/n$. Then, extracting a subsequence, we see that $g_n \in C_c(\mathbf{R}^N), 0 \leq g_n \leq m$, and g_n converges to $\hat{g}$ in $L^1(\mathbf{R}^N)$ and a.e. Define

$$h_n(x) = \begin{cases} g_n(x) & \text{if } x \in B := \{x : \lim_{n\to\infty} g_n(x) = \hat{g}(x)\}, \\ \hat{g}(x) & \text{otherwise.}\end{cases}$$

Then, by (3), we get

$$\begin{aligned}\int_{f(A)} h_n(y)dy &= \int_{f(A)\cap B} g_n(y)dy = \int_{f(A\cap f^{-1}(B))} g_n(y)dy \\ &\leq \int_{A\cap f^{-1}(B)} g_n(f(x))|\det Df(x)|dx \\ &\leq \int_A h_n(f(x))|\det Df(x)|dx.\end{aligned}$$

Thus, letting $n \to \infty$ and then $m, r \to \infty$, we obtain the desired inequality. ∎

Theorem F.3. *Let $f : \mathbf{R}^N \to \mathbf{R}^N$ be Lipschitz and $A \subset \mathbf{R}^N$ Lebesgue measurable. Then*

$$\int_{f(A)} dy \leq \int_A |\det Df(x)|dx.$$

Proof. Fix $R > 0$ and let $\mathcal{F}$ be the class of Lebesgue measurable sets $E \subset B_R := B_R(0)$, for which this inequality holds. We will show that every Lebesgue measurable set $A \subset B_R$ belongs to $\mathcal{F}$.

(1) Let $\eta : \mathbf{R}^N \to \mathbf{R}$ be Lipschitz. Then

$$|\eta(x + tv) - \eta(x + sv)| \leq L_0|t - s|, \quad |v| = 1, \quad t, s \in \mathbf{R}$$

for some constant $L_0 > 0$. By the proof of Rademacher's theorem,

$$|(D\eta(x), v)| = |d\eta(x, v)| \leq L_0, \quad \text{a.e. } x \in \mathbf{R}^N.$$

Thus $\eta \in W^{1,1}(B_{R+1}(0))$. Let $\eta_n, \tilde{\eta}_n \in C^1(\mathbf{R}^N)$ be the mollifications of $\eta, \tilde{\eta}$ for $h = 1/n$, where

$$\tilde{\eta}(x) = \begin{cases} \eta(x) & \text{if } x \in B_{R+1}(0), \\ 0 & \text{if } x \notin B_{R+1}(0).\end{cases}$$

Since $D\tilde{\eta} = D\eta$ a.e. on $B_{R+1}(0)$, we have that $\tilde{\eta} \in W^{1,1}(\mathbf{R}^N)$. By Theorem 5.1.8 (3), we see that

$$\|\tilde{\eta}_n - \tilde{\eta}\|_{W^{1,1}(\mathbf{R}^N)} \to 0,$$

or, equivalently,

$$\|\tilde{\eta}_n - \tilde{\eta}\|_{L^1(\mathbf{R}^N)} + \|(D\tilde{\eta})_n - D\tilde{\eta}\|_{L^1(\mathbf{R}^N)} \to 0 \quad \text{as } n \to \infty.$$

We note that

$$\begin{aligned}(D\tilde{\eta})_n(x) &= \int_{B_1(0)} \rho(z) D\tilde{\eta}(x - \frac{z}{n}) dz \\ &= \int_{B_1(0)} \rho(z) D\eta(x - \frac{z}{n}) dz = D\eta_n(x) \quad \text{a.e. on } \bar{B}_R.\end{aligned}$$

Hence

$$\int_{\bar{B}_R} |D\eta_n(x) - D\eta(x)| dx \to 0 \quad \text{as } n \to \infty.$$

Therefore, extracting a subsequence, we get

$$\lim_{n\to\infty} |D\eta_n(x) - D\eta(x)| = 0, \quad \text{a.e. on } \bar{B}_R,$$

and, furthermore,

$$|D\eta_n(x)| \le \int_{B_1(0)} \rho(z) |D\eta(x - \frac{z}{n})| dz \le L_0\sqrt{N}, \quad \text{a.e.}$$

(2) By Rademacher's theorem, the derivative $Df(x)$ of f exists a.e. Let f_n be the millification of each component of f. By Lemma 5.1.4 (i), f_n converges to f uniformly on each compact set as $n \to \infty$. Let $F \subset B_R$ be compact. Then there is $k \in \mathbf{N}$, for any $\varepsilon > 0$, such that whenever $n \ge k$, we have

$$|f(x) - f_n(x)| \le \sup_{x\in\bar{B}_R} |f(x) - f_n(x)| < \varepsilon/2, \quad x \in F.$$

Also, for each $x \in F$, there is $x_\varepsilon \in G_\varepsilon$ such that for $n \ge 1$

$$\begin{aligned}|f_n(x) - f_n(x_\varepsilon)| &\le \int_{B_1(0)} \rho(z) |f(x - \frac{z}{n}) - f(x_\varepsilon - \frac{z}{n})| dz \\ &\le L|x - x_\varepsilon| < \varepsilon/2,\end{aligned}$$

where $G_\varepsilon = \cup_{y\in F} B_{\varepsilon/2L}(y)$ and $L > 0$ is the Lipschitz constant of f. Thus, whenever $n \ge k$,

$$|f(x) - f_n(x_\varepsilon)| < \varepsilon, \quad x \in F.$$

For $0 < \delta < \varepsilon$, it is clear that $x_\delta \in G_\delta \subset G_\varepsilon$. Extracting a subsequence, we have $x_\delta \to \tilde{x} \in \bar{G}_\varepsilon$ as $\delta \searrow 0$. Hence, for $n \ge k$

$$\begin{aligned}|f(x) - f_n(\tilde{x})| &\le |f(x) - f_n(x_\delta)| + |f_n(x_\delta) - f_n(\tilde{x})| \\ &< \delta + |f_n(x_\delta) - f_n(\tilde{x})| \to 0 \quad \text{as } \delta \searrow 0, \quad x \in F.\end{aligned}$$

This yields that $f(x) = f_n(\tilde{x})$ and then

$$f(F) \subset \bigcap_{\varepsilon>0} \bigcup_{k} \bigcap_{n\geq k} f_n(\bar{G}_\varepsilon).$$

By Theorem F.2, we get

$$\int_{f_k(\bar{G}_\varepsilon)} dy \leq \int_{\bar{G}_\varepsilon} |\det Df_k(x)| dx.$$

Therefore, by (1) and the dominated convergence theorem,

$$\begin{aligned}\int_{f(F)} dy &\leq \lim_{k\to\infty} \int_{\bigcap\limits_{n\geq k} f_n(\bar{G}_\varepsilon)} dy \leq \limsup_{k\to\infty} \int_{f_k(\bar{G}_\varepsilon)} dy \\ &\leq \lim_{k\to\infty} \int_{\bar{G}_\varepsilon} |\det Df_k(x)| dx = \int_{\bar{G}_\varepsilon} |\det Df(x)| dx, \\ &\leq C \int_{\bar{G}_\varepsilon} dx, \quad \forall \varepsilon > 0.\end{aligned}$$

By compactness, $F = \bigcap\limits_{\varepsilon>0} \bar{G}_\varepsilon$. Letting $\varepsilon \to 0$, we deduce that

$$\int_{f(F)} dy \leq \int_F |\det Df(x)| dx \leq C \int_F dx$$

for some constant $C > 0$, independent of F.

(3) By (2) and Theorem F.1 (2), we have

$$\int_{f(G)} dy \leq \int_G |\det Df(x)| dx \leq C \int_G dx$$

for any open set G. Therefore,

$$\mathcal{G} := \{G \subset B_R : \text{open}\} \subset \mathcal{F}.$$

(4) By Theorem F.1 (2), we have

$$\mathcal{G} \subset \mathcal{F}_0 := \{E \subset B_R : E = \bigcup_{n=1}^{\infty} F_n = \bigcap_{n=1}^{\infty} G_n,\ F_n \text{ closed},\ G_n \text{ open}\}.$$

It is easy to see that $\mathcal{F}_0$ is closed under the operations $\cdot^c$ and $\cup$. Thus $\mathcal{F}_0$ is an algebra.

(5) Claim: $\mathcal{F}$ is monotone. Let $E_n \in \mathcal{F}, n = 1, 2, \ldots,$ be a nonincreasing sequence. Then

$$\begin{aligned}\int_{f(\bigcap\limits_n E_n)} dy &\leq \int_{\bigcap\limits_n f(E_n)} dy = \lim_{n\to\infty} \int_{f(E_n)} dy \\ &\leq \lim_{n\to\infty} \int_{E_n} |\det Df(x)| dx = \int_{\bigcap\limits_n E_n} |\det Df(x)| dx,\end{aligned}$$

which implies $\bigcap\limits_n E_n \in \mathcal{F}$. By analogy, we can show that $\mathcal{F}$ is closed under the limit operation for a nondecreasing sequence.

(6) By (4) and (5), we have $\mathcal{F}_0 \subset \mathcal{F}$. A standard result on σ-algebra gives that

$$\{E \subset B_R : \text{ Borel}\} = \sigma(\mathcal{G}) \subset \sigma(\mathcal{F}_0) \subset \mathcal{F}.$$

(7) Let $A \subset B_R$ be Lebesgue measurable. Then there are a Borel set E and a null set M such that $A = E \cup M$. We know by Theorem 4.4.4 (4) that $f(M)$ is null. By the same line as Theorem F.2 (2), taking into account (3), we see that $f(A)$ is Lebesgue measurable. Therefore, by (6),

$$\begin{aligned}\int_{f(A)} dy &= \int_{f(E)\cup f(M)} dy = \int_{f(E)} dy \leq \int_E |\det Df(x)|dx \\ &= \int_A |\det Df(x)|dx,\end{aligned}$$

which implies $A \in \mathcal{F}$. ■

Appendix G

The Brouwer Fixed–Point Theorem

Definition G.1. (a) *$m+1$ points $v_0, v_1, \ldots, v_m$ in $\mathbf{R}^N$ are said to be affinely independent if $v_1 - v_0, \ldots, v_m - v_0$ are linearly independent.*

(b) *For a subset A of $\mathbf{R}^N$, the convex hull $co(A)$ of A is the set of all convex combinations of members of A, that is, the set of all sums $\sum_{i=1}^{n} \lambda_i x_i$, in which $x_i \in A, \lambda_i \geq 0, \sum_{i=1}^{n} \lambda_i = 1$; n is arbitrary.*

(c) *For $m+1$ affinely independent points $v_0, v_1, \ldots, v_m$ in $\mathbf{R}^N (0 \leq m \leq N)$, the set $S := co\{v_0, v_1, \ldots, v_m\}$ is said to be an m-simplex.*

The points v_i are vertices of S and the sets $co\{v_{i_0}, \ldots, v_{i_n}\}(0 \leq n \leq m)$ are faces of S.

(d) *Let S be an m-simplex of $\mathbf{R}^N$. A simplicial subdivision of S is a finite collection $\Sigma = \{S_i : i \in I\}$ of simplexes $S_i \subset S$ such that*

(d_1) $S = \cup_{i \in I} S_i$;

(d_2) $S_i \cap S_j$ is either empty or equal to a common face of S_i and S_j if $i \neq j$;

(d_3) if $S_i \in \Sigma$ and F is a face of S_i, then $F \in \Sigma$.

(e) *Let S be an m-simplex of $\mathbf{R}^N$ and $\Sigma = \{S_i : i \in I\}$ a simplicial subdivision of S. Let $\mathcal{V} = \cup_{i \in I} V(S_i)$ for the set $V(S_i)$ of vertices of S_i.*

A mapping $\sigma : \mathcal{V} \to \{0, \ldots, m\}$ is the label of Σ if

$$\begin{cases} z \in \mathcal{V},\ z \in co\{v_{i_0}, \ldots, v_{i_n}\} \Rightarrow \sigma(z) = \text{one of } i_0, \ldots, i_n, \\ \sigma(v_j) = j, \quad \text{for } j = 0, 1, \ldots, m. \end{cases}$$

Σ is said to be labeled if there is a label σ of Σ.

Theorem G.2 (Sperner's lemma). *Let $S = co\{v_0, v_1, \ldots, v_m\}$ be an m-simplex of $\mathbf{R}^N$ and Σ a simplicial subdivision of S. Assume that Σ is labeled. Then there is an odd number of m-simplexes $T \in \Sigma$ such that $\sigma(V(T)) = \{0, 1, \ldots, m\}$.*

Proof. For any $k_0, k_1, \ldots, k_n \in \{0, 1, \ldots, m\}$ and $0 \leq n \leq m$, we define the counting function $\nu(k_0, k_1, \ldots, k_n)$ by the number of $i \in I$ of $S_i \in \Sigma$ such that $\sigma(V(S_i)) = \{k_0, k_1, \ldots, k_n\}$, $\nu(k_0, k_1, \ldots, k_n) = 0$ if such S_i does not exist, where $\{k_0, k_1, \ldots, k_n\}$ means the set of n points $k_0, k_1, \ldots, k_n$, for instance, $\{0, 0, 1, 2\} = \{2, 0, 1, 0\} \neq \{0, 1, 2\}$.

By induction on m, we will show that $\nu(0, 1, \ldots, m)$ is odd. Obviously, $\nu(0) = 1$.

(1) The case $m = 1$. $S = co\{v_0, v_1\}$ is subdiveded into line segments and Σ contains 0-simplexes S_i such that

$$\sigma(V(S_i)) = \{0\} \quad (\sigma(v_0) = 0, \sigma(v_1) = 1).$$

Consider the sum

$$s_1 := 2\nu(0, 0) + \nu(0, 1).$$

We note that s_1 is the number of faces 0 in the labels of vertices in $\mathcal{V}$. s_1 counts every interior face 0 twice, because every interior face is shared by two shared simplexes. But s_1 counts the single 0 on the boundary of S once. Therefore,

$$2\nu(0, 0) + \nu(0, 1) = 2\nu_i(0) + \nu_b(0),$$

where $\nu_i(0)$ and $\nu_b(0)$ be the numbers of 0 in the interior and on the boundary, respectively. Clearly, $\nu_b(0) = 1$. Thus $\nu(0, 1)$ is odd.

(2) The case $m = 2$. $S = co\{v_0, v_1, v_2\}$ is subdivided into triangles and Σ contains 1-simplexes S_i such that

$$\sigma(V(S_i)) = \{0, 1\} \quad (\sigma(v_0) = 0, \sigma(v_1) = 1, \sigma(v_2) = 2).$$

Consider the sum

$$s_2 := 2[\nu(0, 1, 0) + \nu(0, 1, 1)] + \nu(0, 1, 2),$$

which is the number of faces $(0, 1)$ in the labels of vertexes in $\mathcal{V}$. The s_2 counts every interior face $(0, 1)$ twice, because every interior face is shared by two simplexes. But s_2 counts every face $(0, 1)$ on the boundary of S once, because each face on the boundary belongs to only one simplex. Therefore,

$$2[\nu(0, 1, 0) + \nu(0, 1, 1)] + \nu(0, 1, 2) = 2\nu_i(0, 1) + \nu_b(0, 1),$$

where $\nu_i(0, 1)$ and $\nu_b(0, 1)$ be the numbers of faces $(0, 1)$ in the interior and on the boundary, respectively. By (1), $\nu_b(0, 1)$ is odd. Thus $\nu(0, 1, 2)$ is odd.

(3) The general case $m \geq 3$. Σ contains $(m-1)$-simplexes S_i such that

$$\sigma(V(S_i)) = \{0, 1, \ldots, m-1\} \quad (\sigma(v_0) = 0, \ldots, \sigma(v_m) = m).$$

Consider the sum

$$s_m := 2[\sum_{k=0}^{m-1} \nu(0, 1, \ldots, m-1, k)] + \nu(0, 1, \ldots, m-1, m),$$

which is the number of faces $(0, 1, \ldots, m-1)$ in the labels of vertexes in $\mathcal{V}$. s_m counts every interior face $(0, 1, \ldots, m-1)$ twice, because every interior face is shared by two simplexes. But s_m counts every face $(0, 1, \ldots, m-1)$ on the boundary of S once, because each face on the boundary belongs to only one simplex. Therefore,

$$s_m = 2\nu_i(0, 1, \ldots, m-1) + \nu_b(0, 1, \ldots, m-1),$$

where $\nu_i(0, 1, \ldots, m-1)$ and $\nu_b(0, 1, \ldots, m-1)$ be the numbers of faces $(0, 1, \ldots, m-1)$ in the interior and on the boundary, respectively. By induction hypothesis, $\nu_b(0, 1, \ldots, m-1)$ is odd. Thus $\nu(0, 1, \ldots, m-1, m)$ is odd. ■

Theorem G.3 (Knaster–Kuratowski–Mazurkiewicz lemma). *Let $S = co\{v_0, v_1, \ldots, v_m\}$ be an m-simplex of $\mathbf{R}^N$ and let E_i, $i = 0, 1, \ldots, m$, be closed subsets of $\mathbf{R}^N$. We assume that*

$$co\{v_{i_0}, \ldots, v_{i_n}\} \subset \bigcup_{i \in J} E_i, \qquad \textit{for every} \quad J = \{i_0, \ldots, i_n\} \subset \{0, 1, \ldots, m\}.$$

Then $\cap_{i=0}^m E_i \neq \phi$.

Proof.

(1) Let $U^{(n)} = co\{v_0, v_1, \ldots, v_n\}$ be an n-simplex for $n \leq m$. By induction, we define the barycentric subdivision $T^{(n)}$ of $U^{(n)}$ into $(n+1)!$ smaller n-simplexes $T_{i,j}^{(n)}$ as follows. For $n = 1$, $U^{(1)}$ is divided into two 1-simplexes

$$T_{1,1}^{(1)} := co\{v_0, u_1\}, \quad T_{1,2}^{(1)} := co\{v_1, u_1\},$$

where $u_1 = (v_0 + v_1)/2$ is its barycenter. We set

$$T^{(1)} := \{T_{1,1}^{(1)}, T_{1,2}^{(1)}\}.$$

For $n \geq 2$, let F_i, $i = 1, \ldots, n+1$, be the $(n-1)$ dimensional faces of $U^{(n)}$. By induction hypothesis, each F_i is subdivided into simplexes, denoted by $T_{i,j}^{(n-1)}$, $j = 1, \ldots, n!$. There are $(n+1)! = (n+1)n!$ such $(n-1)$-simplexes $T_{i,j}^{(n-1)}$. Divide $U^{(n)}$ into $(n+1)!$ simplexes $T_{i,j}^{(n)}$ by segments joining the barycenter $u_n = (v_0 + v_1 + \cdots + v_n)/(n+1)$ to $T_{i,j}^{(n-1)}$. More precisely, $T_{i,j}^{(n)}$ is given by

$$T_{i,j}^{(n)} = co(T_{i,j}^{(n-1)} \cup \{u_n\}).$$

We set

$$T^{(n)} := \{T_{i,j}^{(n)} : i = 1, \ldots, n+1, j = 1, \ldots, n!\}.$$

Furthermore, by iteration of this formula, we see that each $T_{i,j}^{(n)}$ can be rewitten as

$$T_{i,j}^{(n)} = co\{u_0, u_1, \ldots, u_n\},$$

where $u_0 = v_{i_0}, u_1 = (v_{i_0} + v_{i_1})/2, \ldots, u_n = (v_{i_0} + v_{i_1} + \cdots + v_{i_n})/(n + 1)$ for a suitable choice of permutation $(i_0, i_1, \ldots, i_n)$ of $(0, 1, \cdots, n)$. In particular, the collection of the barycentric subdivision and their faces for S forms a simplicial subdivision Σ of S.

(2) We claim that

$$\mu(T^{(n)}_{i,j}) \leq \frac{n}{n+1}\mu(U) \leq \frac{m}{m+1}\mu(S),$$

where $\mu(U)$ denotes the maximum distance between two vertices of the simplex $U := U^{(n)}$. Setting $w_j = v_{i_j}$, $j = 0, 1, \ldots, n$, we have

$$|w_j - u_n| = \frac{1}{n+1}|\sum_{i=0}^{n}(w_j - w_i)| \leq \frac{1}{n+1}\sum_{i\neq j}|w_j - w_i| \leq \frac{n}{n+1}\mu(U).$$

Hence, for $u_j = \sum_{i=0}^{n} \lambda_i w_i \in T^{(n)}_{i,j}$,

$$\begin{aligned}|u_j - u_n| &= |\sum_{i=0}^{n}\lambda_i(w_i - u_n)| \leq \sum_{i=0}^{n}\lambda_i\frac{n}{n+1}\mu(U)\\ &= \frac{n}{n+1}\mu(U), \quad 0 \leq j \leq n-1.\end{aligned}$$

Similarly,

$$|u_i - u_j| \leq \frac{j}{j+1}\mu(U) \leq \frac{n}{n+1}\mu(U), \quad 0 \leq i < j \leq n-1.$$

Therefore,

$$\max_{0\leq i,j\leq n}|u_i - u_j| \leq \frac{n}{n+1}\mu(U),$$

which implies the desired inequality.

(3) By iterating this process k-times for each $U^{(n)} \subset S$, we obtain the kth simplicial subdivision $\Sigma^{(k)}$ of S and

$$\mu(T) < (\frac{m}{m+1})^k\mu(S), \qquad \text{for all } T \in \Sigma^{(k)}.$$

(4) Let $\mathcal{V}^{(k)}$ be the set of vertices of all $T \in \Sigma^{(k)}$. For any $z \in \mathcal{V}^{(k)}$, there is $J = \{i_0, \ldots, i_n\} \subset \{0, 1, \ldots, m\}$ such that $z \in co\{v_{i_0}, \ldots, v_{i_n}\}$. Hence $z \in E_{i_j}$ for some $i_j \in J$. Putting $\sigma(z) = i_j$, we see that σ is a label of $\Sigma^{(k)}$.

By Sperner's lemma, there exists an m-simplex $T_k = co\{v_0^{(k)}, \ldots, v_m^{(k)}\} \in \Sigma^{(k)}$ such that $\sigma(V(T_k)) = \{0, 1, \ldots, m\}$. Without loss of generality, we may assume that $\sigma(v_i^{(k)}) = i$ and, by assumption, $v_i^{(k)} \in E_i$, for every $i = 0, 1, \ldots, m$. Since S is compact, extracting a subsequence, we have $v_i^{(k)} \to \bar{v}_i \in E_i$ as $k \to \infty$, for all i. Furthermore, by (3), $\mu(T_k) \to 0$. Thus $\bar{v}_0 = \cdots = \bar{v}_m \in \cap_{i=0}^{m} E_i$. ∎

Theorem G.4 (Brouwer). *Let K be a nonempty convex compact subset of $\mathbf{R}^N$ and $f : K \to K$ a continuous function. Then f has a fixed point $x \in K$, that is, $f(\bar{x}) = \bar{x}$.*

Proof.

(1) Let K be an m-simplex $S = co\{v_0, v_1, \ldots, v_m\}$. For any $x \in S$, we can write

$$x = \sum_{i=0}^{m} \lambda_i(x) v_i, \quad \lambda_i(x) \geq 0, \quad \sum_{i=0}^{m} \lambda_i(x) = 1.$$

We set $\Lambda(x) = (\lambda_0(x), \ldots, \lambda_m(x))$ for any $x \in S$. Let $x_n \in S \to x$. Then, by the Bolzano–Weierstrass theorem, there exists a subsequence $\{x_{n_k}\}$ of $\{x_n\}$ such that $\Lambda(x_{n_k})$ converges to $a = (a_0, \ldots, a_m) \in \mathbf{R}^{m+1}$ as $k \to \infty$. Clearly, $\sum_{i=0}^{m} \lambda_i(x) v_i = \sum_{i=0}^{m} a_i v_i$ and $\sum_{i=0}^{m} (\lambda_i(x) - a_i) = 0$, so that $\lambda_i(x) = a_i$, for all i. This yields that the whole sequence $\Lambda(x_n)$ converges to $\Lambda(x)$. If not, there are infinitely many x_j in $\{x_n\}$ such that $\Lambda(x_j) \notin B_\delta(\Lambda(x))$ for some $\delta > 0$. But, we find a subsequence $\{x_{j_k}\}$ of $\{x_j\}$ such that $\Lambda(x_{j_k}) \to \Lambda(x)$. This is a contradiction. Thus $\lambda_i(x)$ is continuous for all i.

Now we claim that

$$co\{v_i, i \in J\} \subset \bigcup_{i \in J} E_i, \qquad \text{for every } J \subset \{0, 1, \ldots, m\},$$

where E_i is the closed subset of S defined by

$$E_i = \{x \in S : \lambda_i(f(x)) \leq \lambda_i(x)\}, \quad i = 0, 1, \ldots, m.$$

If $x \in co\{v_i, i \in J\}$, then $\sum_{i \in J} \lambda_i(x) = 1$. Suppose $x \notin \cup_{i \in J} E_i$. Then $\lambda_i(f(x)) > \lambda_i(x)$, for all $i \in J$. Thus

$$1 = \sum_{i=0}^{m} \lambda_i(f(x)) \geq \sum_{i \in J} \lambda_i(f(x)) > \sum_{i \in J} \lambda_i(x) = 1,$$

which is a contradiction.

By the Knaster–Kuratowski–Mazurkiewicz lemma, there exists $\bar{x} \in \cap_{i=0}^{m} E_i$, that is, $\lambda_i(f(\bar{x})) \leq \lambda_i(\bar{x})$ for all i. Since

$$\sum_{i=0}^{m} \lambda_i(f(\bar{x})) = \sum_{i=0}^{m} \lambda_i(\bar{x}) = 1,$$

we have $\lambda_i(f(\bar{x})) = \lambda_i(\bar{x})$, for all i. Therefore, $f(\bar{x}) = \bar{x}$.

(2) Let K be convex and compact. By boundedness, K is contained in some sufficiently large simplex S. For each $x \in S$, there exists $y \in K$ such that

$$|x - y| = \min_{z \in K} |x - z|.$$

Let $y_i, i = 1, 2$, be such a y corresponding to $x_i \in S$. By convexity, for $0 < t \leq 1$,

$$|x_1 - y_1|^2 + |x_2 - y_2|^2 \leq |x_1 - (y_1 + t(y_2 - y_1))|^2 + |x_2 - (y_2 + t(y_1 - y_2))|^2.$$

Hence

$$0 \leq -t(x_1 - y_1, y_2 - y_1) - t(x_2 - y_2, y_1 - y_2) + t^2|y_2 - y_1|^2.$$

Dividing by t and letting $t \to 0$, we have

$$|y_2 - y_1|^2 \leq (x_2 - x_1, y_2 - y_1) \leq |x_2 - x_1||y_2 - y_1|.$$

Thus

$$|y_2 - y_1| \leq |x_2 - x_1|.$$

This yields that y is uniquely determined. Define $p(x) = y$ for each $x \in S$. Then $p(x)$ is continuous, and $p(x) = x$ if $x \in K$. By (1), the map $f \circ p : S \to K \subset S$ has a fixed point $\bar{x} \in S$. Hence $\bar{x} = f \circ p(\bar{x}) \in K$. Therefore, $\bar{x} = f(\bar{x})$. ■

Appendix H

The Ascoli–Arzelà Theorem

Theorem H.1 (Ascoli–Arzelà). *Let $C(K)$ be the space of all continuous functions f on a comapct subset K of $\mathbf{R}^N$ with norm $\|f\| = \sup_{x\in K} |f(x)|$. Then a sequence $\{f_n\}$ in $C(K)$ has a subsequence converging to some element of $C(K)$ if the following conditions are satisfied:*

(i) $\{f_n\}$ *is bounded, that is,* $\sup_n \|f_n\| < \infty$,
(ii) $\{f_n\}$ *is equicontinuous, that is,* $\lim_{\delta\to 0} \sup_n \sup_{|x-y|<\delta} |f_n(x) - f_n(y)| = 0$.

Proof.

(1) Let $E = \{x_1, x_2, \dots\}$ be a countable dense subset of K. By (i), $\{f_n(x_1)\}$ is bounded. Hence, by the Bolzano–Weierstrass theorem, there is a subsequence

$$\{f_{11}(x_1),\ f_{12}(x_1), \dots,\ f_{1n}(x_1), \dots\}$$

of $\{f_n(x_1)\}$, which is convergent. Similarly, the sequence $\{f_{1k}(x_2) : k = 1, 2, \dots\}$ has a subsquence

$$\{f_{21}(x_2),\ f_{22}(x_2), \dots,\ f_{2n}(x_2), \dots\},$$

which is convergent. Continuing the process, the sequence $\{f_{(n-1)k}(x_n) : k = 1, 2, \dots\}$ has a subsequence

$$\{f_{n1}(x_n),\ f_{n2}(x_n), \dots,\ f_{nn}(x_n), \dots\},$$

which is convergent. We proceed in this way and then set $g_n = f_{nn}$, so that g_n is the nth function occurring in the nth subsequence. We notice that g_n is obtained by picking out the diagonal of $\{f_{mk}\}$.

From its construction, $\{g_n : n \geq m\}$ is a subsequence of $\{f_{mk} : k = 1, 2, \dots\}$ for each $m \geq 1$. Thus we see that $\{g_n : n \geq 1\}$ converges at each point of E.

(2) We claim that the sequence $\{g_n\}$ converges at each point $x \in K$. For any $\delta > 0$, there is $y_\delta \in E$ such that $|x - y_\delta| < \delta$. Then, we have

$$\begin{aligned}|g_n(x) - g_m(x)| &\le |g_n(x) - g_n(y_\delta)| + |g_n(y_\delta) - g_m(y_\delta)| \\ &\quad + |g_m(y_\delta) - g_m(x)| \\ &\le |g_n(y_\delta) - g_m(y_\delta)| + 2\sup_j \sup_{|x-y|<\delta} |g_j(x) - g_j(y)| \\ &\le |g_n(y_\delta) - g_m(y_\delta)| + 2\sup_k \sup_{|x-y|<\delta} |f_k(x) - f_k(y)|.\end{aligned}$$

Letting $n, m \to \infty$ and then $\delta \to 0$, by (1) and (ii), we get

$$\lim_{n,m\to\infty} |g_n(x) - g_m(x)| = 0.$$

By the Cauchy criterion, this yields that $\{g_n(x)\}$ is convergent.

(3) We claim that $\{g_n\}$ is convergent in $C(K)$. Let $y \in K$. By (2) and (ii), there exist $N = N(y, \varepsilon) \in \mathbf{N}$ and $\delta = \delta(\varepsilon) > 0$, for any $\varepsilon > 0$, such that

$$|g_n(x) - g_m(y)| \le \sup_j |g_j(x) - g_j(y)| + |g_n(y) - g_m(y)| < \varepsilon, \qquad \text{for } x \in B_\delta(y),\ \ n, m \ge N.$$

Since the collection of balls $\{B_\delta(\hat{y}) : \hat{y} \in K\}$ covers K, by compactness, there is a finite covering $\{B_\delta(y_1), \cdots, B_\delta(y_k)\}$ of K. Hence, $y \in B_\delta(y_j)$ for some $1 \le j \le k$ and

$$|g_n(y) - g_m(y_j)| < \varepsilon, \qquad \text{for } n, m \ge N(y_j, \varepsilon).$$

Define $N_0 = \max\{N(y_1, \varepsilon), \ldots, N(y_k, \varepsilon)\}$. Then, we have for $m, n \ge N_0$

$$\begin{aligned}|g_n(y) - g_m(y)| &\le |g_n(y) - g_n(y_j)| + |g_n(y_j) \\ &\quad - g_m(y_j)| + |g_m(y_j) - g_m(y)| < 3\varepsilon.\end{aligned}$$

This yields that $\lim_{n,m\to\infty} \|g_n - g_m\| = 0$. By completeness, we conclude that $\{g_n\}$ converges to some $g \in C(K)$. ∎

Bibliography

[1] R. A. Adams and J. J. F. Fournier. (2003). *Sobolev Spaces*. New York: Academic Press.

[2] R. B. Ash. (1972). *Real Analysis and Probability*. New York: Academic Press.

[3] M. Akian, J. L. Menaldi, and A. Sulem. (1996). On an investment-consumption model with transaction costs, *SIAM J Control Optim.*, 34, 329–364.

[4] R. G. D. Allen. (1967). *Marco-Economic Theory: A Mathematical Treatment*. London: Macmillan.

[5] J. P. Aubin. (1979). *Mathematical Methods of Game and Economic Theory*. Amsterdam: North Holland.

[6] C. Baiocchi and A. Capelo. (1984). *Variational and Quasivariational Inequalities*. New York: John Wiley & Sons.

[7] V. Barbu and G. Da Prato. (1983). *Hamilton-Jacobi Equations in Hilbert Spaces*. Research Notes in Mathematics, vol. 86. Boston: Pitman.

[8] M. Bardi and I. Capuzzo-Dolcetta. (1997). *Optimal Control and Viscosity Solutions of Hamilton–Jacobi–Bellman Equations*. Boston: Birkhäuser.

[9] R. F. Bass. (1998). *Diffusions and Elliptic Operators*. New York: Springer-Verlag.

[10] R. Bellman. (1957). *Dynamic Programming*. Princeton NJ: Princeton University Press.

[11] V. E. Beneš. (1971). Existence of optimal stochastic control law, *SIAM J Control*, 9, 446–472.

[12] V. E. Beneš. (1975). Composition and invariance methods for solving some stochastic control problems, *Adv. Appl. Probab.*, 7, 299–329.

[13] V. E. Beneš, L. A. Shepp, and H. S. Witsenhausen. (1980). Some solvable stochastic control problems, *Stochastics*, 4, 39–83.

[14] V. E. Beneš and I. Karatzas. (1981). Examples of optimal control for partially observable systems: Comparison, classical, and martingale methods, *Stochastics*, 5, 43–64.

[15] A. Bensoussan. (1979). Optimal impulsive control theory, *Lecture Notes in Control and Information Sciences*, 16, 17–41.

[16] A. Bensoussan. (1982). *Stochastic Control by Functional Analysis Methods*. Amsterdam: North Holland.

[17] A. Bensoussan. (1988). *Pertubation Methods in Optimal Control*. New York: John Wiley & Sons.

[18] A. Bensoussan. (1992). *Stochastic Control of Partially Observable Systems*. Cambridge, UK: Cambridge University Press.

[19] A. Bensoussan and J. Frehse. (1992). On Bellman equations of ergodic control in R^n, *J reine angrew. math.*, 429, 125–160.

[20] A. Bensoussan and A. Friedman. (1974). Nonlinear variational inequalities and differential games with stopping times, *J Funct. Analysis*, 16, 305–352.

[21] A. Bensoussan and A. Friedman. (1977). Nonzero-sum stochastic differential games with stopping times and free boundary problems, *Trans. Amer. Math. Soc.*, 231, 275–327.
[22] A. Bensoussan and J. L. Lions. (1978). *Applications des Inéquations Variationnelles en Contrôle Stochastique*. Paris: Dunod.
[23] A. Bensoussan and J. L. Lions. (1982). *Contrôle Impulsionnel et Inéquations Quasi-variationnelles*. Paris: Dunod.
[24] A. Bensoussan, S. P. Sethi, R. Vickson, and N. Derzko. (1984). Stochastic production planning with production constraints, *SIAM J Control Optim.*, 22, 920–935.
[25] J. M. Bismut and B. Skalli. (1977). Temps d'arrêt optimal, théorie générale de processus et processus de Markov, *Z. Wahr. verw. Geb.*, 39, 301–313.
[26] J. M. Bismut. (1979). Contrôle de processus alternants et applications, *Z. Wahr. verw. Geb.*, 47, 241–288.
[27] R. M. Blumenthal and R. K. Getoor. (1968). *Markov Processes and Potential Theory*. New York: Academic Press.
[28] R. Boel, P. P. Varaiya, and E. Wong. (1975). Martingalers on jump processes I, II, *SIAM J Control Optim.*, 13, 999–1061.
[29] V. S. Borkar. (1989). *Optimal Control of Diffusion Processes*, Research Notes in Math., vol. 203, Pitman, Harlow: Longman.
[30] H. Brezis. (1992). *Analyse Fonctionnelle: Théorie et Applications*. Paris: Masson.
[31] P. Cannarsa and C. Sinestrari. (2004). Semiconcave Functions, *Hamilton–Jacobi Equations, and Optimal Control*, Boston: Birkhäuser.
[32] Y. Z. Chen and L. C. Wu. (1998). *Second-Order Elliptic Equations and Elliptic Systems*, Translations of Mathematical Monographs, vol. 174. Providence, RI: American Mathematics Society.
[33] A. C. Chiang. (1992). *Elements of Dynamic Optimization*. New York: McGraw-Hill.
[34] C. W. Clark. (1976). *Mathematical Bioeconomics: The Optimal Exploitation of Renewable Resources*, New York: John Wiley & Sons.
[35] F. H. Clarke. (1983). *Optimization and Nonsmooth Analysis*. New York: John Wiley & Sons.
[36] M. G. Crandall, H. Ishii, and P. L. Lions. (1992). User's guide to viscosity solutions of second-order partial differential equations, *Bull. Amer. Math. Soc.*, 27, 1–67.
[37] G. Da Prato. (1985). Some results on Bellman equation in Hilbert spaces, *SIAM J Control Optim.*, 23, 61–71.
[38] G. Da Prato and A. Ichikawa. (1990). Quadratic control for linear time-varying systems, *SIAM J Control Optim.*, 28, 359–381.
[39] G. Da Prato and J. Zabczyk. (1992). *Stochastic Equations in Infinite Dimensions*. Cambridge, UK: Cambridge University Press.
[40] M. H. A. Davis. (1977). Linear Estimation and Stochastic Control. London: Chapman & Hall.
[41] M. H. A. Davis and A. R. Norman. (1990). Portfolio selection with transaction costs, *Math. Oper. Res.*, 15, 676–713.
[42] M. H. A. Davis and M. Zervos. (1994). A problem of singular stochastic control with discretionary stopping, *Ann. Appl. Prob.*, 4, 226–240.
[43] M. H. A. Davis and P. P. Varaiya. (1973). Dynamic programming conditions for partially observable stochastic systems, *SIAM J Control Optim.*, 11, 226–261.
[44] S. Dayanik and I. Karatzas. (2003). On the optimal stopping problem for one-dimensional diffusions, *Stoch. Proc. Appl.*, 107, 173–212.
[45] C. Dellacherie and P. A. Meyer. (1975/1980). *Probabilités et Potentiel*. Paris: Hermann.
[46] A. K. Dixit and R. S. Pindyck. (1994). *Investment under Uncertainty*. Princeton, NJ: Princeton University Press.
[47] J. L. Doob. (1953). *Stochastic Processes*, New York: John Wiley & Sons.
[48] D. Duffie. (1986). Stochastic equilibria: Existence, spanning number, and the 'no expected financial gain from trade' hypothesis, *Econometrica*, 54, 1161–1183.
[49] D. Duffie and W. Zame. (1989). The consumption-based capital asset pricing model, Econometrica, 57, 1279–1297.

[50] D. Duffie. (1996). *Dynamic Asset Pricing Theory*. 2nd ed. Princeton, NJ: Princeton University Press.

[51] D. Duffie, W. H. Fleming, H. M. Soner, and T. Zariphopoulou. (1977). Hedging in incomplete markets with HARA utility, *J Econ. Dyn. Contr.*, 21, 753–782.

[52] T. D. Duncan and P. P. Varaiya. (1971). On the solutions of a stochastic control system, *SIAM J Control Optim.*, 9, 354–371.

[53] E. B. Dynkin. (1965). *Markov Processes, Vol. I–II*, Berlin: Springer-Verlag.

[54] E. B. Dynkin and A. A. Yushkevich. (1969). *Markov Processes: Theorems and Problems*, New York: Plenum Press.

[55] N. El Karoui. (1981). Les aspects probabilistes du contrôle stochastique, *Lecture Notes in Math.*, 876, Berlin: Springer-Verlag.

[56] N. El Karoui and M. C. Quenez. (1995). Dynamic programming and pricing of contingent claims in an incomplete market, *SIAM J. Control Optim.*, 33, 29–66.

[57] N. El Karoui, S. Peng, and M. C. Quenez. (1997). Backward stochastic differential equations in finance, *Math. Finance*, 7, 1–71.

[58] R. J. Elliott. (1982). *Stochastic Calculus and Applications*. New York: Springer-Verlag.

[59] L. C. Evans. (1998). *Partial Differential Equations*. Providence, RI: American Mathematical Society.

[60] L. C. Evans and A. Friedman. (1979). Optimal stochastic switching and the Dirichlet problem for the Bellman equation, *Trans. Amer. Math. Soc.*, 253, 365–389.

[61] L. C. Evans and R. F. Gariepy. (1992). *Measure Theory and Fine Properties of Functions*. Boca Raton, FL: CRC Press.

[62] A. G. Fakeev. (1970). Optimal stopping rules for stochastic processes with continuous parameter, *Theory Probab. Appl.*, 15, 324–331.

[63] W. H. Fleming and R. W. Rishel. (1975). *Deterministic and Stochastic Optimal Control*. New York: Springer-Verlag.

[64] W. H. Fleming and M. Nisio. (1996). On the existence of optimal stochastic controls, *J Math. Mech.*, 15, 777–794.

[65] W. H. Fleming and H. M. Soner. (1993). *Controlled Markov Processes and Viscosity Solutions*. New York: Springer-Verlag.

[66] W. H. Fleming and T. Zariphopoulou. (1991). An optimal investment/consumption model with borrowing, *Math. Oper. Res.*, 16, 802–822.

[67] J. N. Franklin. (2002). *Methods of Mathematical Economics*. Philadelphia: SIAM.

[68] A. Friedman. (1975, 1976). *Stochastic Differential Equations and Applications, Vols. I–II*. New York: Academic Press.

[69] A. Friedman. (1982). *Foundations of Modern Analysis*. New York: Dover.

[70] M. G. Garroni and J. L. Menaldi. (2002). Second-Order Integro-Differential Problems, Research Note in Mathematics, vol. 430. Boca Raton, FL: Chapman & Hall/CRC.

[71] I. I. Gikhman and A. V. Skorokhod. (1969). *Introduction to the Theory of Random Processes*. Philadelphia: Saunders.

[72] I. I. Gihman and A. V. Skorohod. (1972). *Stochastic Differential Equations*. Berlin: Springer-Verlag.

[73] I. I. Gihman and A. V. Skorohod. (1979). *Controlled Markov Processes*. New York: Springer-Verlag.

[74] D. Gilbarg and N. S. Trudinger. (1983). *Elliptic Partial Differential Equations of Second Order*. 2nd ed. New York: Springer-Verlag.

[75] N. Ikeda and S. Watanabe. (1981). *Stochastic Differential Equations and Diffusion Processes*. Amsterdam: North Holland.

[76] N. Ikeda and S. Watanabe. (1977). A comparison theorem for solutions of stochastic differential equations and its applications, *Osaka J. Math.*, 14, 619–633.

[77] H. Ishii. (1989). On uniqueness and existence of solutions of fully nonlinear second-order elliptic PDEs, *Comm. Pure Appl. Math.*, 42, 14–45.

[78] H. Ishii. (1994). Viscosity solutions to nonlinear partial differential equations, *Sugaku*, 46, 144–157 (in Japanese).

[79] K. Itô and H. P. McKean, Jr. (1965). *Diffusion Processes and Their Sample Paths*. Berlin: Springer-Verlag.

[80] M. I. Kamien and N. L. Schwartz. (1991). *Dynamic Optimization: The Calculus of Variations and Optimal Control in Economics and Management*. 2nd ed. Amsterdam: North Holland.

[81] K. Kamizono and H. Morimoto. (2002). On a variational inequality associated with a stopping game combined with a control, *Stoch. Stoch. Rep.*, 73, 99–123.

[82] I. Karatzas. (1983). A class of singular stochastic control problems, *Adv. Appl. Probab.*, 15, 225–254.

[83] I. Karatzas and S. E. Shreve. (1984). Connections between optimal stopping and singular stochastic control I: Monotone follower problems, *SIAM J Control Optim.*, 22, 856–877.

[84] I. Karatzas and S. E. Shreve. (1985). Connections between optimal stopping and singular stochastic control II: Reflected follower problems, *SIAM J Control Optim.*, 23, 433–451.

[85] I. Karatzas, J. P. Lehoczky, S. P. Sethi, and S. E. Shreve. (1986). Explicit solution of a general consumption/investment problem, *Math. Oper, Res.*, 11, 261–294.

[86] I. Karatzas, J. P. Lehoczky, and S. E. Shreve. (1990). Existence and uniqueness of multi-agent equilibrium in a stochastic, dynamic consumption/investment model, *Math. Oper. Res.*, 15, 80–128.

[87] I. Karatzas and S. E. Shreve. (1991). Brownian Motion and Stochastic Calculus, 2nd ed. New York: Springer-Verlag.

[88] I. Karatzas. (1996). *Lectures on the Mathematics of Finance*. CRM Monographs 8. Providence, RI: American/Mathematical Society.

[89] I. Karatzas and S. E. Shreve. (1998). *Methods of Mathematical Finance*. New York: Springer-Verlag.

[90] I. Karatzas and H. Wang. (2000). A barrier option of American type, *Appl. Math. Optim.*, 42, 259–279.

[91] I. Karatzas and H. Wang. (2000). Utility maximization with discretionary stopping, *SIAM J Control Optim.* 39, 306–329.

[92] I. Karatzas and D. Ocone. (2002). A leavable bounded-velocity stochastic control problem, *Stoch. Proc. Appl.*, 99, 31–51.

[93] T. Kato. (1966). *Functional Analysis*. Tokyo: Kyôritsu-Shuppan. (in Japanese).

[94] K. Kawaguchi and H. Morimoto. (2007). Long-run average welfare in a pollution accumlation model, *J Econ. Dyn. Contr.*, 31, 703–720.

[95] D. Kinderlehrer and G. Stampacchia. (1980). *An Introduction to Variational Inequalities and Their Applications*. New York: Academic Press.

[96] H. Kunita and S. Watanabe. (1967). On square integrable martingales, *Nagoya Math. J*, 30, 209–245.

[97] S. Koike. (2004). A Beginner's Guide to the Theory of Viscosity Solutions. MSJ Memoirs, vol. 13, Japan: Mathematics Society.

[98] S. Koike and H. Morimoto. (2003). On variational inequalities for leavable bounded-velocity control, *Appl. Math. Optim.*, 48, 1–20.

[99] N. Krylov. (1980). *Controlled Diffusion Processes*. Berlin: Springer-Verlag.

[100] H. Kwakernaak and R. Sivan. (1972). *Linear Optimal Control Systems*. New York: John Wiley & Sons.

[101] O. A. Ladyzhenskaya and N. N. Ural'tseva. (1968). Linear and Quasilinear Elliptic Equations. New York: Academic Press.

[102] P. L. Lions. (1982). *Generalized Solutions of Hamilton–Jacobi–Bellman Equations*, Research Notes in Mathematics, vol. 69. Boston: Pitman.

[103] P. L. Lions. (1983). Optimal control of diffusion processes and Hamilton–Jacobi–Bellman equations, Part I: The dynamic programming principle and applications; Part II: Viscosity solutions and uniqueness, *Comm. Partial Differential Equations*, 8, 1101–1174, 1229–1276.

[104] P. L. Lions and A. S. Sznitman. (1984). Stochastic differential equations with reflecting boundary conditions, *Comm. Pure Appl. Math.*, 37, 511–537.

[105] P. L. Lions and P. E. Souganidis. (1988). Viscosity solutions of second-order equations, stochastic control and stochastic diffrential games, IMA Volumes in Mathematics and Its Applications, 10, 293–309.

[106] R. S. Liptser and A. N. Shiryayev. (1977, 1978). *Statistics of Random Processes, Vols. I–II*. New York: Springer-Verlag.

[107] L. Ljungqvist and T. J. Sargent. (2000). *Recursive Macroeconomic Theory*. Cambridge, MA: MIT Press.

[108] J. L. Menaldi. (1980). On the optimal stopping time problem for degenerate diffusions, *SIAM J Control Optim.*, 18, 697–721.

[109] J. L. Menaldi. (1980). On the optimal impulse control problem for degenerate diffusions, *SIAM J Control Optim.*, 18, 722–739.

[110] J. L. Menaldi and M. Robin. (1983). On some cheap control problems for diffusion processes, *Trans. Amer. Math. Soc.*, 278, 771–802.

[111] J. L. Menaldi. (1987). Optimal impulse control problems for degenerate diffusions with jumps, *Acta. Appl. Math.*, 8, 165–198.

[112] R. C. Merton. (1971). Optimal consumption and portfolio rules in a continuous-time model, *J Economic Theory*, 3, 373–413; Erratum 6 (1973), 213–214.

[113] R. C. Merton. (1975). An asymptotic theory of growth under uncertainty, *Review of Economic Studies*, 42, 375–393.

[114] R. C. Merton. (1990). *Continuous-Time Finance*. Oxford and Cambridge, UK: Basil Blackwell.

[115] H. Morimoto. (1982). Optimal stopping and a martingale approach to the penalty method, *Tôhoku Math. J*, 34, 407–416.

[116] H. Morimoto. (1984). Dynkin games and martingale methods, *Stochastics*, 13, 213–228.

[117] H. Morimoto. (1986). Non-zero-sum discrete parameter stochastic games with stopping times, *Probab. Th. Rel. Fields*, 72, 155–160.

[118] H. Morimoto. (1987). On noncooperative n-player cyclic stopping games, *Stochastics*, 20, 27–37.

[119] H. Morimoto. (1987). Optimal switching for alternating processes, *Appl. Math. Optim.*, 16, 1–17.

[120] H. Morimoto. (1991). On average cost stopping time problems, *Probab. Th. Rel. Fields*, 90, 469–490.

[121] H. Morimoto. (1994). Perturbation methods in optimal stopping with average cost criterion, *Stoch. Stoch. Rep.*, 47, 229–245.

[122] H. Morimoto and M. Okada. (1999). Some results on the Bellman equation of ergodic control, *SIAM J Control Optim.*, 38, 159–174.

[123] H. Morimoto and Y. Fujita. (2000). Ergodic control in stochastic manufacturing systems with constant demand, *J Math. Anal. Appl.*, 243, 228–248.

[124] H. Morimoto and K. Kawaguchi. (2002). Optimal exploitation of renewable resources by the viscosity solution method, *Stoch. Anal. Appl.*, 20, 927–946.

[125] H. Morimoto. (2003). Variational inequalities for combined control and stopping, *SIAM J Control Optim.*, 42, 686–708.

[126] H. Morimoto and X. Y. Zhou. (2008). Optimal consumption in a growth model with the Cobb–Douglas production function, *SIAM J Control Optim.*, 47, 2991–3006.

[127] H. Nagai. (1999). *Stochastic Differential Equations*, Tokyo: Kyôritsu-Shuppan (in Japanese).

[128] J. Neveu. (1964). *Bases Mathématique du Calcul des Probabilités*. Paris: Masson.

[129] M. Nisio, (1981). *Stochastic Control Theory*, ISI Lecture Notes 9, MacMillan.

[130] M. Nisio. (1976). On a nonlinear semigroup attached to optimal stochastic control, *Publ. RIMS, Kyoto Univ.*, 12, 513–537.

[131] M. Nisio. (1976). On stochastic optimal controls and envelope of Markovian semigroups, *Proc. Intern. Symp. SDE Kyoto*, edited by K. Ito, Japan: Kinokuniya, 297–325.

[132] B. Øksendal. (2003). *Stochastic Differential Equations*. Berlin: Springer-Verlag.

[133] É. Pardoux. (1996). Backward stochastic differential equations and viscosity solutions of systems of semilinear parabolic and elliptic PDEs of second order, Stochastic Analysis and Related Topics VI, edited by Decreuseford, J. Gjerde, B. Øksendal and A. Süleyman Üstünel. Boston: Birkhauser, 79–127.

[134] D. Revuz and M. Yor. (1994). *Continuous Martingales and Brownian Motion*. 2nd ed. Berlin: Springer-Verlag.

[135] M. Robin. (1983). Long-term average cost control problems for continuous time Markov processes: A survey, *Acta. Appl. Math.*, 1, 281–299.

[136] M. Robin. (1981). On some impulse control problems with long run average cost, *SIAM J Control Optim.*, 19, 333–358.

[137] R. T. Rockafellar and R. J.-B. Wets. (1998). *Variational Analysis*. Berlin: Springer-Verlag.

[138] L. C. G. Rogers and D. Williams. (1979, 1989). Diffusions, Markov Processes, and Martingales, Vols. I–II. New York: John Wiley & Sons.

[139] H. L. Royden. (1968). *Real Analysis*, 2nd ed. New York: Macmillan.

[140] W. Rudin. (1987). Real and Complex Analysis, 3rd ed. New York: McGraw-Hill.

[141] S. P. Sethi and Q. Zhang. (1994). Hierarchical Decision Making in Stochastic Manufacturing Systems, Boston: Birkhäuser.

[142] S. P. Sethi and G. L. Thompson. (2000). *Optimal Control Theory*. Boston: Kluwer.

[143] S. P. Sethi, H. Zhang, and Q. Zhang. (2005). *Average-Cost Control of Stochastic Manufacturing Systems*. New York: Springer-Verlag.

[144] S. E. Shreve and H. M. Soner. (1994). Optimal investment and consumption with transaction costs, *Ann. Appl. Probab.*, 4, 609–692.

[145] S. E. Shreve and G. L. Xu. (1992). A duality method for optimal consumption and investment under short-selling prohibition: I. General market coefficients; and II. Constant market coefficients, *Ann. Appl. Probab.*, 2, 87–112, 314–328.

[146] S. E. Shreve. (2004). Stochastic Calculus for Finance. Vols. I–II. New York: Springer-Verlag.

[147] A. N. Shiryayev. (1978). *Optimal Stopping Rules*. New York: Springer-Verlag.

[148] A. V. Skorokhod. (1965). *Studies in the Theory of Random Processes*. Boston: Addison-Wesley.

[149] H. M. Soner. (1997). Controlled Markov processes, viscosity solutions and applications to mathematical finance, *Lecture Notes in Math.*, 1660, 134–185.

[150] Ł. Stettner and J. Zabczyk. (1981). Strong envelopes of stochastic processes and a penalty method, *Stochastics*, 4, 267–280.

[151] D. W. Stroock. (1993). *Probability Theory: An Analytic View*, Cambridge, UK: Cambridge University Press.

[152] D. W. Stroock and S. R. S. Varadhan. (1979). *Multidimensional Diffusion Processes*. New York: Springer-Verlag.

[153] M. E. Thompson. (1971). Continuous parameter optimal stopping problems, *Z. Wahr. verw. Geb.*, 19, 302–318.

[154] J. H. Van Schuppen and E. Wong. (1974). Transformation of local martingales under a change of law, *Ann. Prob.*, 2, 879–888.

[155] E. Wong and B. Hajek. (1985). *Stochastic Processes in Engineering Systems*. New York: Springer-Verlag.

[156] Z. Wu, J. Yin, and C. Wang. (2006). *Ellitic and Parabolic Equations*. Singapore: World Scientific.

[157] T. Yamada and S. Watanabe. (1971). On the uniqueness of solutions of stochastic differential equations, *J Math. Kyoto, Univ.*, 11, 155–167.

[158] K. Yoshida. (1965). *Functional Analysis*, Berlin: Springer-Verlag.

[159] T. Zariphopoulou. (1994). Consumption-investment models with constraints, *SIAM J Control Optim.*, 32, 59–85.

[160] J. Yong and X. Y. Zhou. (1999). *Stochastic Controls: Hamiltonian Systems and HJB Equations*. New York: Springer-Verlag.

[161] W. P. Ziemer. (1989). Weakly Differentiable Functions, New York: Springer-Verlag.

Index